Contents

Nelson Advanced Modular Science

The Organism and the Environment

SECOND EDITION

ADDS • ERICA LARKCOM • RUTH MILLER

Thomas Nelson and Sons Ltd
Nelson House Mayfield Road
Walton-on-Thames Surrey
KT12 5PL UK

© John Adds, Erica Larkcom, Ruth Miller 1997

First published by Thomas Nelson and Sons Ltd 1997

ISBN 0 17 448274 4
NPN 9 8 7 6 5 4

Printed in China

Publication Team

Acknowledgements:
Acquisitions: Mary Ashby/Chris Coyer
Cover Design: Liam Reardon
Design: Maria Pritchard
Editorial Management: Simon Bell
Editorial: Liz Jones
Marketing: Jane Lewis
Production: Liam Reardon
Typesetting and artwork: Hardlines

Introduction

As modularisation of syllabuses gains momentum, there is a corresponding demand for a modular format in supporting texts. The Nelson Advanced Modular Science series has been written by Chief Examiners and those involved directly with the A level examinations. The books are based on the ULEAC (London) AS and A level modular syllabuses in Biology and Human Biology, Chemistry and Physics. Each module text offers complete and self-contained coverage of all the topics in the module. The texts also include examples outside the prescribed syllabus to broaden your understanding and help to illustrate the principle which is being presented. There are regular review questions and practical investigations while you read about and study the topic. Finally, there are typical examination questions with mark schemes so that you can test yourself and know what the examiners really want in the answers.

The Organism and the Environment is the second book in the biology series. This text gives emphasis to whole organisms, living in interacting communities within ecosystems. A central theme is the flow of energy through ecosystems and the way this is related to different modes of nutrition. You will gain an appreciation of the enormous diversity of living organisms, how they are adapted to their environment and how species have changed over time. In particular, we look at the effects of one species, Homo sapiens, and see how humans have changed the face of the Earth since they appeared, a mere
400 000 years ago. You will see how human activities have influenced ecosystems in the past and you can try to predict the impact of these effects in the future. The authors hope that through your study and understanding of biology you will be able to make informed decisions and show balanced judgement about environmental matters which have now become global issues with implications on an international scale.

Erica Larkcom B.A., M.A., C.Biol., M.I.Biol., Subject Officer for A level Biology, formerly Head of Biology, Great Cornard Upper School, Suffolk

John Adds B.A., C.Biol., M.I.Biol., Dip.Ed., Chief Examiner for A level Biology, Head of Biology, Abbey Tutorial College, London

Ruth Miller B.Sc., C.Biol., M.I.Biol., Chief Examiner for AS and A level Biology, formerly Head of Biology, Sir William Perkins's School, Chertsey

Note to teachers on safety

When practical instructions have been given we have attemted to indicate hazardous substances and operations by using standard symbols and appropriate precautions. Nevertheless you should be aware of your obligations under the Health and Safety at Work Act, Control of Substances Hazardous to Health (COSHH) Regulations and the Management of Health and Safety at Work Regulations. In this respect you should follow the requirments of your employers at all times.

In carrying out practical work, students should be encouraged to carry out their own risk assessments, i.e. they should identify hazards and suitable ways of reducing the risks from them. However, they must be checked by the teacher/lecturer. Students must also know what to do in an emergency, such as a fire.

The teachers/lecturers should be familiar and up to date with current advice from professional bodies.

Acknowledgements

The authors wish to thank the following for their help and encouragement during the preparation of this book: David Hartley, English Nature, The Field Studies Council, The Worldwide Fund for Nature.

Christine Grey-Wilson provided sketches for figures in Chapters 6 and 7; Adrian Warn produced the original versions of Figures 5.21, 7.19 and 7.22; John Schollar at The National Centre for Biotechnology Education provided reference for Figures 9.15, 9.17 and 9.18.

The following Figures are based on illustrations in Jones, S., Martin, R. and Pilbeam D. (eds) The Cambridge Encyclopedia of Human Evolution: Figures 9.6, 9.8, 9.9, 9.10, 9.11, 9.12, 9.13, 9.14, 9.19, 9.25, 9.26, 9.28, 9.31 to 9.34 and 9.42. The authors are grateful to the publishers and authors of this work for permission to use copyright material.

The authors and publishers would like to thank the following for permission to reproduce photographs:

Brian and Cherry Alexander: 10.1(b)

Science Photo Library: Figure 1.9, Figure 2.10(b), Figure 2.15(a), Figure 2.17(a)

Biophoto Associates/Science Photo Library: Figure 10.12

Andy Crump, TDR, WHO/Science Photo Library: Figure 11.13, Figure 11.23

John Reader/Science Photo Library: Figures 9.20(a) and (b)

Vaughan Fleming/Science Photo Library: Figure 7.1(a)

Phillippe Plailly/Eurelios/Science Photo Library: Figure 9.29

Gamma/Harrison-API: Figure 10.2

Mike Birkhead/Oxford Scientific Films: Figure 9.1

Planet Earth Photo Library: Figure 1.18(a), Figure 2.1(b), Figure 2.1(d), Figure 3.1, Figure 7.1(b), Figure 7.2, Figure 8.6

Peter Scoones: Figure 1.18(b)

Heather Angel/Biofotos: Figure 2.1(c), Figure 4.1(a and b), Figure 4.14

Image Select/Ann Ronan Picture Library: Figure 6.3(a), Figure 7.26(a and b), Figure 8.1, Figure 9.3

Image Select/Jean-Paul Thomas: Figure 6.3(b)

James Balog/Tony Stone Images: Figure 10.11

Paul Chesley/Tony Stone Images: Figure 11.28(h)

David Hiser/Tony Stone Images: Figure 11.30(b)

James Martin/Tony Stone Images: Figure 11.30(a)

Art Wolfe/Tony Stone Images: Figures 9.5(a) and (b), Figure 11.28(g)

Jon Riley/Tony Stone Images: Figure 11.29

Manoj Shah/Tony Stone Images: Figure 9.5(c)

Jess Stock/Tony Stone Images: Figure 10.14

George Chan/Tony Stone Images: Figure 11.10(b)

Daniel J. Cox/Tony Stone Images: Figure 9.5(d)

David Levy/Tony Stone Images: Figure 10.15(c)

Lorne Resnick/Tony Stone Images: Figure 10.1(a), Figure 10.8

Trip/K Cardwell: Figure 11.15(b)

Werner Forman Archive: Figure 9.37

John Adds: Figure 3.9, Figure 3.11, Figure 3.13, Figure 5.1

Judith Clark: Figure 11.32

Erica Larkcom: Figure 3.16, Figure 6.2(a), Figure 6.2(b), Figure 6.3(c), Figure 6.4(a), Figure 6.4(b), Figure 6.5(a, b and c), Figure 6.10, Figure 6.12, Figure 6.13, Figure 6.21, Figure 6.22(a), Figure 6.22(b), Figure 7.5, Figure 7.8, Figure 9.41(b), Figure 10.9, Figure 10.15(a) and (b), Figure 10.16(a), Figure 11.10(a), 11.15(b), 11.22, 11.26(a) and (b), 11.28(a), (b), (c), (d), (e) and (f)

Joy Larkcom: Figure 7.9

Anne Stephens: Figure 9.41(a), 10.23(a) and (b), 11.25

Ruth Miller: Figure 5.23

The examination questions and mark schemes on pages 263–279 appear by permission of Edexcel (London Examinations).

Autotrophic nutrition in flowering plants

All living organisms need energy for growth and maintenance. **Autotrophic** organisms are able to use external sources of energy in the synthesis of their organic food materials, whereas **heterotrophic** organisms must be supplied with ready-made organic compounds from which to derive their energy. Algae, green plants and certain prokaryotes can obtain the energy for synthesis directly from the sun's radiation. It is then used to build up essential organic compounds from inorganic molecules. Such organisms are called **photosynthetic** and possess special pigments which can absorb the necessary light energy. A few specialised prokaryotes are able to use energy derived from certain types of chemical reaction in the synthesis of organic molecules from inorganic ones. These organisms are called **chemosynthetic** and include the nitrifying bacteria, *Nitrosomonas* and *Nitrobacter,* which are important in the nitrogen cycle (see Chapter 5). All other organisms are heterotrophic and are dependent on autotrophic organisms for their energy supplies, so must feed on plants, or on other animals which have eaten plants. Photosynthesis is the source of energy and organic materials for other organisms besides plants. So it can be seen that the ultimate source of all metabolic energy is the Sun and photosynthesis is responsible for the maintenance of life on Earth.

Photosynthesis

Photosynthesis in green plants is the process in which energy from the sun is transformed into chemical bond energy in organic molecules. It is a process in which energy is transduced from one form to another and results in the inorganic molecules, carbon dioxide and water, being built up into organic molecules. Oxygen is produced as a waste product. In green plants, the first stable organic molecules to be formed in photosynthesis are simple sugars, which can be used as a source of energy or used in the synthesis of other organic molecules. The general equation for photosynthesis can be written as:

$$CO_2 \quad + \quad 2H_2O \quad \xrightarrow[\text{chlorophyll}]{\text{light}} \quad (CH_2O) \quad + \quad O_2 \quad + \quad H_2O$$

$$\text{carbon dioxide} \quad + \quad \text{water} \quad \rightarrow \quad \text{carbohydrate} \quad + \quad \text{oxygen} \quad + \quad \text{water}$$

NB: Water is a product as well as a reactant in the process.

Photosynthesis occurs in two stages:
- the **light-dependent stage**, which requires light energy and results in the production of **ATP** (adenosine triphosphate) and **NADPH** (reduced nicotinamide adenine dinucleotide phosphate)
- the **light-independent stage** in which the NADPH is used to reduce carbon dioxide to carbohydrate. ATP is required in this stage.

AUTOTROPHIC NUTRITION IN FLOWERING PLANTS

From the direct products of photosynthesis, green plants can synthesise proteins, polysaccharides, lipids and nucleic acids, all of which contribute to the structure and functioning of cells and organelles. In addition, the sugars and the polysaccharide starch form energy stores. During respiration, the sugars are oxidised to carbon dioxide and water, releasing energy which is stored in molecules of ATP ready for use in cellular activities such as the synthesis of new protoplasm (growth), active transport and movement of protoplasm. Figure 1.1 summarises these activities.

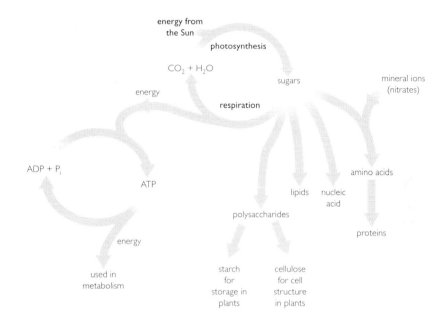

Figure 1.1 Diagram to show how the products of photosynthesis may be used

In addition to the raw materials carbon dioxide and water, a supply of mineral ions is necessary for amino acid and, subsequently, protein formation. Carbon dioxide is obtained from the atmosphere and enters the plants through the stomata on the aerial parts, especially the leaves. The water and mineral ions are obtained from the soil and are transported through the plant in the xylem tissue from the roots to the leaves. Water is necessary for all living processes and there are no special ways in which plants can ensure a supply of water specifically for photosynthesis. Water is continually taken up by green plants to maintain the turgidity of the tissues and to replace that lost in transpiration. The amount of water needed in photosynthesis is small compared with that taken up and lost through transpiration.

Net photosynthesis

In order for growth to occur in a plant, there must be a net gain of energy, so photosynthesis must exceed respiration. The total amount of carbon dioxide 'fixed' as carbohydrate is referred to as **gross photosynthesis** (**GP**). A certain amount of carbon dioxide will be released as a result of respiration (**R**). So growth will depend on the difference between gross photosynthesis and respiration, which is known as **net photosynthesis** (**NP**).

$$NP = GP - R$$

The site of photosynthesis

The leaf as a photosynthetic organ

The main photosynthetic organ of a green plant is the **leaf** (Figure 1.2), although the cortex of green herbaceous stems may contain cells with large numbers of chloroplasts, enabling photosynthesis to occur. In most plants, the leaf provides a large, thin, flat surface, which traps the light effectively. Leaves are particularly well-adapted as photosynthetic organs because:

- they provide a large surface area over which light can be absorbed
- they are thin, so the diffusion paths of the gases during gaseous exchange are short: carbon dioxide diffuses in and oxygen diffuses out
- the midrib and extensive network of veins provide good support for the thin lamina or **blade**
- the extensive network of veins enables efficient transport of materials to and from the photosynthesising cells. None of the photosynthesising cells are far away from the xylem – transporting water and mineral ions needed in the process, or from the phloem – transporting the products of photosynthesis away
- the mesophyll tissue, made up of palisade cells and spongy cells, contains large numbers of chloroplasts. The chloroplasts contain the green pigment **chlorophyll**, which absorbs the light energy.

An examination of its internal structure reveals how the leaf is further adapted to the function of photosynthesis. The organisation of the different tissues can be seen in the transverse section of a dicotyledon leaf in Figure 1.3 and the functions of these tissues are summarised in Table 1.1.

The palisade tissue of the leaf has the greatest concentration of chloroplasts in its cells, so it is here that most photosynthesis takes place. The structure of a palisade cell has been described in Module 1, but it is relevant here to consider

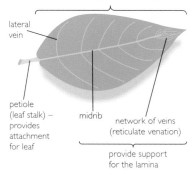

(a)   lamina (leaf blade) – large, broad, flat; large surface area for efficient light absorption and gaseous exchange

lateral vein

petiole (leaf stalk) – provides attachment for leaf

midrib

network of veins (reticulate venation)

provide support for the lamina

Dicotyledonous leaf (e.g. privet). Definite upper and lower surface (dorsiventral) with most stomata on the lower surface

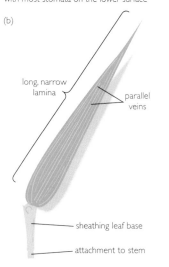

(b)

long, narrow lamina

parallel veins

sheathing leaf base

attachment to stem

Monocotyledonous leaf (e.g. grasses). No distinct difference between upper and lower surfaces (isobilateral) with equal distribution of stomata

Figure 1.2 External features of leaves: (a) dicotyledonous, (b) monocotyledonous

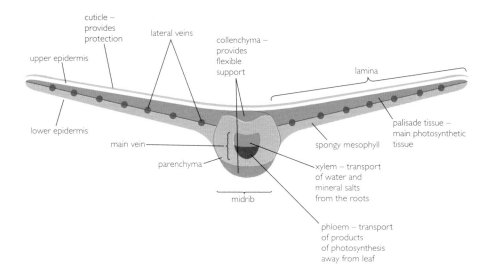

cuticle – provides protection

lateral veins

collenchyma – provides flexible support

lamina

upper epidermis

lower epidermis

main vein

parenchyma

midrib

palisade tissue – main photosynthetic tissue

spongy mesophyll

xylem – transport of water and mineral salts from the roots

phloem – transport of products of photosynthesis away from leaf

Figure 1.3 Tissue plan of transverse section through a dicotyledonous leaf

AUTOTROPHIC NUTRITION IN FLOWERING PLANTS

Table 1.1 *Leaf tissues and their functions*

Tissue	Structure	Function
upper epidermis	layer 1 cell thick, covered by cuticle; cells flattened; no chloroplasts; usually no stomata present, but if present, fewer than on lower epidermis	cuticle protects and prevents evaporation of water from epidermal cells; transparent so light allowed through to palisade mesophyll
palisade mesophyll	densely packed layer of column-shaped cells; thin cell walls; large numbers of chloroplasts	most photosynthesis occurs in this layer; shape of cells and dense packing means high concentration of chloroplasts; turgidity contributes to support
spongy mesophyll	irregular shaped cells containing fewer chloroplasts; form loose network with large air spaces	important for the diffusion of gases; air spaces connect up with stomata; some photosynthesis; turgidity contributes to support
vascular tissue in midrib and veins	xylem: lignified; vessels and tracheids	xylem conducts water and mineral ions to photosynthesising cells; provides support for the lamina
	phloem: sieve tubes and companion cells	phloem removes the products of photosynthesis (in the form of sucrose)
collenchyma (often found in midrib region)	living cells with extra cellulose thickening in the cell walls	provides flexible support in the leaf, particularly in the midrib region; turgidity also contributes to support
sclerenchyma (often found in midrib region)	lignified cells, fibres; no living contents	provides support in the midrib region
lower epidermis	layer 1 cell thick; covered by cuticle; cells similar to upper epidermis; many stomata present	important for the presence of stomata allowing gaseous exchange

how it is adapted to its function. Because of their columnar shape and orientation, the palisade cells appear to be tightly packed in a layer just below the upper epidermis. Careful observation shows that substantial parts of the long sides of these cells are exposed to the intercellular air spaces, so enabling rapid uptake of available carbon dioxide (Figure 1.4). As the upper epidermis is transparent, the palisade cells are in a position to receive the maximum amount of available light. The cell walls are thin and there is only a thin layer of cytoplasm, allowing rapid diffusion of materials into the chloroplasts.

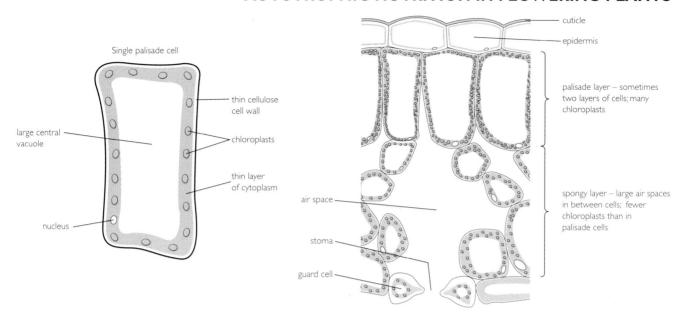

Figure 1.4 Diagram showing cellular structure of leaf tissue with detail of a palisade cell (magnification c.× 500)

Chloroplasts and chloroplast pigments

Chloroplasts are saucer-shaped plastids, usually between 4 and 10 μm in diameter and 1 μm thick, surrounded by a double membrane forming the chloroplast envelope. Inside the chloroplast, is a complex system of internal membranes called **lamellae**, or **thylakoids**, which develop from extensions of the inner membrane of the envelope. They are embedded in a colourless matrix called the **stroma**. In some regions, the thylakoids are arranged in stacks, forming **grana**, which are connected to each other by **stroma thylakoids**. A simplified model of the ultrastructure of a chloroplast is illustrated in Figure 1.5. The membranes which make up the thylakoids, in which the chlorophyll molecules are embedded, consist of approximately equal proportions of protein and lipid. Conversion of light energy to chemical energy takes place in the thylakoids and the reactions of the Calvin cycle, involving the reduction of carbon dioxide to carbohydrates, take place in the stroma.

The chloroplasts contain small 70 S ribosomes and circular DNA, and are similar in structure to some photosynthetic prokaryotes. The presence of these structures has led to the suggestion that the chloroplasts of green plants are the descendants of cyanobacteria, which became incorporated into eukaryotic cells at an early stage in the evolution of green plants. This theory is known as the **endosymbiont theory** and applies to mitochondria as well as chloroplasts.

All photosynthetic organisms possess at least one pigment which can absorb light and start the reactions of photosynthesis. These pigments can be extracted using organic solvents and separated from each other by chromatography. Three different groups of pigments are found in photosynthetic organisms:

- **chlorophylls**
- **carotenoids**
- **phycobilins.**

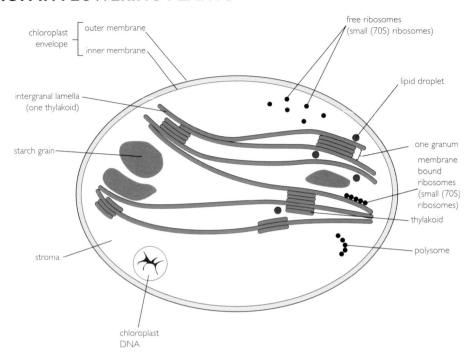

Figure 1.5 Chloroplast ultrastructure (simplified; magnification × 15000)

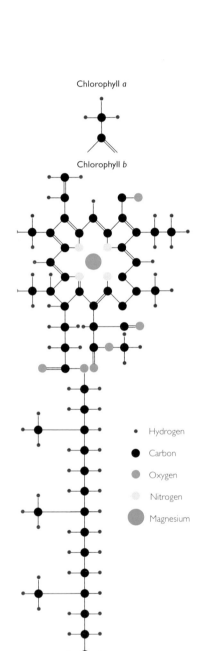

Chlorophyll *a*

Chlorophyll *b*

- · Hydrogen
- ● Carbon
- ● Oxygen
- ○ Nitrogen
- ● Magnesium

Figure 1.6 Atomic structure of chlorophyll molecules a *and* b

The chlorophylls give plants their green colour as they absorb red and blue-violet light, reflecting green light. The molecules have a porphyrin 'head' attached to a long hydrocarbon (phytol) 'tail'. The porphyrin head is polar and consists of a tetrapyrrole ring with a magnesium ion (Mg^{2+}) in its centre, whereas the hydrocarbon tail is non-polar. The porphyrin part of the molecule is bound to the protein of the chloroplast lamellae and the hydrocarbon chain extends into the lipid layer. The chemical structure is shown, for interest, in Figure 1.6.

There are several forms of chlorophyll, differing slightly in colour, chemical structure and absorption peaks. The most abundant is **chlorophyll *a***, which plays a central role in the absorption of light energy and is present in all photosynthetic organisms where oxygen is evolved. **Chlorophyll *b*** is also found in the leaves of green plants and in the green algae. Other forms of chlorophyll are characteristic of some groups of algae. Present in all photosynthetic organisms, the **carotenoids** are yellow or orange pigments, which absorb blue-violet light. They are usually masked by the chlorophylls, but their colour shows up in the autumn, when the chlorophylls break down. They are hydrocarbons, situated quite close to the chlorophyll molecules in the chloroplast lamellae. The light energy which they absorb is transferred to chlorophyll *a*. The most widespread of the carotenoids is *β*-**carotene**, present in all green plants and algae. **Fucoxanthin** is a yellow-brown pigment, related to the carotenoids, which gives the characteristic colour to brown algae.

The **phycobilins**, found in cyanobacteria and red marine algae, are structurally similar to chlorophyll *a*, but do not contain a magnesium atom. They absorb light in the middle of the visible spectrum, which enables the red marine algae to photosynthesise in dim light conditions under water.

Table 1.2 summarises the various photosynthetic pigments, their occurrence and the range of their light absorption.

Table 1.2 *Characteristics of photosynthetic pigments*

Pigment	Colour	Occurrence	Wavelengths of light absorbed
chlorophyll *a*	blue-green	green plants, algae	peaks at 420 nm and 660 nm
chlorophyll *b*	yellowish-green	green plants, green algae	peaks at 435 nm and 643 nm
β-carotene	orange-yellow	green plants, algae	between 425 and 480 nm
fucoxanthin	yellow-brown	brown algae (sea-weeds)	between 425 and 475 nm
phycobilins	red	red algae, cyanobacteria	between 490 and 576 nm

Absorption and action spectra

An **absorption spectrum** is a graph of the relative absorbance of different wavelengths of light by a photosynthetic pigment. It is obtained by subjecting solutions of each pigment to different wavelengths of light and measuring how much light is absorbed.

As shown in Figure 1.7, chlorophyll *a* absorbs blue-violet light (420 nm) and red light (660 nm). Very little absorbance occurs between 500 and 600 nm, so light of these wavelengths will be reflected. The absorption spectrum for chlorophyll *b* is very similar.

An **action spectrum** is a graph showing the amount of photosynthesis at different wavelengths of light. It can be obtained by allowing plants, such as Canadian pondweed, to photosynthesise for a stated time in light of each wavelength in turn and measuring the volume of gas evolved. A graph is then plotted of rate of photosynthesis against wavelength of light.

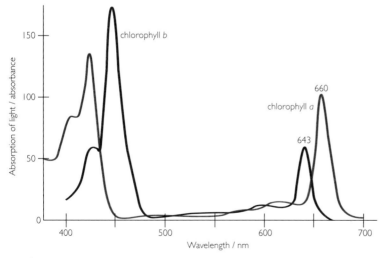

Figure 1.7 Absorption spectra for chlorophylls a *and* b

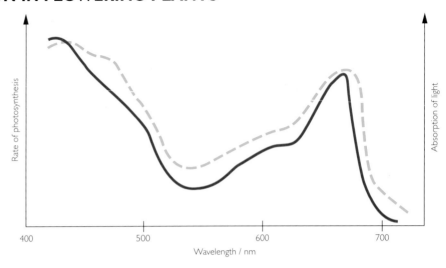

Figure 1.8 Graph comparing absorption and action spectra

If an action spectrum and the absorption spectrum for an extract of chlorophyll pigments are plotted on the same graph, as shown in Figure 1.8, it can be seen that they resemble each other closely, indicating that there is good evidence for these pigments being involved in the absorption of light for photosynthesis.

Light-dependent reaction

In this stage, energy is transferred from light into ATP and reduced coenzyme, NADPH, is formed. In order to appreciate how these events take place, it is necessary to understand how the chlorophyll molecules are arranged and their association with the thylakoid membranes.

The photosynthetic pigment molecules are organised into two photosystems, **Photosystem I (PSI)** and **Photosystem II (PSII)**. These photosystems appear on the thylakoid membranes as different sized particles, which were first discovered when chloroplasts were subjected to a technique known as 'freeze-fracturing' before being viewed using electron microscopy. It appeared that the larger particles, now known to be PSII, were associated with the granal thylakoids and the smaller particles, PSI, with the stromal thylakoids (Figure 1.9).

Before we can see how each photosystem works, we need to recognise that there are two types of photosynthetic pigments: **primary** pigments and **accessory** pigments. Two primary pigments are known, both of which are specialised forms of chlorophyll *a*. They are designated **P680**, with an absorption peak at 680 nm, and **P700**, with an absorption peak at 700 nm. The accessory pigments include all the other forms of chlorophyll (other forms of chlorophyll *a* with different absorption peaks and chlorophyll *b*) and the carotenoids. They pass their energy on to the primary pigments. The primary pigments then emit the electrons which cause the light-dependent reactions of photosynthesis.

Each photosystem consists of a large number of accessory pigment molecules which absorb light energy and transfer it to a **reaction centre** consisting of one of the primary pigment molecules. In **PSI**, the primary pigment molecule is **P700** and in **PSII**, the primary pigment molecule is **P680**.

Chlorophyll molecules contain electrons that can be excited by light energy to become **high energy electrons**. When light is absorbed, these electrons are emitted and are taken up by other molecules, known as **electron acceptors** or **electron carriers**, which pass them on to another molecule. The chlorophyll molecule, having lost electrons, is left in an oxidised state, the electron carrier is reduced. A word equation for the effect of light on chlorophyll can be written as follows:

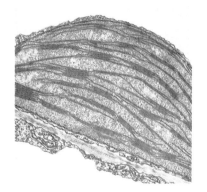

Figure 1.9 EM of chloroplast

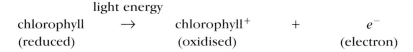

$$\text{chlorophyll} \xrightarrow{\text{light energy}} \text{chlorophyll}^+ + e^-$$
$$\text{(reduced)} \qquad \text{(oxidised)} \qquad \text{(electron)}$$

The electrons are passed along a chain of electron carriers in a series of oxidation–reduction (**redox**) reactions. Each carrier in the series is at a lower energy level than the one preceding it and as the electrons are passed along, sufficient energy is released to build up molecules of ATP from ADP and inorganic phosphate. As this process uses light energy, it is referred to as **photophosphorylation** to distinguish it from **oxidative phosphorylation** which occurs as a result of respiration.

In the light-dependent stage, light energy absorbed by the chlorophyll molecules in PSI and PSII is transferred to the P700 and P680 chlorophyll *a* molecules respectively. High energy electrons are emitted from both P700 and P680, leaving them both oxidised. In order for more light energy to be absorbed, the electrons need to be replaced. Associated with PSII is an enzyme which catalyses the splitting of water into hydrogen ions, electrons and oxygen.

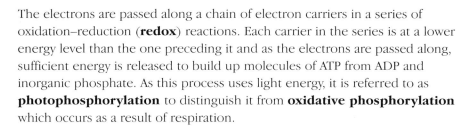

$$H_2O \rightarrow 2H^+ + \tfrac{1}{2}O_2 + 2e^-$$
$$\text{water} \quad \text{hydrogen} \quad \text{oxygen} \quad \text{electrons}$$
$$\text{ions} \qquad \text{ions}$$

The electrons released from the water replace the electrons emitted from P680, the hydrogen ions are released into the lumen of the thylakoid and the oxygen is released as a waste product. The two electrons emitted from P680 are picked up by electron acceptor A and passed along a chain of carriers. In this process, some of the energy from these electrons is used to move hydrogen ions across the thylakoid membrane from the stroma. Eventually they replace the electrons emitted from P700, thus restoring stability to the P700 chlorophyll molecule.

The electrons emitted from P700 are passed to electron acceptor B and used to reduce NADP in the stroma. This reaction also removes hydrogen ions from the stroma.

The thylakoid membranes are impermeable to hydrogen ions and there is a pH difference between the lumen of the thylakoid and the stroma, with a high concentration of hydrogen ions inside the lumen and a low concentration of hydrogen ions in the stroma. On the thylakoid membranes are complex protein molecules made up of two protein parts, one acting as a pore and the other acting as ATP synthetase (ATPase). As the hydrogen ions move from the lumen of the thylakoid through the channel created by the pore, energy is transferred from the hydrogen ions and used to build up ATP from ADP and inorganic phosphate, catalysed by ATP synthetase. The hydrogen ion concentration gradient across the thylakoid membrane is maintained by:

- the splitting of water
- the transport of electrons from PSII along the chain of carriers
- the formation of NADPH.

The events described above and illustrated by Figure 1.10 and Figure 1.11 are referred to as **non-cyclic photophosphorylation**, and result in the production of NADPH and ATP, both of which are used in the light-independent stage.

Cyclic photophosphorylation, in which only ATP is formed, involves PSI. Light absorbed by P700 causes electrons to be emitted, which are picked up by a different electron carrier. The energy in the electrons is used to move hydrogen ions across the thylakoid membrane into the lumen. The electrons are then passed back into P700 to replace the electrons that were emitted. ATP is synthesised as described above. Figure 1.12 illustrates this.

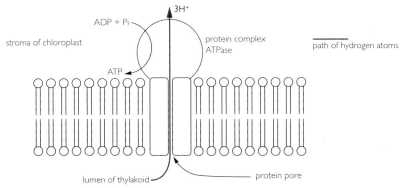

One ATP molecule is built up for every 3 H⁺ ions passing through the pore down the gradient

Figure 1.10 Diagram of thylakoid membrane with protein parts to illustrate ATP formation

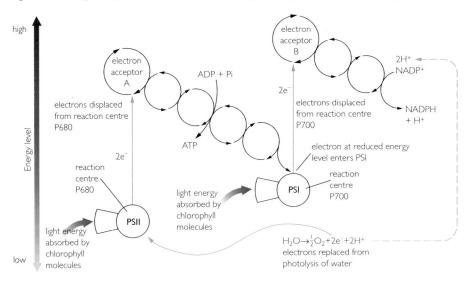

Figure 1.11 Non-cyclic photophosphorylation

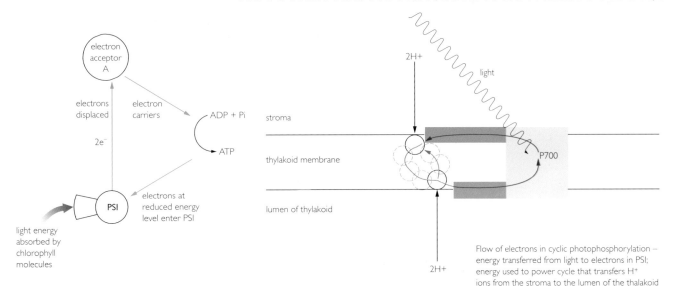

Figure 1.12 Cyclic photophosphorylation

Light-independent reaction

This stage of photosynthesis takes place in the stroma of the chloroplast. **NADPH** from the light-dependent stage is used to reduce **carbon dioxide** to **carbohydrate** and the **ATP** is used to provide the necessary energy. The sequence of enzyme-controlled reactions in this stage was determined by **Calvin** and his associates, using the alga *Chlorella* and the radioactive isotope ^{14}C. His experiment is summarised in Figure 1.13.

Calvin identified the first product of photosynthesis as **glycerate-3-phosphate** (**phosphoglyceric acid**), a 3C organic acid, which is reduced to **glyceraldehyde-3-phosphate** and from which other carbohydrates can be synthesised.

Carbon dioxide diffuses into the cells of the palisade tissue and into the chloroplasts, along a concentration gradient. In the stroma of the chloroplast, a molecule of carbon dioxide combines with a **carbon dioxide acceptor molecule**, a 5C pentose, **ribulose 1,3 bisphosphate (RuBP)**. Two molecules of glycerate-3-phosphate are formed. This is a **carboxylation** (addition of CO_2) and the enzyme responsible for catalysing the reaction is the **carboxylase, ribulose bisphosphate carboxylase (RuBisCO)**. The glycerate-3-phosphate is first phosphorylated, using ATP, then reduced to **glyceraldehyde-3-phosphate**, using NADPH. This process is summarised in Figure 1.14.

Glyceraldehyde-3-phosphate is used for two purposes:
- to regenerate the CO_2 acceptor molecule RuBP so that the fixation of carbon dioxide can continue
- to build up more complex organic compounds such as sugars, starch and amino acids.

The regeneration of RuBP involves a complex series of reactions in which 3, 4, 5, 6 and 7 carbon sugar phosphates are formed. For every six molecules of glyceraldehyde-3-phosphate formed during the Calvin cycle, only one can be

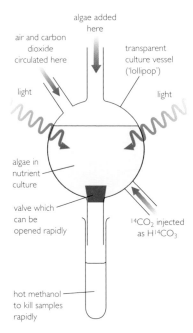

Figure 1.13 Diagram illustrating Calvin's 'lollipop' apparatus to determine the path taken by carbon dioxide in photosynthesis. Carbon dioxide containing radioactive carbon is bubbled through the algal suspension in the flask

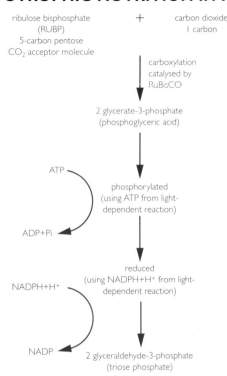

Figure 1.14 Equation summarising the formation of glyceraldehyde-3-phosphate

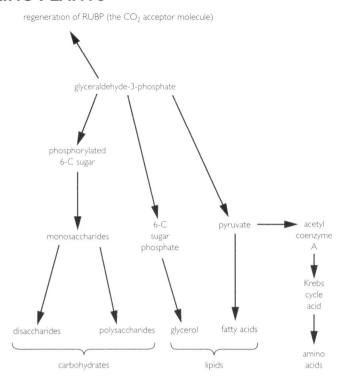

Figure 1.16 Summarising the fate of glyceraldehyde-3-phosphate

used for the synthesis of product as five are needed to regenerate a molecule of RuBP. Some starch may be built up in the chloroplast, but the synthesis of sucrose takes place in the cytosol. The glyceraldehyde-3-phosphate is able to leave the chloroplast via a specific transport protein in the inner chloroplast membrane. It may then be built up into hexose phosphates and eventually sucrose or starch, or it may be used in the synthesis of fatty acids or amino acids. Figure 1.15 summarises the fate of GALP. Recent research has indicated that different end products are made under different conditions. When light intensity and carbon dioxide concentration are high, then sugars and starch are formed, but in low light intensities amino acids are formed.

Factors affecting the rate of photosynthesis

Photosynthesis involves a series of reactions and the overall rate at which the process occurs will be dependent on the rate of the slowest of these reactions. If the light intensity is low, then the rate at which NADPH and ATP are being produced in the light-dependent stage will have an effect on the rate of the reactions in the light-independent stage. In these circumstances, light is said to be a **limiting factor**. If the light intensity is increased, then the rate of photosynthesis will increase until another factor, such as temperature or carbon dioxide concentration, becomes limiting (Figure 1.16).

The chief external factors which affect the rate of photosynthesis are **light intensity, carbon dioxide concentration** and **temperature**.

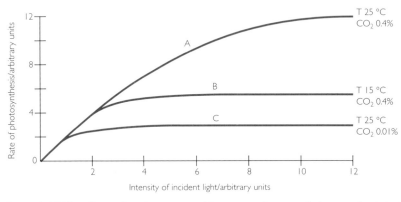

Figure 1.16 The effects of various external factors on the rate of photosynthesis in Chlorella

Light intensity

No photosynthesis occurs in the absence of light, but respiration continues and the net gas exchange of a green plant will show uptake of oxygen and release of carbon dioxide. It could also be shown that carbohydrates are being used up. At very low light intensities, some photosynthesis will occur using the carbon dioxide released by respiration, so the net gas exchange will still show uptake of oxygen and release of carbon dioxide. As the light intensity increases, so the rate of photosynthesis increases until the amount of carbon dioxide released from respiration is equal to the amount used up in photosynthesis. At this light intensity, known as the **light compensation point** (Figure 1.17), the rate of carbon dioxide production during respiration is equal to the rate at which carbon dioxide is taken up for photosynthesis. At light intensities higher than this, there will be a net uptake of carbon dioxide and release of oxygen and the amount of carbohydrate in the plant will increase.

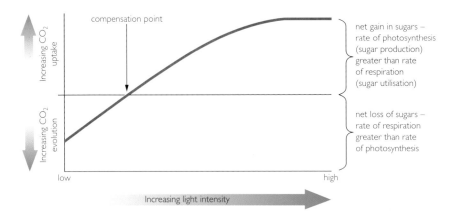

Figure 1.17 Light intensity and the compensation point

As the light intensity increases further, the rate decreases and then reaches a plateau, where further increase in light intensity has no effect on the rate of photosynthesis. At this stage, either another factor has become limiting or light saturation has been reached.

Prolonged exposure to strong sunlight may cause damage to the chloroplast pigments. If this happens, the leaves appear bleached and eventually die.

13

AUTOTROPHIC NUTRITION IN FLOWERING PLANTS

Figure 1.18 (top) shade plant, wood sorrel (Oxalis); (bottom) sun plant, rose bay willow herb (Epilobium)

Sun and shade plants and leaves

The compensation points of plants adapted to grow in shady situations, such as the wood sorrel (*Oxalis*), are quite different from plants such as rose bay willow herb (*Epilobium*), which grow in more open situations. The compensation point of the 'shade' plants is reached at lower light intensities than for the 'sun' plants. In addition to their compensation points, these plants may show differences in leaf anatomy, similar to the differences found in 'sun' and 'shade' leaves on the same plant.

Table 1.4 *Table of differences between 'sun' and 'shade' leaves*

'Sun' leaves	'Shade' leaves
thick leaves	thin leaves
palisade tissue often two or three cells thick	palisade tissue consists of a single layer of cells
most of the chloroplasts in the palisade tissue	chloroplasts more evenly distributed in palisade and spongy mesophyll tissue
efficient absorption of light at high light intensities	efficient absorption of light at low light intensity; at higher light intensities light passes through

Carbon dioxide concentration

Carbon dioxide is needed in the light-independent reactions, where it is involved in the formation of carbohydrates. The carbon dioxide concentration of the atmosphere is about 0.035 per cent, or 350 parts per million, by volumé, and under normal conditions this is the factor which limits the rate of photosynthesis. It can be shown that increasing the concentration of carbon dioxide will increase the rate of photosynthesis and it is possible to increase the growth of glasshouse crops, such as tomatoes and lettuces, by enriching the atmosphere with carbon dioxide.

Temperature

Temperature affects the rate of chemical reactions and is important in the light-independent stage of the photosynthetic process because the reactions are enzyme-controlled. In temperate climates, the optimum temperature for photosynthesis is about 25 °C, but an increase of 10 °C will double the rate, provided no other factor becomes limiting.

Photorespiration

Oxygen is a waste product of the light-dependent stage of photosynthesis. When light intensity and temperature are high, a great deal of oxygen is produced as the rate of photosynthesis speeds up. High concentrations of oxygen have an inhibiting effect on photosynthesis because the enzyme ribulose bisphosphate carboxylase, which catalyses the combination of carbon dioxide with the acceptor molecule, can also act as an oxygenase. If oxygen is present, it will compete with the carbon dioxide for the active site of the enzyme. This is an example of **competitive inhibition** and the degree of inhibition depends on the relative concentrations of the two gases.

When oxygen combines with RuBP, only one molecule of glycerate-3-phosphate is formed, together with one molecule of a 2 carbon compound, **phosphoglycolate**. The phosphoglycolate cannot be used directly in the synthesis of carbohydrates, but undergoes a series of reactions in which it is converted to a molecule of phosphoglyceric acid. During this pathway, which is termed **photorespiration**, oxygen and energy are used up and carbon dioxide is released. Under these conditions, there is a 30 to 40 per cent reduction in yield from photosynthesis. It should be noted that photorespiration is not related to the respiratory process described in the Module 1 text *Cell Biology and Genetics*.

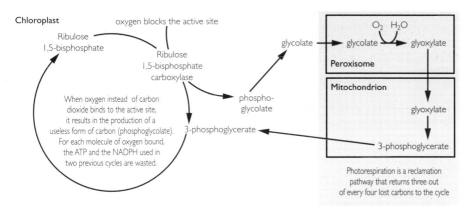

Figure 1.19 Diagram to show uptake of oxygen by RuBP and photorespiration

C_3 and C_4 plants

Plants in which the first products of photosynthesis are 3 carbon compounds, such as phosphoglyceric acid, are termed **C_3 plants**. For some plants which grow in tropical climates, where light intensity and temperatures are high, the first products of photosynthesis have been shown to be 4 carbon compounds. Such plants, which include sugar cane and maize, are termed **C_4 plants**. Further investigation of these plants has shown that the carbon dioxide is initially taken up by a different acceptor molecule in the mesophyll cells and then moved to the cells around the vascular bundle (bundle sheath cells), where the reactions of the Calvin cycle take place. The internal anatomy of the leaves is different from C_3 plants as the chloroplasts in the mesophyll cells have large grana, whereas those in the bundle sheath cells have very few grana. The series of reactions involved was described by Hatch and Slack in the 1960s and is now known as the **Hatch–Slack pathway** (Figure 1.20).

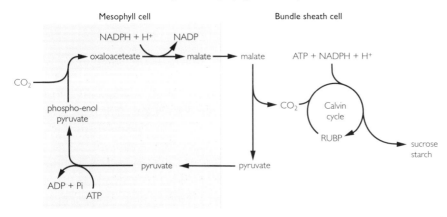

Figure 1.20 The Hatch–Slack pathway in mesophyll and bundle sheath cell

Table 1.5 *Differences between C_3 and C_4 plants*

C_3 plants	C_4 plants
carbon dioxide acceptor is ribulose bisphosphate and the enzyme is RUBP carboxylase	first carbon dioxide acceptor is phosphoenol pyruvate in the mesophyll cells and the enzyme is PEP carboxylase RUBP present in bundle sheath cells where CO_2 concentration is very high
photorespiration occurs because RUBP carboxylase can accept oxygen as well as carbon dioxide	photorespiration does not occur; PEP carboxylase does not accept oxygen and CO_2 concentration high in bundle sheath cells
less tolerant of dry conditions	more tolerant of dry conditions
reach maximum rate of photosynthesis at lower light intensities than C_4 plants	reach maximum rate of photosynthesis at much higher light intensities than C_3 plants
increase in temperature above 25 °C does not increase rate of photosynthesis	increase in temperature above 25 °C greatly increases rate of photosynthesis
less energy consumed in process, but overall less efficient than C_4 plants; usually lower yields	consumes more energy, but more efficient and yields usually higher than C_3 plants
examples: cereals, tobacco, beans	examples: maize, sugar

In the C_4 plants, carbon dioxide combines with **phospho-enol pyruvate** (**PEP**), a 3 carbon compound, to form a 4 carbon compound, **oxaloacetate**. This reaction is catalysed by the enzyme **phospho-enol pyruvate carboxylase** (**PEP carboxylase**), which is very efficient at accepting carbon dioxide when the concentration of oxygen is high. The oxaloacetate is converted to another 4 carbon compound, **malate**, and moved from the mesophyll cells to the bundle sheath cells. Carbon dioxide is released and enters the Calvin cycle by combining with RuBP. In this situation, the carbon dioxide concentration is high and the oxygen concentration is low, so there is no photorespiration. Photosynthesis is more efficient as there is less carbon dioxide lost, although the pathway does use more energy. C_4 plants are able to exploit the higher temperatures of tropical climates and also appear to be more tolerant of dry conditions.

Mineral nutrition

In addition to carbon, hydrogen and oxygen, which are obtained from carbon dioxide and water, green plants need at least thirteen essential elements in order to produce new tissues and maintain their correct functioning. **Nitrogen** and **sulphur** are essential constituents of amino acids, **phosphorus** is needed for the synthesis of nucleic acids and phospholipids and **potassium** is an important constituent of cell sap. These elements are taken up by the roots as inorganic ions from the soil water; nitrogen as nitrate ions (NO_3^-), sulphur as sulphate ions (SO_4^{2-}), phosphorus as phosphate ions (PO_4^{3-}) and potassium as potassium ions (K^+). Some of the functions of nitrates, phosphates and magnesium are shown in Table 1.6.

Most absorption occurs in the root hair region of the roots, situated a few centimetres from the root apex. In this region, the cells in the outermost layer, called the piliferous layer, have long hair-like extensions which penetrate between the mineral particles of the soil. These mineral particles are surrounded by films of water in which the mineral ions are dissolved. The cellulose cell walls of the root hair cells are freely permeable, so water and mineral ions diffuse through and are then taken up into the root hair cells, either by diffusion or by active transport.

Table 1.6 *Functions of nitrate, phosphate and magnesium ions*

Mineral ion	Functions
nitrate [NO_3^-]	essential for the synthesis of amino acids, proteins, nucleic acids, pigment molecules, coenzymes
phosphate [PO_4^{3-}]	required for synthesis of nucleic acids, phospholipids; component of nucleotides (ATP); phosphate group involved in energy transduction; phosphorylated intermediates in metabolism
magnesium [Mg^{2+}]	constituent of chlorophyll molecule; required as an activator for many dehydrogenase and phosphate transfer enzymes

The movement of ions across membranes is affected by a combination of the concentration gradient and an electrical gradient known as the **electrochemical gradient**. The ions will be attracted to areas of opposite charge and will move away from areas of similar charge. Where the internal concentration of an ion is less than its concentration in the soil water, it will diffuse into the root along the electrochemical gradient. In many cases, however, the uptake of ions is a process of accumulation, where movement takes place from a lower concentration in the soil water to a higher concentration in the plant tissues. This is an active process and requires energy from respiration. It has been found that the relative concentrations of different ions inside cells differs from their relative concentrations in the soil water, providing evidence for selective uptake. It appears that the cells will take up those ions they require in preference to others. The cell membranes are very effective diffusion barriers and the active transport of ions appears to be associated with specific ion pumps located on them.

Once inside the root, the ions may be transported across the root in the cytoplasm of the cortex cells, the **symplast pathway.** They pass from cell to cell by means of the **plasmodesmata**, which are fine cytoplasmic connections between neighbouring cells. When the xylem is reached, it is thought that the ions are actively secreted into this tissue, from where they are transported, in the **transpiration stream**, to the aerial parts of the plant. Alternatively, the ions may be transported by the **apoplast pathway**, through the cell walls, partly by diffusion and partly with the flow of water due to transpiration. Water uptake by the roots of plants is described in greater detail in Chapter 3 and illustrated in Figure 3.10.

AUTOTROPHIC NUTRITION IN FLOWERING PLANTS

Chromatography of chloroplast pigments

Introduction

Chloroplasts contain a number of different pigments, including chlorophyll *a*, chlorophyll *b*, β-carotene and xanthophylls such as lutein. These can be extracted using organic solvents and separated using the technique of chromatography. Thin layer chromatography, using small silica gel plates, gives excellent and rapid separation.

CAUTION: the solvents used in this practical are highly flammable and harmful. Use of a fume cupboard is recommended.

HIGHLY FLAMMABLE
propanone
petroleum ether
hexane

Materials

- Plastic-backed silica gel chromatography plates, cut to the size of microscope slides
- Fresh leaf material, such as spinach
- Mortar and pestle
- Fine sand
- Scissors
- Stoppered test tube
- Propanone
- Distilled water
- Petroleum ether (40 °C – 60 °C fraction)
- Pipette
- Small sample bottle
- Fine glass tube, such as a melting point tube
- Glass beaker covered with suitable lid
- Solvent: hexane-diethyl ether-propanone in the proportions 60 : 30 : 20 by volume (see **Note** in the margin)

EXTREMELY FLAMMABLE
diethyl ether

HARMFUL
diethyl ether
(intoxicating vapour)
hexane
petroleum ether

Method

1. Use scissors to cut up about 1 g of fresh leaf tissue and grind thoroughly in a mortar with some fine sand.
2. Transfer the ground up material to a stoppered test tube, add 4 cm³ of propanone, shake thoroughly then leave to stand for ten minutes.
3. Add 3 cm³ of distilled water and shake.
4. Now add 3 cm³ of petroleum ether, shake gently, then allow the solvents to separate. The pigments will collect in the top layer of petroleum ether which can be removed with a pipette and transferred to a small sample bottle.
5. Prepare a silica gel chromatography plate by carefully drawing a pencil line origin about 1.5 cm from the bottom. Avoid touching the surface of the plate with your fingers.
6. Use a fine glass tube to transfer the extract to the origin. Briefly touch the surface of the plate with the end of the tube, allow the petroleum ether to evaporate and repeat this process several times to produce a small, concentrated spot of pigment (Figure 1.21).

Note: If no fume cupboard is available it is recommended that propanone (90%) – petroleum ether (80–100°C fraction) in the proportions 1:9 by volume is used as the solvent

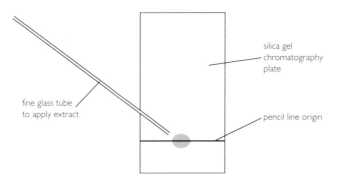

silica gel chromatography plate

fine glass tube to apply extract

pencil line origin

Figure 1.21

7 Stand the plate in a small, covered beaker containing solvent (hexane-diethyl ether-propanone). The level of solvent **must** be below the level of the pigment spot. Allow the solvent to rise until the solvent front reaches the top 1–2 cm of the plate. Do not allow the solvent to reach the top edge.

8 Remove your completed chromatogram and allow the solvent to evaporate. Make a careful drawing to show the positions and colours of the separated pigment spots.

Results and discussion

The separated pigments may be identified by their **Rf values**. The Rf value is defined as the distance moved by the spot (measured from the origin to the centre of the spot) divided by the distance moved by the solvent (measured from the origin to the solvent front). Rf values for some chloroplast pigments are given Table 1.3.

Table 1.3

Pigments	Rf values
β-carotene	0.98
phaeophytin	0.53
chlorophyll *a*	0.42
chlorophyll *b*	0.34
xanthophylls	0.30 – 0.07

1 Were all the leaf pigments removed by this technique?
2 How many pigments were you able to identify in your extract?

The effect of light intensity on the rate of photosynthesis

Introduction

A convenient way of measuring the rate of photosynthesis is to find the rate of production of oxygen bubbles from a submerged aquatic plant, such as *Elodea canadensis* (Canadian pond weed). A freshly cut shoot of *Elodea*, supported by tying carefully to a glass rod using cotton, can be inverted in a large beaker full of water. The cut end should be no more than about 1 cm below the surface of the water. When illuminated, a stream of bubbles should emerge from the cut end of the shoot. A tiny amount of detergent added to the water will allow the bubbles to escape freely. These bubbles are not pure oxygen (nitrogen and carbon dioxide are also present), but the rate of bubbling will give a good indication of the rate of photosynthesis.

A more quantitative method involves collecting the bubbles produced in a standard time and measuring their total volume. The apparatus shown in Figure 1.22 (known as a photosynthometer) can be used for such quantitative studies.

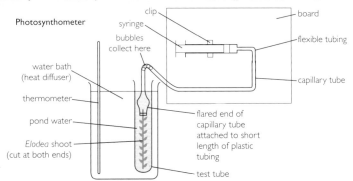

Figure 1.22

AUTOTROPHIC NUTRITION IN FLOWERING PLANTS

Materials

- Photosynthometer
- 60 or 100 W bench lamp. If available, a slide projector with a heat shield provides a high intensity light source
- Fresh *Elodea*. If possible, this should be kept in dilute sodium hydrogencarbonate solution and brightly illuminated before the experiment
- Stop clock
- Metre rule
- Sodium hydrogencarbonate

Method

1 Remove the plunger from the syringe and completely fill the apparatus with water from a tap. Replace the plunger.
2 Assemble the apparatus as shown in the Figure 1.22. Push in the plunger almost to the end of the syringe and ensure that there are no air bubbles in the tubing.
3 Add a pinch of sodium hydrogencarbonate to the test tube containing the *Elodea* and stir to dissolve. This will provide carbon dioxide and ensure that it is not a limiting factor.
4 When ready to start the experiment, darken the laboratory and start with the bench lamp close to the *Elodea*. Allow time for the *Elodea* to equilibrate, then collect the oxygen given off for a suitable period of time.
5 Using the plunger, carefully draw the bubbles into the capillary tube so that the volume or length of the bubble can be measured.
6 Take two more readings at this distance.
7 Repeat this procedure with the lamp at different, measured distances from the *Elodea*, such as 10, 15, 20, 30, 40 and 80 cm. In each case allow time for the *Elodea* to equilibrate.
8 Record the distances and corresponding volume of oxygen produced. Check the thermometer to ensure that the temperature of the water remains constant (why?).

Results and discussion

1 The light intensity is inversely proportional to the square of the distance (d) between the light source and the *Elodea*. You can calculate the relative light intensity by $1/d^2$. Since the figures for light intensity are small, each one can be multiplied by 10 000. This makes them easier to plot.
2 Make a table to show the mean rate of photosynthesis (volume of oxygen produced per unit time) at each light intensity.
3 Plot a graph to show the relationship betweeen the rate of photosynthesis and light intensity. Comment on the shape of your graph.

Further work

The same apparatus can be used to investigate the effect of other factors, such as carbon dioxide concentration or temperature, on the rate of photosynthesis. Carbon dioxide concentration can be varied by using a range of sodium hydrogencarbonate solutions, for example, 0.1, 0.2, 0.3 0.4 and 0.5%. Careful experimental design is necessary to keep other factors constant.

Heterotrophic nutrition

The different types of heterotrophic nutrition

Heterotrophic organisms need to be supplied with food consisting of complex organic molecules, which they use to obtain energy for metabolism and the materials required for growth, repair and replacement of tissues. These organic molecules are obtained either directly from green plants or from organisms that have fed on green plants. The way in which heterotrophic organisms obtain their food varies greatly, but once obtained, the complex organic compounds have to be broken down into simpler, soluble molecules before they can be absorbed.

Figure 2.1 All these organisms are heterotrophic: (a) lionesses feeding at a kill; (b) parasitic bracket fungi; (c) hermit crab with anemone attached to its shell

Within an ecosystem, heterotrophic organisms are the **consumers** in food chains. Green plants, known as **primary producers**, are autotrophic and able to synthesise the organic molecules which are required by consumers. Primary consumers are **herbivores**, feeding directly on green plants. Secondary consumers are **carnivores** and feed on the herbivores. Tertiary consumers are also carnivores as they feed on the secondary consumers. Heterotrophic organisms that feed on both plant and animal material are known as **omnivores**.

Four main types of heterotrophic nutrition are recognised:
- **holozoic** which is characteristic of higher animals
- **saprobiontic/saprotrophic** in which organisms feed on dead organic

remains of other organisms
- **parasitic** where an organism obtains food from another living organism called the **host**
- **mutualism**, a form of **symbiosis**, in which there is a close association between two organisms, each contributing to and benefiting from the relationship.

Holozoic nutrition

Holozoic nutrition is characteristic of higher animals, including humans, and consists of five stages:
- **ingestion** – food is taken into the body through the mouth
- **digestion** – the food is first mechanically broken down by the teeth in the mouth and then chemically broken down by **hydrolysing enzymes** in the **stomach, duodenum** and **ileum**
- **absorption** – the smaller soluble molecules are taken up into the **bloodstream** from the **duodenum** and **ileum**
- **assimilation** – the absorbed products of digestion are incorporated and used by the body
- **egestion** – the undigested parts of the food are eliminated from the body through the **anus** during **defaecation.**

Most of these stages take place in a long, tubular digestive tract, known as the **alimentary canal** or **gut** (see Figure 2.3). In some animals, such as the earthworm, the gut is a simple straight tube which extends from the mouth to the anus, but in others it is coiled and varies in thickness along its length. In mammals, including humans, there are other organs associated with it, such as the liver and the pancreas.

Components of food

In order to maintain good health, the food which is taken in, known as the **diet**, should contain the following:
- energy-providing foods such as **carbohydrates** and **lipids**
- growth-promoting foods such as **proteins**
- **vitamins** and **minerals**
- **water.**

These compounds should be in sufficient quantities and in the correct proportions to give an adequate, balanced diet. A balanced diet for any individual person will depend on their age, sex, body size and occupation. Carbohydrates, lipids and proteins are needed in larger quantities than the other components of the diet, as these compounds supply the energy for metabolism and the materials for growth and maintenance. Only small quantities of vitamins and minerals are needed, and the amount of water will vary according to the activity of the individual and the prevailing climatic conditions.

Carbohydrates

Carbohydrates are the main sources of energy in the diet. The hexose sugar, glucose, is the main respiratory substrate within body cells, but hexose sugars are interconvertible, so no carbohydrates are considered essential in the same way as are some amino acids and fatty acids. Carbohydrate in the human diet comprises sugars, starch and cellulose. Monosaccharide sugars can be absorbed straight away as they do not need to be broken down, but disaccharides and starch have to undergo chemical digestion before they can be absorbed. Cellulose cannot be digested by humans, but it has a function in the diet as it provides bulk and helps to keep the food moving along the alimentary canal. Some carbohydrates and their sources are summarised in Table 2.1.

Table 2.1 *Sources of some carbohydrates in the human diet*

Carbohydrate	Class	Source in the diet
glucose	monosaccharide (hexose)	ripe fruits, honey
fructose	monosaccharide (hexose)	ripe fruits, honey
sucrose	disaccharide	sugar cane, sugar beet, refined sugar
lactose	disaccharide	milk
starch	polysaccharide	cereal grains (wheat flour), potato tubers and other plant storage organs
cellulose	polysaccharide	plant cell walls

Proteins

Proteins are growth-promoting foods, providing a source of amino acids from which new tissues can be built up and cell components replaced. In the body, plasma proteins, enzymes and hormones are used up or become worn out, so need to be replaced, together with structures such as nails, hair and skin cells which are being lost continually. It has been estimated that about 4 per cent of the total body protein needs to be resynthesised each day, so an adult needs to take in about 40 g of protein as part of the daily diet, just for repair and replacement of tissues. This protein intake must include the nine **essential amino acids**, which humans cannot synthesise, otherwise other amino acids cannot be made and growth will be inhibited. Amino acids other than the essential ones can be made by converting one to another in a process known as **transamination**. Protein from animal sources contains all of these essential amino acids, but most plant protein does not. For example, potatoes are deficient in methionine and maize lacks tryptophan, however soya beans contain most of the essential amino acids. Table 2.2 summarises the sources of protein in the human diet.

Table 2.2 *Sources of protein in the human diet*

Type of protein	Source
animal	muscle tissue, liver, milk and milk products, cheese, yoghurt, eggs
plant	cereal grains, pulses and other legumes

Lipids

Fats and oils are energy-providing foods in the human diet. They can be used directly as respiratory substrates in some tissues. On oxidation, they yield twice as much energy as the same mass of carbohydrate. They are essential in the diet as constituents of cell membranes and of the myelin sheath surrounding nerve fibres. Excess dietary fat can be stored in special cells which form **adipose tissue**. This is located beneath the skin and around the body organs, where it acts as a protective cushion and insulates the body, as well as providing an energy store. Fats and oils are **triglycerides**, composed of one molecule of glycerol and three fatty acid molecules. Fats from animal sources are solid at room temperatures and contain a high proportion of **saturated fatty acids**. Oils from plant sources, such as olive oil and groundnut oil, are liquid at room temperatures and have a high proportion of **unsaturated fatty acids.** While most fatty acids can be synthesised by humans, a few, known as **essential fatty acids**, can not, so must be taken in as part of a balanced diet.

Water

Water is the main constituent of blood, lymph and other body fluids. It is needed for chemical reactions to take place and it is used to transport materials around the body. Water is lost from the body each day through excretion, in the faeces, by evaporation from the lungs and skin and through sweating. This water must be replaced either by drinking or by eating foods which contain water. A certain amount of water is gained from metabolic reactions such as respiration and from condensation reactions which are involved in the building up of fats, proteins and carbohydrates. Table 2.3 shows the typical daily gain and loss in an adult.

Table 2.3 *Balance of water gain and loss during a 24 hour period*

Water gain/dm^3 per 24 hours		Water loss/dm^3 per 24 hours	
from drinking	1.2	in urine	1.5
from food	1.0	in faeces	0.1
from metabolic reactions	0.35	evaporation from skin and lungs	0.9
		as sweat	0.05
total	2.55	total	2.55

Vitamins as essential nutrients

Vitamins are organic molecules, needed in small quantities to maintain metabolic functions. They cannot be synthesised by the body, so must be obtained from food and absorbed in the small intestine or, in the case of vitamin K, obtained from the activities of intestinal bacteria. Some vitamins can be synthesised from precursors, known as **provitamins**, which need to be supplied by the diet. An example of such a provitamin is **carotene**, the orange pigment found in carrots, from which **retinol** (vitamin A) is built up. The vitamin requirements of mammals differ from species to species. For example, humans need a supply of **ascorbic acid** (vitamin C) in the diet, but most other mammals are able to synthesise this vitamin.

If a particular vitamin is missing from the diet, or not present in sufficient quantities, then metabolic activity is affected, resulting in a deficiency disease. The symptoms of many **deficiency diseases** were described and linked with restricted diets long before vitamins were discovered. In the days of sailing ships, the symptoms of **scurvy** were common amongst sailors on long sea voyages, due to the lack of fresh vegetables and fruit.

Vitamins may be either **fat-soluble** (A, D and K) or **water-soluble** (B group and C). Fat-soluble vitamins are absorbed in the ileum with fat from the diet. They can be stored in the body, particularly in the liver, and do not have to be taken in every day. Water-soluble vitamins dissolve in the body fluids and any in excess are removed from the body in the urine, so they need to be obtained every day. Table 2.4 summarises the major vitamins needed in the human diet, their sources, metabolic roles and deficiency symptoms.

Table 2.4 *Sources, roles and deficiency symptoms of the major vitamins needed in the human diet*

Vitamin	Major source	Role	Deficiency symptoms
A retinol (fat soluble)	fish liver oil, liver, dairy products; green vegetables contain carotene which is converted to retinol in the body	needed for healthy growth and maintenance of epithelial tissues especially mucus membranes; essential for vision	night blindness
D calciferol (fat soluble)	fish liver oil, butter, eggs; made by action of sunlight on the skin	needed to maintain levels of calcium and phosphorus in the body; enhances calcium absorption in the intestine	leads to rickets in children in which bones fail to calcify; in adults bones may fracture easily
K phylloquinone (fat soluble)	green leafy vegetables; made by action of bacteria in the gut	needed for the normal clotting of the blood; involved in the formation of prothrombin in the liver	delay in blood clotting
B_1 thiamine (water soluble)	meat, wholemeal bread, vegetables	coenzyme in release of energy from carbohydrates	beri-beri, in which nerves degenerate
B_2 riboflavine (water soluble)	most foods, including milk	coenzyme in electron transport; needed for release of energy from food	rarely deficient
B_3 nicotinic acid [niacin] (water soluble)	meat, wholemeal bread, potatoes, yeast extract	needed for the release of energy from food	skin disease and diarrhoea, pellagra
B_{12} cobalt-containing compounds (water soluble)	liver, yeast extract	needed for nucleic acid synthesis in dividing cells	pernicious anaemia
C ascorbic acid (water soluble)	potatoes, green vegetables, fruits (particularly citrus fruits)	needed for the maintenance of healthy connective tissues	bleeding gums, failure of wounds to heal; scurvy

Mineral ions

Mineral ions, like vitamins, are required in small quantities and must be obtained from the diet. They are needed because they:

- are components of body tissues, for example **calcium** and **phosphate** in bone
- help to maintain the correct pH of the body fluids, for example **sodium**, **potassium** and **chloride**
- are cofactors for some enzymes, for example **magnesium**
- are enzyme activators, for example **calcium** and **iron**.

Ions such as calcium and phosphate, termed **macronutrients**, are needed in larger quantities than iron, which is referred to as a **trace element**, or **micronutrient**. Mineral ions also resemble vitamins in that their absence from the diet results in deficiency diseases. Table 2.5 summarises the sources, roles and deficiency symptoms of major mineral ions needed in the human diet.

Table 2.5 *Major mineral ions needed in the human diet*

Mineral ion	Source	Role	Deficiency
calcium (Ca^{2+})	cheese, bread, milk	needed for formation of bones and teeth, muscle contraction, nerve function, bloodclotting	poor skeletal growth; soft bones (rickets); muscular spasm
phosphorus (PO_4^{3-})	all foods have some	needed for formation of bones and teeth; component of nucleic acids and ATP	not often deficient
potassium (K^+)	vegetables, fruit	acts with sodium in the transmission of nerve impulses; anion–cation balance in cells	not often deficient
sodium (Na^+)	all foods have some	acts with potassium in the transmission of nerve impulses; anion–cation balance in cells	muscular cramp
chlorine (Cl^-)	all foods have some	solute concentration and anion–cation balance in cells; activity of nerve cells	muscular cramp
iron (Fe^{2+}, Fe^{3+})	meat, liver, red wine	constituent of haemoglobin; enzyme activator	anaemia
iodine (trace)	sea foods	constituent of thyroxine	cretinism (stunted growth and mental retardation) in children; simple goitre in adults
fluorine (trace)	drinking water	contributes to hardness of bones and teeth	dental caries especially in children

Digestion

Digestion is the process in which complex food molecules are broken down into simpler molecules. It can be divided into
- **mechanical** digestion
- **chemical** digestion.

In mechanical digestion, the large particles of ingested food may be sliced, crushed or otherwise broken up by the teeth. This reduces the food to smaller lumps which are easier to swallow and have a larger surface area, so that subsequent enzyme action during chemical digestion is more effective.

Chemical digestion involves the action of **hydrolases** (hydrolytic enzymes) on the food constituents. The chemical breakdown of the complex organic molecules takes place progressively in stages until simple, smaller, soluble molecules are formed, which can then be absorbed. There are three major groups of digestive enzymes involved:
- **carbohydrases** hydrolyse the glycosidic bonds in carbohydrates
- **proteases** hydrolyse the peptide bonds in proteins and polypeptides
- **lipases** hydrolyse the ester bonds in triglycerides.

The processes which take place are the reverse of the condensation reactions which occur when carbohydrates, proteins and lipids are formed.

Carbohydrases

Some monosaccharide sugars, such as glucose and fructose, may be ingested, but the bulk of carbohydrate in the human diet is in the form of polysaccharides and as such requires digestion. Starch from plants and glycogen from animal sources are split into smaller units, usually disaccharides, by **amylases**. The disaccharides are then hydrolysed to monosaccharides which are small enough to be absorbed (Figure 2.2).

Carbohydrate digestion occurs in the mouth, duodenum and ileum under alkaline conditions. Carbohydrate digestion in humans is summarised in Table 2.6.

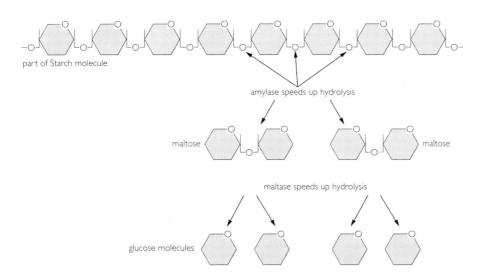

part of Starch molecule

amylase speeds up hydrolysis

maltose   maltose

maltase speeds up hydrolysis

glucose molecules

Figure 2.2 Hydrolysis of starch to maltose and glucose catalysed by amylase and maltase

HETEROTROPHIC NUTRITION

Table 2.6 *Carbohydrate digestion in humans*

Location in alimentary canal	Optimum pH	Enzymes involved	Products of digestion
mouth and buccal cavity	7.0	salivary amylase	dextrins (short chains of glucose residues), some maltose; food does not remain here long enough for much digestion
duodenum	7.0	pancreatic amylase from pancreatic juice	starch broken down to maltose
ileum (bound to membranes of microvilli on the epithelial mucosa)	8.5	maltase lactase sucrase	maltose to glucose; lactose to glucose and galactose; sucrose to glucose and fructose

In the mouth, saliva is produced from three pairs of salivary glands (humans normally secrete about 1.5 dm^3 each day). Saliva consists mostly of water, together with enzymes, mineral ions and mucus. The constituents of saliva and their roles are summarised in Table 2.7.

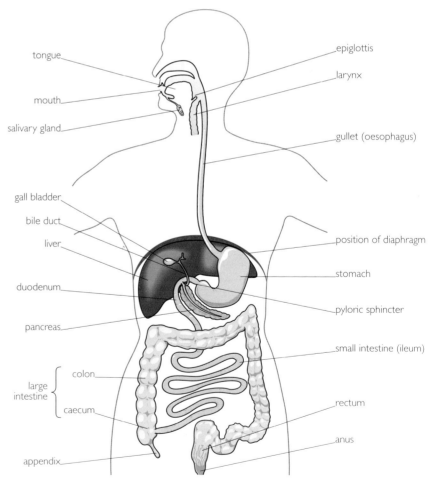

Figure 2.3 *Human alimentary canal and associated organs*

Table 2.7 *Constituents of saliva and their roles*

Constituent	Role
salivary amylase	initiates the digestion of cooked starch to maltose
lysozyme	catalyses the breakdown of cell walls of some pathogenic bacteria
mucus	moistens and lubricates the food; helps in bolus formation
chloride ions	activate salivary amylase
other mineral ions (phosphates, hydrogen carbonate)	help to maintain the correct pH (about pH 7)

Very little starch digestion takes place in the mouth, because the food does not remain there long enough. When a bolus of food is swallowed, it is transported down the oesophagus to the stomach, where the extremely acid conditions inhibit the action of salivary amylase. No carbohydrases are present in the gastric juice from the gastric glands in the mucosa of the stomach wall.

When the food reaches the duodenum, pancreatic juice containing amylase enters through the pancreatic duct. In addition to proteases and lipases, pancreatic juice also contains alkaline salts which help to neutralise the acid from the stomach. Bile from the liver is also added via the bile duct. It contains hydrogencarbonate ions which contribute to the creation of alkaline conditions in which the enzymes from the pancreas and from the intestinal juice work most effectively. Pancreatic amylase hydrolyses any remaining starch to maltose.

The enzymes involved in the hydrolysis of disaccharides, such as maltose, lactose and sucrose, are located on the membranes of the microvilli of the epithelial mucosa. When hydrolysis is completed, the monosaccharides can then be absorbed.

Proteases

There are two groups of proteases:
- **endopeptidases** speed up the hydrolysis of peptide bonds within the protein molecules
- **exopeptidases** act on terminal peptide bonds (those at the ends of the polypeptide chains).

Hydrolysis involving endopeptidases results in proteins being broken down into short polypeptide chains. Pepsin, trypsin and chymotrypsin are endopeptidases, each only capable of hydrolysing specific peptide bonds. For example, trypsin catalyses the hydrolysis of peptide bonds which involve the amino acids lysine or arginine. Pepsin, trypsin and chymotrypsin are secreted into the alimentary canal in their inactive forms: pepsin as pepsinogen, trypsin as trypsinogen and chymotrypsin as chymotrypsinogen. This ensures that these enzymes are only activated when there is food requiring digestion in the alimentary canal and prevents the enzymes damaging the cells in which they are produced.

Pepsinogen is converted to pepsin by the action of hydrochloric acid in the stomach. Once some pepsin has been formed, it will bring about the conversion of more pepsinogen to pepsin. Trypsinogen is converted to trypsin by the action of the enzyme **enterokinase**, which is secreted in the ileum, and chymotrypsinogen is activated by trypsin.

It is also worth noting that some young mammals produce the enzyme **rennin**, secreted as pro-rennin, in the stomach. It is activated by hydrochloric acid and its function is to coagulate the soluble milk protein, caseinogen, to the insoluble calcium salt of casein, which is then hydrolysed by pepsin.

The action of exopeptidases results in the breakdown of the short polypeptide chains by the removal of amino acids. There are two kinds of exopeptidases:
- **aminopeptidases** hydrolyse peptide bonds at the amino end of a polypeptide chain
- **carboxypeptidases** hydrolyse peptide bonds at the carboxyl end of a polypeptide chain.

Protein digestion begins in the stomach, where the optimum pH for hydrolysis by pepsin is 1.5 to 2.0. Here the proteins and long polypeptide chains are broken down to shorter polypeptides. In the duodenum and ileum, where the pH is alkaline, trypsin and chymotrypsin from the pancreatic juice hydrolyse the proteins to shorter polypeptides. Carboxypeptidases are present in the pancreatic juice and their action results in the production of amino acids. Aminopeptidases are present on the microvilli of the epithelial mucosa.

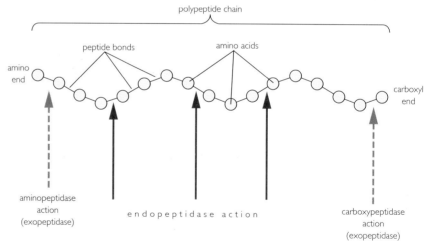

Figure 2.4 Digestion of protein involving endo- and exopeptidases

Lipases
Most of the lipids in a human diet are triglycerides and, before they can be digested, they need to undergo emulsification into droplets. Bile salts from the liver, mostly **sodium glycocholate** and **sodium taurocholate**, are released into the duodenum via the bile duct from the gall bladder, where they have been stored. These salts lower the surface tension between the oil globules and water, bringing about emulsification and providing a larger surface area for the action of lipases.

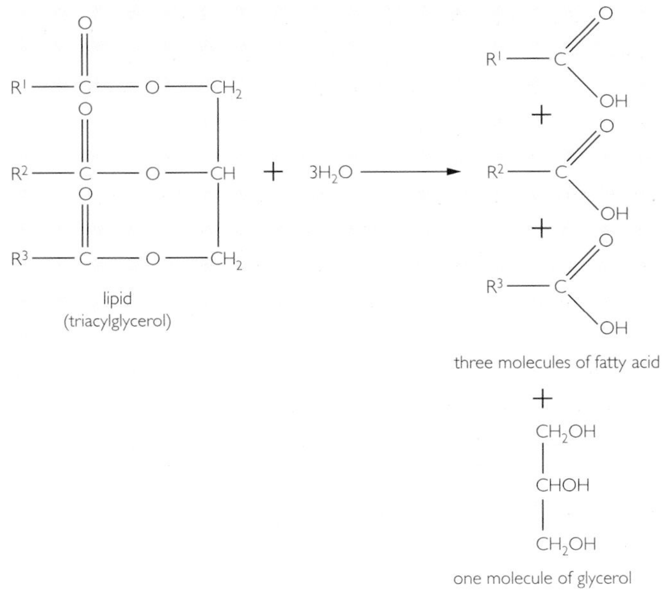

Figure 2.5 Hydrolysis of lipids catalysed by lipase

Lipases are secreted by the pancreas and released into the duodenum in the pancreatic juice. They catalyse the hydrolysis of triglycerides into monoglycerides, fatty acids and glycerol.

Structure of the alimentary canal in relation to digestion and absorption

The alimentary canal, or gut, has a basic common structure along its length, although it is specialised in certain regions to carry out its various roles. It extends as a tube from the **mouth** to the **anus** and along its length the wall is composed of four layers. These are:

- the **mucosa**, the innermost layer surrounding the **lumen**, made up of glandular epithelium and connective tissue containing blood vessels and lymph vessels
- a layer of connective tissue, the **submucosa**, containing nerves, blood vessels and lymph vessels, together with elastic fibres and collagen
- the **muscularis externa**, composed of circular and longitudinal layers of smooth muscle fibres
- the outermost layer, the **serosa**, made up of loose connective tissue.

Along the whole length of the gut, the glandular epithelium of the mucosa contains **goblet cells** which secrete **mucus**. The mucus lubricates the passage of food along the gut and also protects it from the digestive action of enzymes. In the mucosa of the stomach, there are simple, tubular **gastric glands** which secrete **gastric juice**. In the mucosa of the duodenum and ileum, the **intestinal glands** in the **crypts of Lieberkuhn** secrete **intestinal juice** which contains mucus and enterokinase.

The mucosa is separated from the submucosa by a thin layer of smooth muscle, the **muscularis mucosa**. In the submucosa of the duodenum, **Brunner's glands** produce an alkaline solution containing mucus, but no enzymes.

External to the serosa is the **peritoneum**, which is a double-layered membrane surrounding the gut and lining the abdominal cavity. The **mesenteries**, which are extensions of the peritoneum, hold the gut canal in place by anchoring it to the abdominal wall. The surfaces of the cells of the peritoneum are moist, preventing frictional damage as the gut moves against other organs in the abdominal cavity.

At certain regions along the gut are ducts from external glands, such as the salivary glands, the liver and the pancreas. Figure 2.6 gives a general plan of the gut wall with detail of specialised areas.

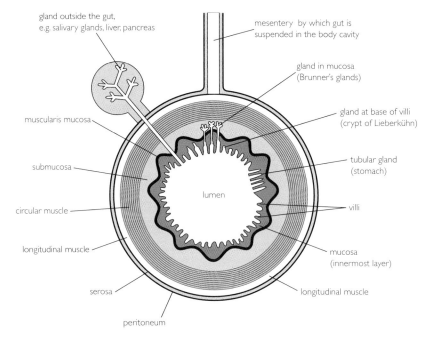

Figure 2.6 General structure of the gut wall with detail of specialised regions

Mastication and movement of food along the alimentary canal

Mastication

Mechanical digestion of ingested food is referred to as **mastication** and is achieved by means of teeth in the mouth, or buccal cavity. Teeth are present in the upper and lower jaws and in an adult human there are 32 permanent teeth. These consist, in each jaw, of: four **incisors**, two **canines**, four **premolars** and six **molars**, arranged as shown in Figure 2.7.

The exposed part of each tooth, the **crown**, projects above the gum and is covered with **enamel**, which is very hard and resistant to decay. The **root** is embedded in the jawbone and is held in place by **periodontal fibres**, connected to the jawbone at one end and to the **cement** surrounding the outside of the root at the other. The **neck** is the part which is surrounded by the gum, but not embedded in the jawbone.

Most of the tooth consists of **dentine**, a hard, bone-like substance, which has many tiny channels extending through it. These channels, known as **canaliculi**, contain strands of cytoplasm from the dentine-producing cells in

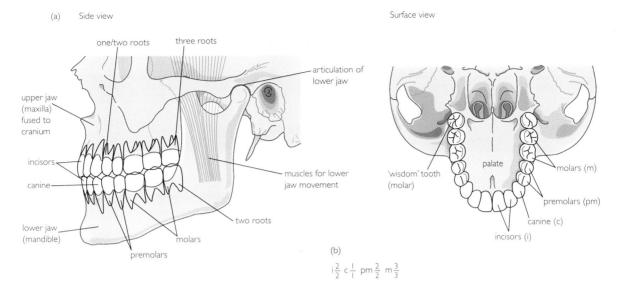

(a)   Side view

one/two roots   three roots

articulation of
lower jaw

upper jaw
(maxilla)
fused to
cranium

incisors

canine

muscles for lower
jaw movement

lower jaw
(mandible)

two roots

molars

premolars

Surface view

'wisdom' tooth
(molar)

palate

molars (m)

premolars (pm)

canine (c)

incisors (i)

(b)

$i\frac{2}{2}\ c\frac{1}{1}\ pm\frac{2}{2}\ m\frac{3}{3}$

Figure 2.7 (a) Arrangement of adult human teeth; (b) Dental formula of permanent dentition

the pulp cavity, located in the central part of the tooth. In addition to the dentine-producing cells, the pulp cavity contains nerve endings and blood vessels (Figure 2.8).

The incisors have flattened crowns with sharp, chisel-like edges, which are used to bite off lumps of food. The canines have conical, pointed crowns and assist in the biting process. In carnivorous animals, the canines are much bigger and more pointed and are used to pierce and hold live prey. Both the incisors and the canines have single roots. The cheek teeth, premolars and molars, have rounded projections, called **cusps**, on their crowns. These assist in crushing and grinding food during the process of chewing. Typically the cheek teeth each have more than one root, the molars on the upper jaw possessing three or four, so giving greater anchorage in the jawbone.

Crushing and grinding of the food not only breaks up larger lumps into smaller pieces, but also mixes it with saliva, which moistens and lubricates it. The tongue and cheek muscles also help to form the moistened food into a mass called a **bolus**, which is manipulated to the back of the buccal cavity before being swallowed.

Movement of food along the alimentary canal

Associated with the muscularis externa are networks of nerves, one of which controls the contraction and relaxation of the muscles which bring about the movement of food along the alimentary canal. This movement, known as **peristalsis**, is initiated by the act of swallowing (Figure 2.9). The bolus stretches the wall and the circular muscle behind the bolus contracts, pushing the food forwards. The circular muscle in the region surrounding the bolus is relaxed, increasing the diameter of the lumen and enabling the food to be pushed forward. Peristaltic movements occur all the way along the gut. In the small intestine, the alternate contraction and relaxation of the muscles brings about **segmentation movements** which help to bring the contents of the lumen of the gut in contact with the epithelium of the wall where absorption occurs.

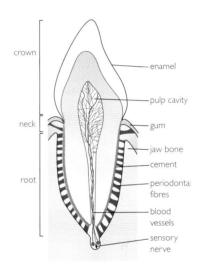

crown

enamel

pulp cavity

neck

gum

jaw bone

cement

root

periodontal
fibres

blood
vessels

sensory
nerve

Figure 2.8 LS human canine tooth

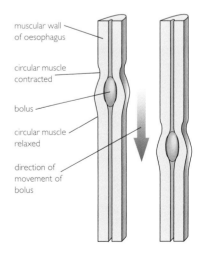

muscular wall
of oesophagus

circular muscle
contracted

bolus

circular muscle
relaxed

direction of
movement of
bolus

Figure 2.9 Diagram to illustrate peristalsis

At certain points along the alimentary canal, the circular muscle is thicker and forms rings of muscle called **sphincters**, which control the passage of food from one region to another. The **cardiac sphincter**, found between the oesophagus and the stomach, controls the entry of food into the stomach. At the other end of the stomach, the **pyloric sphincter** relaxes to allow the passage of food from the stomach into the duodenum when it has reached the right consistency. There are additional sphincters at the junction of the ileum and the caecum and at the anus.

Absorption

Most absorption occurs in the small and large intestines, although it has been shown that water and alcohol can be absorbed from the stomach. The digested food is absorbed in the small intestine and water is mostly taken up from the large intestine.

The first 20 cm of the small intestine is known as the **duodenum** and it is here that the secretions from the liver and the pancreas are added. The rest of the small intestine is called the **ileum** and is about 5 m long. As has already been described, the final stages of digestion of carbohydrates, lipids and proteins take place in the duodenum and ileum, whilst at the same time the process of absorption is occurring. There are several features of this region of the gut which contribute to the efficiency of absorption.

- It is long, providing a large surface area over which absorption can occur.
- There are large numbers of finger-like projections, the **villi**, in the mucosa, increasing the surface area for absorption.
- The villi possess smooth muscle fibres, which contract and relax, mixing up the contents and bringing the columnar epithelial cells of the absorptive surface into greater contact with the digested food.
- The columnar epithelial cells possess **microvilli**, which further increase the surface area available for absorption.
- Each villus has an extensive capillary network so that the absorbed food is transported away quickly, maintaining concentration gradients.
- There is a lacteal in each villus into which absorbed fats pass.

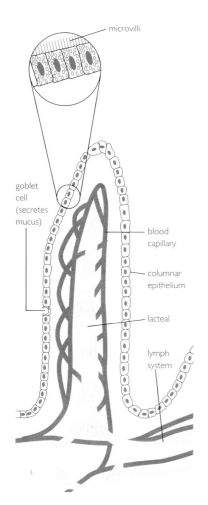

microvilli

goblet cell (secretes mucus)

blood capillary

columnar epithelium

lacteal

lymph system

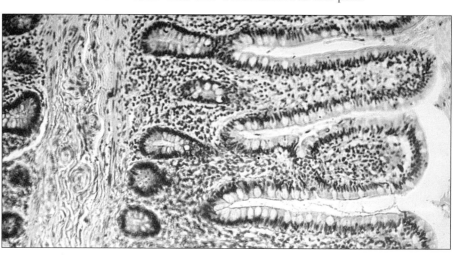

Figure 2.10 (a) Diagram to illustrate VS through a villus; (b) Photomicrograph of a section through the ileum showing villi

Carbohydrates, in the form of monosaccharides, and amino acids are absorbed partly by diffusion and partly by active transport. Soon after a meal, there will be a higher concentration of monosaccharides and amino acids in the ileum, so a concentration gradient will exist and diffusion of these molecules will occur across the mucosal epithelium into the blood capillaries. This process is rather slow and will not take up all the digested food, so it is supplemented by active transport involving a sodium–potassium pump (Figure 2.11).

In the membrane of the epithelial cells is a **glucose transporter protein**, which has binding sites for both glucose molecules and sodium ions. The sodium–potassium pump actively transports sodium ions out of the cells against the electrochemical gradient. Glucose molecules and sodium ions bind to the transporter proteins and the sodium ions diffuse into the cells along their electrochemical gradient, carrying the glucose molecules with them. Inside the cells, the glucose molecules and sodium ions dissociate from the transporter protein, the glucose concentration of the cell increases and glucose moves into the blood by facilitated diffusion. Similar mechanisms exist for the active uptake of dipeptides and amino acids.

The absorption of the products of fat digestion, namely fatty acids and glycerol, involves a different mechanism. If the fatty acids are short, they can diffuse directly into the blood from the epithelial cells. The longer chain fatty acids, monoglycerides and glycerol diffuse into the epithelial cells of the mucosa, where they recombine to form fats. The fatty acids and monoglycerides have polar heads and non-polar tails, so they tend to clump together to form spherical structures, called **micelles**, in the lumen of the gut. Within a micelle, the non-polar tails point towards the middle and the polar heads are on the outside. Micelles can diffuse into the mucosal epithelial cells, but do so more slowly than single fatty acids. The molecules of fat diffuse into the lacteals, where they become coated by proteins present in the lymph to form droplets of lipoprotein called **chylomicrons**. These droplets are passed into the bloodstream from the lymph, where they are hydrolysed back into fatty acids and glycerol by an enzyme present in the blood plasma. They are then transported in the blood, from where they may be taken up and used as respiratory substrates by cells or stored as fat.

Mineral ions, vitamins and water are also absorbed from the contents of the duodenum and the ileum. Any remaining mineral ions and large amounts of water are absorbed from the food residues as they pass through the large intestine. Present also in this region are large numbers of bacteria which synthesise amino acids and vitamins, some of which are of use to humans and can be taken up and absorbed into the bloodstream.

Histology of the ileum wall

The structure of the ileum wall is similar to the general structure of the alimentary canal shown in Figure 2.6. More detail, including the structure of a villus, is shown in Figure 2.10. In the ileum region, the surface layer of the mucosa consists of columnar epithelial cells with large numbers of mucus-secreting goblet cells and the submucosa contains large amounts of lymphoid tissue.

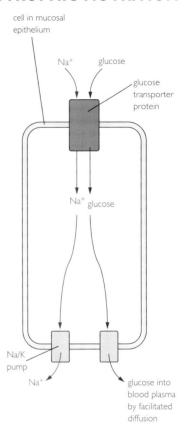

Figure 2.11 Diagram to illustrate glucose transporter protein and sodium–potassium pump in operation is a mucosal epithelial cell

It is relevant to note that the constant passage of food through the ileum damages the tips of the villi and cells are lost from the mucosa. These cells are replaced by cells from the bottom of the crypts of Lieberkuhn situated at the base of the villi. These cells migrate to the surface and gradually move up the villus, eventually being shed from the tip. New cells are produced by mitotic divisions of the crypt cells. It has been estimated that between 50 and 200 g of the intestinal mucosa is renewed every day in an adult human. It takes from 5 to 7 days for a cell from the bottom of the crypt to move up to the tip of a villus.

Hormonal and nervous control

The secretion of digestive enzymes is not a continuous process. It is controlled so that it only occurs in a particular region of the gut when there is food there to be digested. This control is achieved partly by the nervous system and partly by the secretion of hormones. In the initial stages of digestion, the production of saliva, containing salivary amylase, is a reflex action controlled by the nervous system; in a later stage the release of pancreatic juice into the duodenum is controlled by hormones.

Nervous control of the production of saliva in response to food

Small quantities of saliva are secreted continuously in order to keep the mouth moist. Increased saliva production is stimulated by the presence of food in the mouth and also in response to the sight or smell of food. The production of saliva is an involuntary action, referred to as a **reflex action**, over which we have no control. It involves a part of the **autonomic nervous system** known as the **parasympathetic system** and the responses are brought about by stimulation of the **vagus nerve**.

Reflex actions enable rapid, automatic responses to stimuli to be made by muscles or glands, without requiring any thought (the response does not have to be initiated by the brain). The stimulus is detected by a sensory cell or organ, known as the **receptor**. Nerve impulses are generated and transmitted into the central nervous system along **sensory** (receptor) neurones. The impulses are then passed, via **synapses**, to **relay** (intermediate) neurones, which pass the impulses to **effector** (motor) neurones. The effector neurones transmit the impulses to a muscle or a gland, known as the **effector**, which responds by contracting (muscle) or secreting a hormone (gland). A simple reflex arc is illustrated in Figure 2.12.

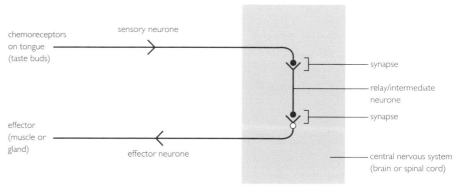

Figure 2.12 Representation of a simple reflex arc

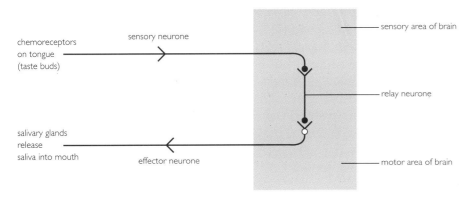

Figure 2.13   Cranial reflex of saliva production

Increased saliva production stimulated by the presence of food in the mouth is an example of an **unconditional cranial reflex** (Figure 2.13), where the stimulus and the response are related. The **taste buds** in the tongue, which are **chemoreceptors**, are stimulated and generate nerve impulses. These are transmitted via sensory neurones to the sensory area of the brain, where a relay neurone transmits them to the motor area. The relay neurone synapses with an effector neurone, which transmits the impulses to the salivary glands. The salivary glands respond by releasing more saliva into the mouth.

The reflex action which involves saliva production at the sight or smell of food is a **conditioned reflex**. In this case, a different, unrelated stimulus produces a response. **Photoreceptors** in the **eye** (**rods** and **cones**), are stimulated by the sight of food. Impulses are transmitted along sensory neurones to the visual cortex in the brain. Impulses from the visual cortex are first passed to an **association centre** and then to the motor area, from which they are transmitted to the salivary glands. Similarly, with the smell of food, sensory **olfactory epithelial cells** in the lining of the **nasal cavity** are stimulated and impulses are transmitted to the **olfactory centre**, before being passed to the association centre. Both these pathways are summarised in Figure 2.14.

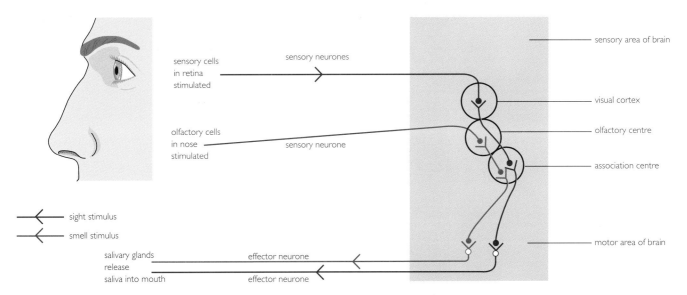

Figure 2.14 Conditioned reflexes involving the eyes and nose

Principles of control by hormones

Hormones are chemical substances secreted into the bloodstream by glands of the **endocrine system**. Hormones circulate in the blood but usually only exert a particular effect on a specific organ, tissue or group of cells. The release of a hormone from an endocrine gland is triggered by the presence of a specific substance or another hormone in the blood, or as a result of stimulation by the **autonomic nervous system**. The amount of hormone released and the duration of its secretion may be regulated by **feedback mechanisms**. The regulation of the blood glucose level by **insulin** and **glucagon** illustrates the principles of hormonal control.

Both insulin and glucagon are secreted from groups of cells in the **pancreas**, called the **islets of Langerhans** (Figure 2.15). Most of the pancreas is composed of **exocrine** tissue, secreting digestive enzymes in the **pancreatic juice**, but interspersed amongst this tissue are clusters of cells with an **endocrine** function. Each group contains a large number of β (**beta**) **cells** which synthesise and secrete **insulin**, a few α (**alpha**) **cells** which synthesise and produce **glucagon**, together with blood capillaries and other secretory cells.

An increase in the level of glucose in the blood flowing through the pancreas stimulates the release of insulin. Insulin receptors have been shown to be present on the cells of the liver, muscle and adipose tissues. Insulin binds to these receptors and has the following effects:
- it changes the permeability of the cell membranes to glucose so that the diffusion of glucose is accelerated
- it lowers the level of glucose in the blood by stimulating **glycogenesis** (the building up of glycogen from glucose) in liver and muscle tissue
- it inhibits **gluconeogenesis** (the breakdown of fats and proteins to carbohydrates) in the liver, muscle and adipose cells
- it promotes the uptake of amino acids into cells and hence protein synthesis.

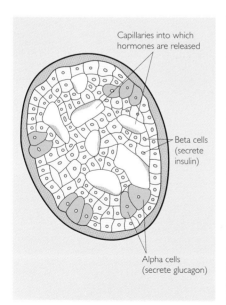

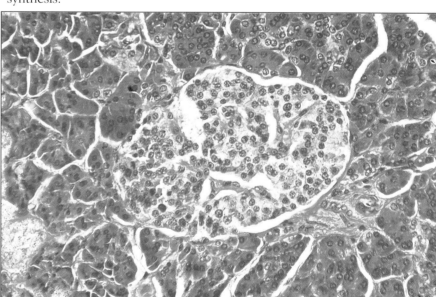

Figure 2.15 (left) Diagram to illustrate structure of islets of Langerhans
(right) Photomicrograph of islets of Langerhans in the pancreas;

During glycogenesis, glucose is first phosphorylated and glycogen is then formed by a condensation process. The effect of insulin is to activate the phosphorylase and **glycogen synthetase** in the liver and muscle cells. Overall, insulin promotes the building up of proteins, lipids and glycogen.

It is interesting to note that glucose is not the only stimulus for insulin release. The presence of food in the stomach stimulates the **vagus nerve**, which results in insulin release before the blood glucose level increases after a meal.

Insulin release inhibits the release of glucagon, but if there is no insulin in the blood then glucagon is released. It is also released in conditions of starvation, exercise and stress. The effects of glucagon are directly opposite to those of insulin and lead to an increase in the level of glucose in the blood. The major site of the action of glucagon is in the liver cells where it promotes the breakdown of glycogen to glucose by activating the enzyme **glycogen phosphorylase** and blocking the synthesis of glycogen by deactivating glycogen synthetase. If there is no insulin in the blood, lipids and proteins break down to give fatty acids, glycerol and amino acids. Glucagon stimulates the uptake of amino acids and glycerol by the liver cells, where they are converted to glucose by gluconeogenesis.

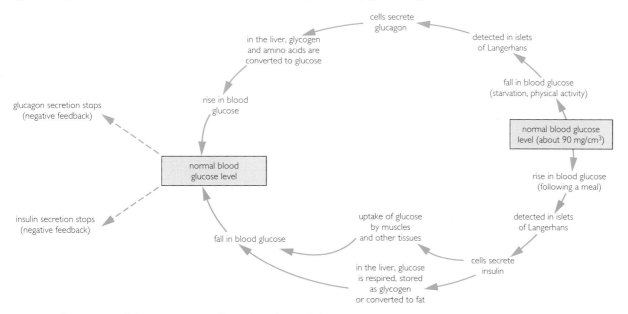

Figure 2.16 Summary of the antagonistic effects of insulin and glucagon

Saprobiontic nutrition

So far in this chapter we have been considering in detail the type of heterotrophic nutrition shown by most animals, namely **holozoic** nutrition. **Saprobiontic**, or **saprotrophic** organisms obtain their nutrition from dead or decaying organic remains of plants and animals. They are **primary consumers** in **detritus food chains** and are often referred to as **decomposers**. Saprobionts secrete enzymes on to the organic matter and absorb the soluble products of this **extracellular** digestion. Any substances released by this digestion which are not taken up by the saprobionts, are made available for uptake by plants, so contributing to the circulation of nutrients. Many bacteria and fungi are saprobiontic.

Rhizopus – *a saprobiontic organism*

Rhizopus stolonifera is a member of the Zygomycota, which are fungi lacking cross-walls in their hyphae. *Rhizopus* is commonly found growing on damp, wholemeal bread, where it can be seen as cotton-like white threads on the surface. After a few days, masses of tiny black sporangia appear, resembling pin heads and giving the fungus its common name 'pin mould'. Closer observation, using a microscope, reveals that the fungus consists of a **mycelium** made up of much-branched **hyphae**, which are **aseptate** (lack cross-walls). Aerial hyphae, called **stolons**, spread over the food substrate and produce tufts of branched hyphae or **rhizoids**, where they touch down. The rhizoids penetrate the substrate, secreting enzymes which digest the complex substances in the food. The soluble products of this digestion are absorbed by the rhizoids and either used in metabolic activities or stored. Three main groups of enzymes are secreted as in other types of heterotrophic organisms: carbohydrases, proteases and lipases.

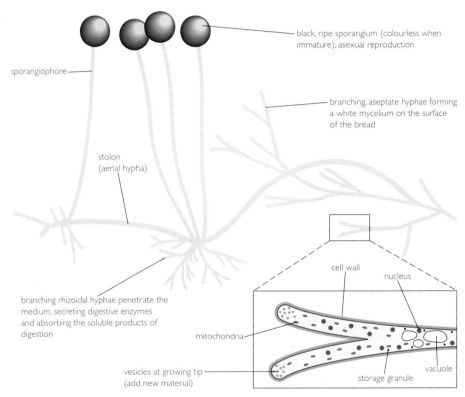

black, ripe sporangium (colourless when immature); asexual reproduction

sporangiophore

branching, aseptate hyphae forming a white mycelium on the surface of the bread

stolon (aerial hypha)

cell wall

nucleus

mitochondria

branching rhizoidal hyphae penetrate the medium, secreting digestive enzymes and absorbing the soluble products of digestion

vacuole

storage granule

vesicles at growing tip (add new material)

Figure 2.17 (a) Rhizopus *on bread, showing sporangia; (b) Diagram of* Rhizopus *showing mycelial structure*

Parasitic nutrition

A **parasite** is an organism which lives in close association with another living organism, the **host**. In this relationship, the parasite is dependent on the host for its food and usually causes the host some degree of harm. **Ectoparasites** live on the outside of their hosts while **endoparasites** live inside their host. Most endoparasites, such as gut parasites, spend their entire lives within their host, but some ectoparasites, such as the tsetse fly, only become attached during feeding. Most parasites are highly adapted to their particular mode of life.

Taenia – *a parasitic organism*

Taenia solium, the pork **tapeworm**, is an endoparasite with two hosts. Its primary host is man, where the adult stage is found attached to the wall of the small intestine of an infected individual. The secondary host is the pig, in which the larval stage develops. The adult stage consists of a flattened, ribbon-like body, made up of a large number of 'segments' called **proglottides** (sing. **proglottis**). It must be emphasised that the proglottides are not true segments, such as those found in earthworms or in the Chordates. The tapeworm can be up to 3.5 m long and is approximately 6 mm wide and 1.5 mm thick. At its anterior end, it has a tiny knob known as the **scolex** which has a double row of hooks and four suckers. Just behind the scolex is a region called the **proliferation zone** where new proglottides form. The rest of the organism consists of proglottides containing both male and female reproductive structures. At the posterior end, after self-fertilisation has occurred, these proglottides contain a greatly enlarged uterus full of fertilised eggs, the rest of the structures having been reabsorbed. These proglottides become detached from the organism and pass out with the faeces of the host. If ingested by the secondary host, a pig, the life cycle proceeds to the next stage, giving rise to a larval stage in muscular tissue.

Taenia has no mouth or alimentary canal, as it absorbs the digested food of its host all over its body surface. Simple, soluble products of digestion of the host's food will be present in the small intestine, so the tapeworm can absorb the food it requires and has no need of a digestive system of its own, nor does it need to secrete digestive enzymes. It has a thick outer covering, known as the **tegument**, consisting of protein and chitin, which protects it from the digestive enzymes of its host. Other adaptations to its mode of life are:

- its ability to live in the low oxygen concentrations present in the human gut
- a reduction in the nervous system and a lack of sense organs
- the possession of suckers and hooks for attachment to the gut wall of the host
- the production of very large numbers of offspring.

If the human host is healthy, little damage is caused by the tapeworm apart from depriving the host of some of its food. The long thin shape of the tapeworm does not block the small intestine. In children and adults already debilitated by other diseases, it can cause problems. Individuals may become less resistant to other diseases and suffer from abdominal pain, vomiting, constipation and loss of appetite.

Mutualism

The term **mutualism** describes a close association between two organisms in which both contribute and where both benefit. This type of association can be illustrated by reference to the relationship between the bacterium ***Rhizobium*** and members of the flowering plant family ***Papilionaceae***.

Rhizobium is a nitrogen-fixing bacterium. In this process, hydrogen ions from carbohydrates, such as glucose, are combined with nitrogen to form ammonia. The ammonia then combines with glutamate to form the amino acid glutamine,

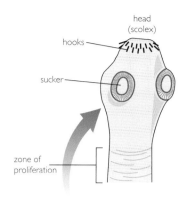

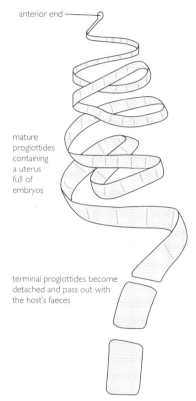

Figure 2.18 Taenia, *adult stage with detail of anterior region*

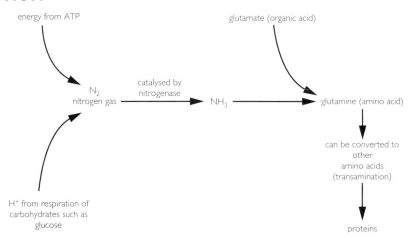

Figure 2.19 Stages in nitrogen fixation

from which other amino acids can be synthesised. The fixation reaction takes place in anaerobic conditions in the cytoplasm of the bacterial cells, catalysed by the enzyme **nitrogenase**.

Rhizobium bacteria are present in the soil. Those near the roots of leguminous plants, such as clover, peas or beans, are attracted towards the roots and are able to penetrate the root hair cells. This attraction has been found to occur in response to a hormone secreted by the roots of the plant. The bacteria also secrete substances which cause the root hair cells to bend around, possibly encouraging penetration. Once inside the root, the bacteria move to the cortex and stimulate the production of auxins and cytokinins by the root tissue. This causes cell division to occur and a nodule of tissue is formed in the cortex, composed of cells containing large numbers of the bacteria. The bacteria become Y-shaped and have a banded appearance. These forms of the bacteria are called **bacteroids**. The bacteroids are able to fix nitrogen in the nodules, where the anaerobic environment is suitable for the efficient functioning of their nitrogenase enzymes. Oxygen molecules are absorbed by a special pigment, **leghaemoglobin**, which surrounds the bacteroids and gives a pinkish colour to the tissues of the nodule. Figure 5.24 shows root nodules in the bean, *Phaseolus multiflorus* and a diagram of a TS through a nodule can be seen in Figure 2.20.

The bacteria benefit from this association by obtaining their supplies of carbohydrate from the photosynthetic activities of the plant. The plant benefits by receiving a supply of ammonia from the bacteria. This enables the leguminous plants to grow in nitrogen-deficient soils.

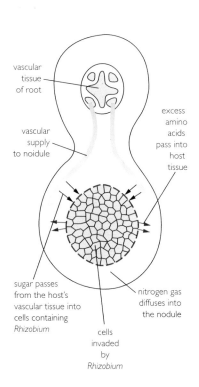

Figure 2.20 TS through a root nodule

Regulation of the internal environment

Regulators and non-regulators

Many organisms live in a relatively unchanging environment, such as sea water, where factors, including the concentration of dissolved salts, pH, temperature and dissolved oxygen content, do not vary widely. These organisms are therefore protected against changes and are not normally faced with the problems of coping with extremes of temperature, or of desiccation. All organisms have evolved from aquatic ancestors and the transition from an aquatic to a terrestrial mode of life requires a number of adaptations if the organism is to survive. For example, air is very much less dense than water so organisms which may have been neutrally buoyant in water need to be able to support their body weight to prevent it collapsing on land. Organisms are also faced with the problem of avoiding desiccation if they are to survive on land.

The ability of an organism to control its internal environment is referred to as **homeostasis**. The concept of homeostasis was first described by Claude Bernard, a French physiologist, in 1857. Bernard realised that it is the ability of mammals, in particular, to regulate their internal environment (that is, the composition of their body fluids) which makes it possible for them to survive in widely fluctuating external conditions. Factors such as body temperature, body water content and blood glucose concentration are controlled by physiological mechanisms so that they stay precisely within narrow limits.

Algae and cnidarians (such as jelly fish) have little ability to control their internal environments and as a result are restricted in their distribution. Marine cnidarians are known as **osmoconformers** because the osmotic concentration of their body fluids is the same as that of the surrounding sea water. This does not mean, however, that their body fluids have exactly the same composition as sea water. They are able to maintain concentrations of ions which are different from their concentrations in sea water, so they do have some ability to regulate their internal environment.

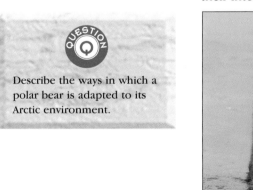

Describe the ways in which a polar bear is adapted to its Arctic environment.

Figure 3.1 Polar bears are adapted to survive in the harsh conditions of the Arctic

REGULATION OF THE INTERNAL ENVIRONMENT

If we contrast the distribution of algae and cnidarians with the distribution of flowering plants and mammals, it is clear that flowering plants and mammals are widely distributed in both aquatic and terrestrial habitats, from the tropics to the Arctic and Antarctic regions. Their ability to survive in harsh conditions is entirely due to the regulation of their internal environments.

Osmoregulation

Organisms are faced with the problem of maintaining their body fluids at the correct concentrations when these are almost always different from their environment. The ability of organisms to control their body water volume and solute concentrations is referred to as **osmoregulation**. The uptake or loss of water by cells occurs by the process of **osmosis**. This is the movement of water between two solutions when the solutions are separated by a partially permeable membrane. Consider what happens if red blood cells, which are unable to osmoregulate, are placed into very dilute salt solution (for example, 0.05%). Water will enter the cells, which will swell up and burst, or **lyse**. Conversely, if the cells are placed into a more concentrated salt solution (such as 1.0%), they will lose water, shrink, and become **crenated** (Figure 3.2).

Osmoregulation in protozoa

Protozoa (Kingdom Protoctista) are unicellular organisms which occur in a wide variety of habitats including sea water, freshwater, soil and as parasites within animals. Protozoa are very diverse in their structure; some have undulipodia or numerous cilia for locomotion. Other protozoans, such as *Amoeba proteus*, are able to move by means of pseudopodia, which are temporary protrusions (Figure 3.3).

Amoeba proteus is a freshwater organism and its internal solute concentration is higher than that of fresh water. Because its cell surface membrane is permeable to water, it will tend to swell as a result of the osmotic uptake of water. Why does the cell not lyse? The water is removed from the cell by means of a **contractile vacuole**. When viewed using a light microscope, this appears as a clear, circular area in the cytoplasm. The contractile vacuole is enclosed in a membrane and surrounded by a layer of small vesicles which empty into the vacuole. Around this structure is a layer of mitochondria, which can be seen using an electron microscope (Figure 3.4).

cell lysis in 0.05% salt

normal red blood

crenated cell in 1.0% salt

Figure 3.2 Osmotic behaviour of a red blood cell

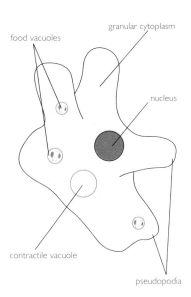

granular cytoplasm

food vacuoles

nucleus

contractile vacuole

pseudopodia

Figure 3.3 Amoeba proteus

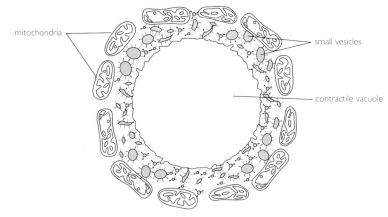

mitochondria

small vesicles

contractile vacuole

Figure 3.4 Contractile vacuole of Amoeba proteus *based on electronmicrographs*

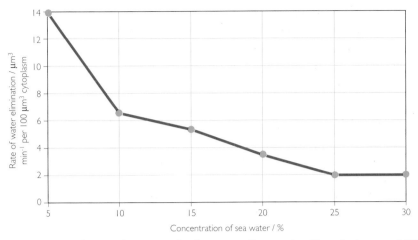

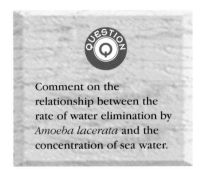

Comment on the relationship between the rate of water elimination by *Amoeba lacerata* and the concentration of sea water.

Figure 3.5 Rate of water elimination by the contractile vacuole of Amoeba lacerata. *100% sea water is pure sea water, 0% is pure water.*

The vacuole takes in water and gradually enlarges, then suddenly decreases in size as the water is expelled to the outside. Contractile vacuoles are absent from most marine protozoa, where the internal and external solute concentrations are almost identical.

Amoeba lacerata is a species of protozoan which can tolerate high salt concentrations and can adapt to living in 50% sea water. The rate of emptying of the contractile vacuole of this species varies inversely with the concentration of the medium in which it is placed. This can be seen graphically in Figure 3.5.

Water relations of plant cells

Plant cells are surrounded by both a plasma membrane and a thick, relatively inelastic cellulose cell wall. The cell wall confers special properties on plant cells as it will resist the osmotic uptake of water. A plant cell placed in distilled water will not swell up and burst, like a red blood cell, but will take up water until the pressure exerted by the cell wall prevents any further expansion. A plant cell in this condition is said to be **fully turgid**. Turgor is important for maintaining mechanical support in plants. If its cells lose water, the plant may wilt.

Water passes through a partially permeable membrane by **osmosis**. We use the term **water potential** to describe the force acting on water molecules in a solution, when separated from pure water by a membrane which is permeable to water only. Consider the situation in Figure 3.6, where a solution of sucrose (the solute) is separated from distilled water by a membrane which will allow water molecules to pass through it, but not sucrose molecules.

We say that the distilled water has a higher water potential than the sucrose solution. Water always tends to move, by osmosis, from a region of high water potential to a region of low water potential, in other words, down a water potential gradient. In the example above, there will be a net movement of water molecules through the membrane into the sucrose solution.

Water potential is given the symbol Ψ (Greek psi) and is measured in units of pressure, kilopascals (kPa). By definition, the water potential of pure water is zero. Adding a solute, such as sucrose, to pure water will decrease the water

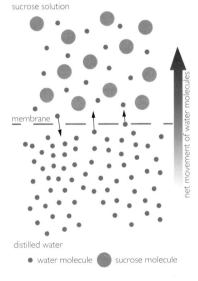

Figure 3.6 Osmosis

potential, that is, it becomes negative. The more solute molecules present, the lower (more negative) the water potential becomes. This change in water potential due to the presence of a solute is referred to as the **solute potential** and is given the symbol Ψ_s. As examples, the solute potential of 0.5 molar sucrose solution is -1450 kPa and the solute potential of 0.7 molar sucrose solution is -2180 kPa.

What would happen if these two sucrose solutions were separated by a membrane permeable to water molecules only? Since water tends to move down a water potential gradient, it will move from the 0.5 molar solution to the 0.7 molar solution, because -2180 kPa is lower than -1450 kPa.

Plant cells contain various solutes, such as sugars, which will exert a solute potential. This is why a plant cell placed in distilled water will tend to take up water. Remember though that this uptake can be opposed by the pressure exerted by the cell wall. This pressure is known as the **pressure potential**, Ψ_p, and because it opposes the solute potential, it usually has a positive value. The overall water relations for a plant cell can be described by the following equation:

$$\Psi = \Psi_s + \Psi_p$$
water potential = solute potential + pressure potential

In a fully turgid cell, the overall water potential of the cell is zero because the values of Ψ_s and Ψ_p are equal and opposite, so they cancel each other out. What will happen if a turgid cell is placed in a strong sucrose solution, such as 0.9 molar? Water will leave the cell because the solute potential of 0.9 molar sucrose (-3020 kPa) is much lower than the water potential of the cell. As the cell loses water, the volume of the cell will decrease and eventually the plasma membrane may lose contact with the cell wall. In this condition the cell is said to be **plasmolysed**. The point at which the plasma membrane is just about to lose contact with the cell wall is known as the point of incipient plasmolysis. At incipient plasmolysis, the pressure potential is zero, so the water potential of the cell is equal to its solute potential, that is $\Psi = \Psi_s$. Figure 3.7 illustrates a plant cell in a state of plasmolysis.

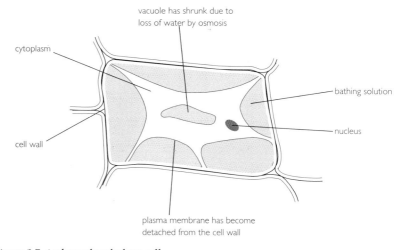

cytoplasm

vacuole has shrunk due to loss of water by osmosis

bathing solution

nucleus

cell wall

plasma membrane has become detached from the cell wall

Figure 3.7 A plasmolysed plant cell

Plasmolysis can be induced experimentally by placing suitable cells, such as small pieces of epidermis from a rhubarb petiole, into 0.9 molar sucrose solution for a few minutes.

Plants and water

In this section, we will consider the uptake of water by roots, its transport to the leaves and evaporation into the atmosphere. This topic is developed in more detail in Module B3, *Biological Systems and their Maintenance*.

Before describing the process of water uptake, it is important to understand the structure of the root and the pathways involved in water transport. A transverse section of a young, dicotyledonous root is shown in Figure 3.8.

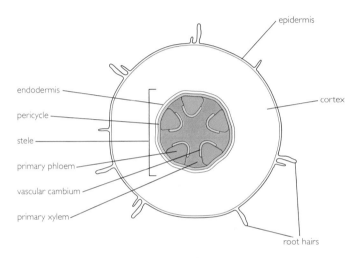

Figure 3.8 Transverse section of young Ranunculus *root*

The layer of cells around the outside is the **epidermis**. Some epidermal cells develop projections called **root hairs**. The central part of the root contains the conducting tissues of the plant, **xylem** and **phloem**. The xylem tissue provides a continuous system for the transport of water and dissolved mineral salts from the root, through the stem to the leaves. The function of phloem is the transport of organic solutes, including sucrose which is formed in photosynthesis. Xylem and phloem are surrounded by a layer of cells known as the **pericycle**. The vascular tissues and their surrounding pericycle form a cylinder of conducting tissue called the **stele**. Just around the outside of the stele is a layer of cells called the **endodermis** which, as described later, has an important role in water movement in the plant. A photomicrograph of the stele of *Ranunculus* can be seen in Figure 3.9.

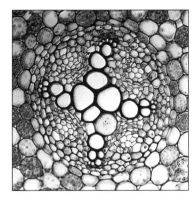

Figure 3.9 Photomicrograph of the stele of Ranunculus *(magnification × 200)*

Between the endodermis and the epidermis there are several layers of relatively large, thin-walled cells forming the **cortex.** The cell walls of cortical cells are highly permeable to water and dissolved solutes. There are also air spaces in the cortex which are important to allow oxygen to diffuse into the root for cell respiration.

Water is taken up mainly by the younger parts of the roots, in the region of root hairs. These are long projections (up to 15 mm) from the epidermal cells which extend among soil particles and serve to increase greatly the surface area for water uptake. The concentration of solutes, such as ions, is very low in the soil water of most soils, so the water potential of this water is close to zero. The solute potentials of plant cells are usually between about -500 to -3000 kPa, so there is a water potential gradient between root hair cells and soil water and water will be taken up by osmosis.

There are two main ways in which water moves across the cortex of the root from the epidermis to the central tissues. These are known as the **apoplast** and the **symplast** pathways. In the symplast pathway, water moves through the cytoplasm, from cell to cell. The cytoplasm of adjacent cells is in contact via **plasmodesmata**, which are fine channels through the cell walls. In the apoplast pathway, water passes through the continuous system of adjacent cell walls. However, when water reaches the endodermal cells its movement is stopped by a waterproof layer in the cell walls, called the **Casparian strip**. This is impregnated with suberin, a waxy compound, which is impermeable to water. Water is therefore prevented from passing around the endodermal cells through the cell walls but instead it must pass through the plasma membrane and cell contents. It is believed that, in this way, endodermal cells are able to regulate the movement of water and dissolved mineral salts, from the soil into the xylem.

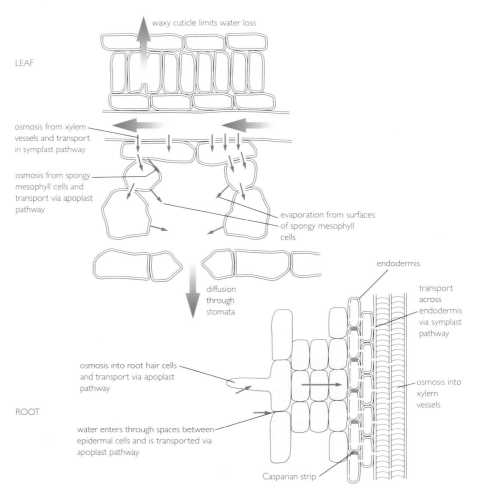

Figure 3.10 The transpiration stream showing uptake of water in a root and loss by evaporation from the leaf

Water is transported to the leaves in xylem vessels. In the leaf, water passes through the walls of xylem vessels to mesophyll cells, and from cell to cell by both the symplast and apoplast pathways. A small percentage of the water taken up will be used in the reactions of photosynthesis (Chapter 1), but most of it evaporates from cell surfaces into the air spaces in the leaf, and into the atmosphere. The evaporation of water from a plant is called **transpiration**, and the flow of water through the plant from roots, through the stem to the leaves in xylem vessels is referred to as the **transpiration stream**.

About 90 per cent of the water lost in transpiration evaporates through pores in the leaf epidermis called **stomata**. Typical stomata consist of two curved **guard cells** on either side of an opening through which gas exchange occurs. A photomicrograph of leaf epidermis is shown in Figure 3.11.

Stomata are able to open and close due to changes in the turgor of guard cells. Stomata may close, for example, at night when carbon dioxide uptake is unnecessary, or in response to some pollutant gases. The rate of transpiration depends on a number of factors, including the width of the stomatal pore (stomatal aperture), temperature, humidity and wind speed. The effects of changes in stomatal aperture and air currents on the rate of transpiration are shown in Figure 3.12.

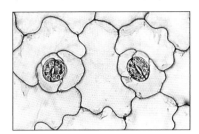

Figure 3.11 Lower epidermis of Kalanchoe, *showing two stomata (magnification × 100)*

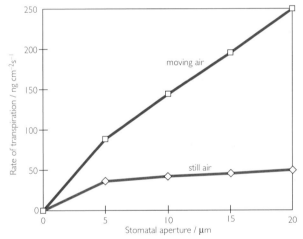

Figure 3.12 Relationship between stomatal aperture and rate of transpiration in still and moving air

Role of transpiration

In 1974, an experiment was conducted by John Hanks at Utah State University to determine the mass of water required to grow a crop of maize. He showed that 600 kg of water were transpired by the maize plants for every 1 kg of dry maize grain produced. This large amount of water lost in transpiration is typical, but what does transpiration achieve?

It is possible for plants to grow in conditions of very high relative humidity, where transpiration will be very low. Indeed, some plants actually grow better in atmospheres of high humidity and submerged aquatic plants do not transpire at all. Nevertheless, transpiration does confer certain benefits on plants. These are listed below.

- **Mineral transport**: transpiration aids the uptake and transport of mineral ions which move through the plant in the transpiration stream.
- **Maintenance of turgidity**: plant cells have an optimum turgidity and some cell functions are less efficient above and below this level. It seems that if transpiration does not occur, plants can become over-turgid and do not grow as well as they do in conditions of slight water stress.
- **Heat loss**: a leaf absorbs both visible light and infra red radiation from its surroundings. If the leaf absorbs more radiant energy than it loses, the temperature of the leaf will rise. This excess heat energy can be lost by transpiration as the evaporation of water has a cooling effect. Each gram of water transpired absorbs between 2.4 and 2.5 kJ from the leaf and its environment.

Water stress

Plants are sometimes classified according to their responses to water. **Hydrophytes** grow in conditions where water is always available, such as in a pond. **Mesophytes** grow in conditions where water availability is intermediate, and **xerophytes** grow in conditions where water is scarce. All plants which live in deserts, for example, are xerophytes and show a number of features, known as xeromorphic adaptations, which help to reduce transpiration and therefore conserve water.

A good example of a xerophyte is marram grass (*Ammophila arenaria*). This species colonises sand dunes around the coasts of western Europe. It is important in helping to stabilise sand dunes because it has an extensive network of rhizomes which help to bind sand, and the tufts of leaves increase the deposition of wind-blown sand. The leaves of *Ammophila* are able to roll up into a cylindrical shape in conditions of water stress. This reduces the surface area of the leaf and encloses the upper (adaxial) epidermis on the inside. The lower (abaxial) epidermis has a thick, waxy cuticle and no stomata; these are confined to the inner surface. Stiff, interlocking hairs help to reduce transpiration further by trapping air within the leaf. Large epidermal cells, known as **hinge cells**, at the base of each furrow, shrink rapidly when the rate of transpiration is high and cause the leaf to roll into a tubular shape. Figure 3.13 shows part of a leaf of *Ammophila* to illustrate these xeromorphic features.

Figure 3.13 Part of a leaf of Ammophila arenaria *(magnification ×50)*

Thermoregulation

The term thermoregulation refers to the regulation of body temperature. Animals can be divided into two groups on this basis: **endotherms**, which maintain nearly constant body temperatures, and **ectotherms**, whose body temperatures vary with the surrounding temperature.

Endotherms

Mammals and birds are able to maintain a constant, high body temperature, more or less irrespective of changes in the environmental temperature. Heat is produced by metabolism and serves to maintain body temperature. These animals are described as endotherms and use physiological and behavioural mechanisms to maintain their body temperature. For an organism to keep a constant body temperature, heat loss must equal heat gain, otherwise the temperature will alter. It has been calculated that if there was no heat loss, metabolic heat alone would raise human body temperature by about 1 °C per hour. Clearly, mammals and birds need to balance heat gains against losses to keep their body temperature constant.

When we speak of a constant body temperature, strictly speaking this term refers to the 'core' temperature of the body, that is, the temperature inside the head, thorax and abdomen. The peripheral temperature may be considerably lower than the core temperature. Figure 3.14 shows the body temperature distribution in a man at a room temperature of 20 °C and 35 °C. The shaded area indicates the core temperature which, at 20 °C room temperature is restricted to the head and trunk but extends into the arms and legs at 35 °C room temperature.

The core temperature of mammals and birds undergoes regular daily fluctuations. Over a 24 hour period, these fluctuations are usually between 1 and 2 °C. This pattern of body temperature differs in diurnal animals (which are active during the day) and nocturnal animals. Diurnal animals show a temperature peak during the day and a minimum at night, but the reverse pattern is seen in nocturnal animals. It is interesting to note that these variations in core body temperature persist even if animals are kept in continuous, constant light. This indicates that these rhythms of body temperature are inherent, or endogenous, in the organisms.

The normal core temperature, disregarding small variations, is almost the same within each of the major groups of endotherms. Most birds maintain their body temperature at 40 ± 2 °C and mammals at 38 ± 2 °C.

Thermoregulation in mammals

We have already stated that for body temperature to remain constant, heat loss must equal heat gain. In cold conditions, heat production can be increased by muscular activity and exercise, shivering (involuntary muscle contractions) and non-shivering thermogenesis (an increase in metabolic rate without noticeable muscle contractions). Heat is lost from the body by radiation, convection and the evaporation of sweat. Evaporative heat loss is due to the loss of heat which occurs when water evaporates. At rest, humans lose about 30 g of water per hour by evaporation, although this can rise to about 1 kg in desert conditions. The evaporation of water requires a considerable amount of heat. To transfer 1 g of water at room temperature to water vapour at the same temperature requires 2.44 kJ. To put this figure into perspective, it is five times the amount of heat required to raise 1 g of water from freezing point to boiling point. In biology, we usually use the figure 2.42 kJ per gram of water, which is an approximation of the value for the evaporation of water at the skin temperature of a sweating person, about 35 °C.

When a person is exposed to hot surroundings, cooling of the body occurs by evaporation of sweat from the body surface. Several large, furry mammals, including cattle, antelopes and camels, also depend mainly on sweating for evaporative heat loss. Although camels have thick fur (which helps to insulate

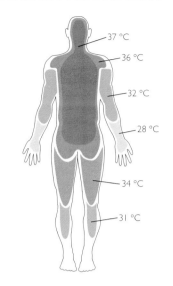

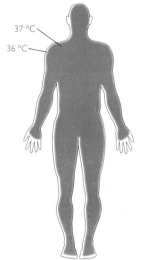

Figure 3.14 *Temperature distribution in the body of a man at room temperatures of 20 °C (top) and 35 °C (bottom)*

them from the sun's heat), evaporation of sweat occurs rapidly because the air in a desert is very dry. Sheep, goats and dogs evaporate most of the water from their respiratory tracts by panting. Some mammals, including kangaroos and some rodents, use a third method for increasing evaporation. They spread saliva over their fur and lick their limbs, which achieves cooling by evaporation.

Heat exchanges due to radiation and convection are determined by physical factors. These include the temperature gradient from the surface of the body to the environment, the amount of air movement and, in humans, the amount of clothing worn. Normally heat is lost by these means but, if the environment is hotter than the body, heat is gained. In this situation, evaporative heat loss is the only method of maintaining body temperature.

Body temperature in endotherms is regulated by a combination of physiological and behavioural responses. In humans, behavioural responses (such as putting on more clothes, or moving into the shade), are more important than physiological responses which act as a means of fine control. The **hypothalamus** (part of the forebrain) plays an essential role in thermoregulation. Sensory input comes from nerve endings in the skin and from nerve cells, which are sensitive to temperature, in the hypothalamus itself. The hypothalamus compares the body temperature with the 'set point' (normal body temperature) and, if these temperatures differ, it initiates appropriate thermoregulatory responses. If the core temperature is greater than the set temperature, mechanisms to increase heat loss come into action or, if the core temperature is lower than the set temperature, heat conservation and increased heat production return the body temperature to its set point.

Role of the skin in thermoregulation
Almost all the heat lost from the body is via the skin, which has a number of important structures and functions concerned with helping to maintain body temperature. Before describing these functions, it is important to understand the general structure of skin. Figure 3.15 shows the structure of human skin.

Heat loss through the skin occurs by **conduction, radiation** and **evaporation** of sweat. Heat loss by conduction and radiation can be varied by altering the blood flow to superficial capillaries. The diameter of arterioles in the skin is controlled by sympathetic nerves originating in the hypothalamus. Low external temperature results in a decrease in the diameter of these vessels (**vasoconstriction**) and the blood flow to capillaries is reduced. This reduces the loss of heat. Conversely, in warm conditions, the diameter of skin arterioles increases (**vasodilatation**), resulting in increased blood flow in the peripheral circulation and a corresponding increase in heat loss.

When the external or body temperature rises, thermal sweating occurs. Sweat glands secrete a dilute solution containing sodium chloride, urea and lactic acid. Evaporation of water leads to the loss of heat by the body. Heavy sweating involves rapid loss of water and salt from the body; dehydration and salt deprivation will ultimately occur unless enough water is drunk and adequate amounts of salt are taken.

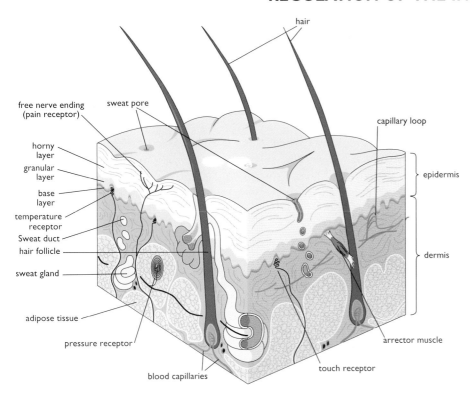

Figure 3.15 Structure of human skin showing structures involved in thermoregulation

In cold conditions, smooth muscles attached to hair follicles in the skin contract, raising the hairs. In humans this has little or no effect, but is important in many other mammals as it helps to trap a layer of air in their fur. Air is a poor conductor of heat so this helps to insulate the body against excessive heat loss. The layer of adipose tissue in the hypodermis (under the dermis) is also an important means of insulation as fat is a poor conductor of heat. This layer of fat tissue is of particular importance in aquatic mammals such as seals and whales, where it forms a thick layer of **blubber** and greatly reduces heat loss to the environment.

Thermoregulation in birds and mammals enables them to survive in the harshest conditions. Studies on the body temperature of the Arctic fox, for example, have shown that it is able to maintain its body temperature within normal mammalian limits even when exposed to an air temperature of −80 °C.

Thermoregulation in reptiles

Reptiles are examples of ectotherms and, although they do use metabolic heat production as a means of increasing heat gain, they largely depend on external sources of heat to raise their body temperature. Solar radiation is used especially by reptiles, which show a number of behavioural responses to control their body temperature. Many reptiles are able to change their skin colour and a dark skin increases the amount of solar energy absorbed. Increasing the area exposed to solar radiation also increases heat gain. This can be achieved by turning the body so that it is at right angles to the sun's rays, spreading the legs and flattening the body. In this way, a lizard can achieve a body temperature considerably higher than that of the surrounding air. When a suitable body temperature has been reached, a lizard may move into the shade, where it is cooler. The repetitive movement of a lizard from a warm environment, where it increases its body temperature, to a cooler environment, where it loses heat, is referred to as **shuttling**. By alternately exposing itself to the sun's rays and seeking shade, a reptile is able to maintain its preferred body temperature, which is also known as the **eccritic** temperature.

Liolaemus, the Peruvian mountain lizard, lives at an altitude of 4000 m where the air temperature is low even in summer. In an investigation, the air temperature and the body temperature of *Liolaemus* were measured during the morning. Between 7.00 a.m. and 8.00 a.m. the air temperature rose from −2.0 °C to 1.5 °C, but the body temperature of the lizard rose from 2.5 °C to 33.0 °C. The lizard's body temperature was thus over 30 °C higher than the air temperature, enabling it to be active in the almost freezing environment.

Figure 3.16 Phrynocephalus luteoguttatus, *a small lizard found in sandy desert regions of southern Afghanistan. When the sand is hot it raises its body above the surface as shown in the photograph. It can bury itself in the sand in a few seconds by sideways movements.*

QUESTION

What are advantages and disadvantages of endothermy?

Many reptiles, including tortoises, snakes and crocodiles, bask in the sun to increase their body temperature. Metabolic heat can also be used to achieve a high body temperature. Some female snakes maintain a high body temperature while incubating their eggs. As an example, a brooding python was observed to maintain her body temperature between 4 and 5 °C above the air temperature by strongly contracting the body muscles. This was accompanied by a corresponding increase in oxygen consumption as her metabolic rate increased and her body weight decreased from 14.3 kg to 10.3 kg over the 30 days incubation period.

Adaptation to the environment

Structural adaptation

In this chapter, we shall consider some of the physical factors affecting the distribution of plants and animals in their habitats. Emphasis is placed on the external features of the organisms for the purposes of identification, so a knowledge of the main external differences between the major groups is desirable. Observations of the species associated with an aquatic or with a terrestrial environment will illustrate that organisms clearly show structural adaptations to their specific habitats.

In an aquatic habitat, such as a pond or stream, the organisms need to maintain their position. If the water is flowing, this might mean that some organisms must be attached so that they remain in the same position, or some may need to swim or move so that they remain in the same conditions. There are many structural adaptations associated with the effects of movements and currents in water. These include streamlining and the possession of fins in fish, flexible stems and variations in leaf form in plants and the flattened shape of crustaceans. Consideration of other physical factors in an aquatic habitat, such as light, oxygen concentration and salinity, will reveal other structural adaptations in the organisms present.

In terrestrial habitats, light, temperature, availability of water and soil type are important factors. In sand dunes, for example, xerophytic conditions exist initially and colonising organisms have to be adapted to lack of water and unstable soil, as well as to exposure to wind. The first colonisers are plants which are salt-tolerant and able to develop extensive root systems, thus maintaining their position. Typical of such situations is marram grass (*Ammophila arenaria*), which shows adaptations to the shifting soil, the lack of water (see Chapter 3) and the exposure.

When carrying out the practical work associated with this module, the relationship of the external features of organisms found in particular habitats to the prevailing physical factors can be studied in greater detail. A knowledge of the relevant distinguishing external features of the major groups is helpful, so that initial identification of organisms can be made.

Classification

An understanding of classification and the principles on which it is based is of great importance in studying and describing the distribution of organisms in their habitats. Classification enables us to link organisms with similar characteristics into groups for easy reference, so it is important that a suitable, sensible system for naming organisms exists and that such a system should be recognised internationally. The use of common names for organisms, such as 'frog', 'buttercup' and 'mushroom', may be widely used in Britain, but these names lack precision. There are many different types of frogs and buttercups and the term 'mushroom' is commonly used for any edible member of the group Basidiomycota.

Figure 4.1 Animals and plants possess structural adaptations which enable them to survive in their particular habitat

ADAPTATION TO THE ENVIRONMENT

In order to avoid confusion, the **binomial system of nomenclature**, based on a scheme devised in the 18th Century by the Swedish naturalist, Carl von Linné (otherwise known as **Linnaeus**) is used. In this system, each organism is given a name consisting of two parts:

- a **generic** name, which states the **genus** and is common to a group of closely related organisms
- a **specific** name, stating the **species**, which is unique to a particular organism and often descriptive of one of its characteristics.

Both names are given in Latin, the generic name beginning with a capital letter and the specific with a lower case letter. It is the convention to use italics when printing the names, but when handwritten it is usual to underline them. For example, in a particular habitat, we might find *Ranunculus bulbosus* (the bulbous buttercup), together with *Primula vulgaris* (the common primrose), *Lumbricus terrestris* (the earthworm) and *Capsella bursa-pastoris* (the shepherd's purse. Latin was used by Linnaeus because it was understood by most educated people in the 18th Century and its use persists in modern times. It enables worldwide recognition of names and any new species which are discovered must be named according to this binomial system, with a detailed written description of the organism.

Many different schemes of classification have been devised, from Aristotle in the 3rd Century BC, dividing animals into those with red blood and those without, to the proposal by the American biologist R H Whittaker in 1959 of the **Five Kingdom System**. The currently recommended scheme is based on Margulis and Schwartz's modification of Whittaker's proposals and recognises the following five kingdoms of living organisms:

- **Prokaryotae**
- **Protoctista**
- **Fungi**
- **Plantae**
- **Animalia.**

The main distinguishing features of these kingdoms are outlined in Table 4.1, together with examples of the groups included in them.

Taxonomy

Taxonomy is the study of the organisation of groups of organisms, called **taxa** (sing. **taxon**), into hierarchies, which attempt to take into account their supposed evolutionary descent. A taxon is a group of organisms which share common features. There are seven major levels of taxon, of which the highest is the kingdom and the lowest is the species.

A **species** is defined as a group of individuals with a large number of features in common. Members of a species can interbreed and produce fertile offspring, but members of one species do not normally breed with members of another species. This definition is restricted to sexually reproducing organisms and can not apply where reproductive behaviour has not been observed, as is the case with fossils, organisms that only reproduce asexually or parthenogenetic forms. A slightly different definition, which specifies a group of organisms showing a

Table 4.1 *The Five Kingdoms*

Kingdom	Characteristics	Representative groups
Prokaryotae	organisms lack nuclei organised within membranes; lack envelope-bound organelles; lack 9 + 2 microtubules	non-photosynthetic: *Escherichia coli*, photosynthetic: cyanobacteria *Anabaena*
Protoctista	eukaryotic organisms with organised envelope-bound nuclei which are neither fungi, plants nor animals; often unicells or assemblages of similar cells	green algae: Chlorophyta brown algae: Phaeophyta Protozoa
Fungi	non-photosynthetic eukaryotic organisms with a protective, non-cellulose wall; absorptive methods of nutrition; usually multinucleate hyphae; spores without flagella	Zygomycota: *Mucor* Ascomycota: *Saccharomyces* Basidiomycota: *Agaricus*
Plantae	multicellular, photosynthetic, eukaryotic organisms; cell walls contain cellulose	Bryophyta: mosses Filicinophyta: ferns Angiospermophyta: flowering plants
Animalia	multicellular, non-photosynthetic eukaryotic organisms with nervous coordination	Cnidarians Annelids Arthropods Molluscs Chordates

close similarity in morphological, biochemical, ecological and life history characters, is more generally used and nearly all species are established on this basis. Some species are widely distributed and show local or regional differences, which justify recognition and result in the taxon being split into a number of **sub-species**, **varieties** or **races**. Such differences, although small, are thought to be significant, but do not prevent interbreeding.

Between the lowest taxon: the species, and the highest: the kingdom, there are five other taxa: **genus**, **family**, **order**, **class** and **phylum**. The relationships between these taxa are summarised in Table 4.2. As the classification proceeds from the kingdom to the species, the number of organisms in each taxon decreases, but the number of shared features increases. Some taxa, such as the phylum **Chordata**, are very large and have been subdivided into sub-phyla for convenience.

In any study of living organisms in their habitats, it is useful to be able to place them in their taxonomic groups. For this to be achieved with any accuracy, a knowledge of the distinguishing external features of the different taxa is required and for correct identification to genus and species level, it is necessary to use **keys**.

ADAPTATION TO THE ENVIRONMENT

Table 4.2 *Taxa used in the classification of organisms*

Taxon	Description	Bladderwrack	Garlic	Human
kingdom	largest group of organisms sharing common features	Protoctista	Plantae	Animalia
phylum	major subdivision of a kingdom	Phaeophyta	Angio -spermophyta	Chordata
class	a group of related orders; subdivision of a phylum	Phaeophyceae	Mono -cotyledonae	Mammalia
order	a group of related families; subdivision of a class	Fucales	Liliales	Primates
family	a group of closely related genera; subdivision of an order	Fucaceae	Liliaceae	Hominidae
genus	a group of related species; subdivision of a family	Fucus	Allium	Homo
species	a group of organisms capable of interbreeding and producing fertile offspring	F.vesiculosus	A.sativum	H.sapiens

A variety of different methods for the identification of organisms is available. Museum specimens, pictures and helpful experts can contribute, but, with some knowledge of the terminology, it is possible to use a suitable key. The easiest keys to follow are **dichotomous** keys. Where observable, external features are used to divide organisms into two groups at each stage. Such keys are referred to as **single access keys** as each stage should be followed through in the correct sequence until the identification is made. An example of a dichotomous key is shown in Fig 4.2

This key combines descriptions of the features as well as illustrations. It should be started at 1 and then followed through until an identification is made. This key is to the orders of insects and once an order has been established, reference is made to the appropriate section of the book, where there will be a further key for identification to species or genus level.

Sometimes, students are required to construct their own keys to separate groups, individual organisms or parts of organisms, such as vertebrae, leaves or

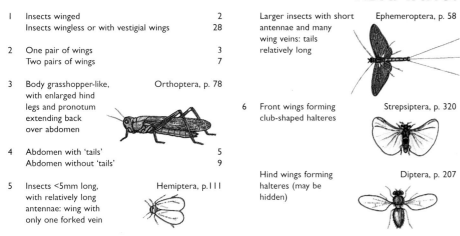

I	Insects winged	2
	Insects wingless or with vestigial wings	28
2	One pair of wings	3
	Two pairs of wings	7
3	Body grasshopper-like, with enlarged hind legs and pronotum extending back over abdomen	Orthoptera, p. 78
4	Abdomen with 'tails'	5
	Abdomen without 'tails'	9
5	Insects <5mm long, with relatively long antennae: wing with only one forked vein	Hemiptera, p.111
	Larger insects with short antennae and many wing veins: tails relatively long	Ephemeroptera, p. 58
6	Front wings forming club-shaped halteres	Strepsiptera, p. 320
	Hind wings forming halteres (may be hidden)	Diptera, p. 207

Figure 4.2 Part of a key to the orders of European insects

flowers. There are several general principles that should be observed, both when constructing and when using keys.

- only use visible, external features
- use permanent, adult features. If the features apply only to larval or immature forms, or to organisms of one sex only, this is usually made clear in the key
- avoid the use of terms such as 'large' and 'small'. Most keys will either use comparative terms such as 'wider' or 'longer than body length'. If specific measurements are given, a range will be included, for example 'legs from 10 to 15 mm'.

Table 4.3 *Characteristics of some major groups of organisms*

Phylum/ major group	Characteristics	Examples
Chlorophyta (green algae)	photosynthetic; pigments similar to green plants; cellulose cell walls; starch stored; may be unicellular, filamentous or colonial; some thalloid forms	*Chlorella* (unicellular); *Spirogyra* (filamentous); *Ulva* (sea lettuce – thalloid form)
Phaeophyta (brown algae)	photosynthetic; chlorophyll pigments and brown pigment fucoxanthin; multicellular; may show alternation of generations	*Fucus vesiculosus* (bladderwrack)
Fungi	non-photosynthetic; consist of hyphae forming a mycelium; spores lack flagella	Phylum Zygomycota: *Rhizopus* (bread mould) Phylum Ascomycota: *Saccharomyces* (yeast) Phylum Basidiomycota: *Agaricus* (field mushroom)
Bryophyta (mosses)	photosynthetic; definite alternation of generations with conspicuous gametophyte; no true roots but multicellular rhizoids; spirally arranged leaves on stem	*Sphagnum* (bog moss)
Filicinophyta (ferns)	photosynthetic; true roots, leaves and stems; young leaves tightly coiled in bud; sporangia in clusters on lower surface of leaves	*Pteridium* (bracken)
Angiospermophyta (angiosperms)	photosynthetic; seed-bearing plants with true flowers; seeds enclosed in fruit formed from the ovary	Class Monocotyledoneae: *Anthoxanthum* (sweet vernal grass) Class Dicotyledoneae: *Ranunculus* (buttercup)

ADAPTATION TO THE ENVIRONMENT

Phylum/ major group	Characteristics	Examples
Cnidarians	animals with two cell layers, ectoderm and endoderm, separated by a mesogloea; ectoderm contains nematoblasts; show radial symmetry with tentacles; many show polymorphism with polyp and medusa forms	Class Hydrozoa; dominant polyp form, medusa reduced or absent: *Hydra* Class Scyphozoa; dominant medusa and reduced polyp: *Aurelia* (jellyfish) Class Anthozoa; polyp forms only: *Actinia* (sea anemone)
Annelids	worm-like animals showing clear segmentation	Class Polychaeta; polychaetes: *Nereis* (ragworm) Class Oligochaeta; earthworms: *Lumbricus* (earthworm)
Arthropods	animals with a chitinous exoskeleton; segmented; jointed appendages	Class Crustacea; crustaceans: *Cancer* (crab) Class Chilopoda; centipedes: *Lithobius* (garden centipede) Class Insecta; insects: *Apis* (honey bee) Class Arachnida; spiders: *Araneus* (garden spider)
Molluscs	unsegmented animals with a head, muscular foot and visceral mass or hump; many species possess calcareous shell secreted by the mantle	Class Gastropoda; gastropods: *Helix* (garden snail) Class Pelycypoda (formerly bivalvia); bivalves: *Mytilus* (mussel)
chordates	animals which have, at some stage during their development, a notochord, a hollow dorsal nerve cord, visceral clefts or gill slits, a post-anal tail	Class Osteichthyes; bony fish: *Clupea* (herring) Class Amphibia; amphibians: *Rana* (frog) Class Reptilia; reptiles: *Lacerta* (lizard) Class Aves; birds: *Colomba* (pigeon) Class Mammalia; mammals: *Homo* (Man)

Physical factors affecting distribution

A number of physical factors affect the distribution of organisms in their habitats. These physical factors are often referred to as **abiotic**, to distinguish them from the **biotic** factors, which involve the effects of other living organisms, including humans, on the distribution of species. Some physical factors such as light, have widespread effects and are important in a range of different situations, both aquatic and terrestrial. Others, such as wave action, are only of significance in aquatic environments.

The physical factors can be divided into
- **climatic** – temperature, light, wind and water availability
- **soil** – often referred to as **edaphic** factors
- **topographic** – altitude, aspect (whether north-facing or south-facing) and inclination (steepness of slope)
- others, such as wave action, which are relevant in specific situations.

Temperature

Most living organisms have an optimum temperature range within which they can survive, thus variations in temperature will affect the rate at which they grow. The main source of heat is the sun's radiation and the temperature of a habitat will depend on its latitude, the season of the year, the time of day and its aspect. The temperature of any environment can also be affected by the presence or absence of vegetation.

Temperatures below freezing will cause physical damage to the living cells as ice crystals will form. If the temperature becomes too high, then enzymes are denatured, causing growth and metabolism to be disrupted. Aquatic habitats tend to have more stable temperature conditions than terrestrial ones, due to the high specific heat of water. It takes a great deal of heat energy to bring about a significant increase in the temperature of a large body of water, so there is less fluctuation in the temperature of an aquatic habitat than there is in a terrestrial one.

Light

Light is the source of energy for photosynthesis, so it influences primary productivity. As the consumers are directly or indirectly dependent on the primary producers, light is therefore essential for the maintenance of life. Light has a number of other roles, one of which, the length of the photoperiod, has considerable influence on the behaviour of both plants and animals. Many activities, such as flowering and germination in plants and reproductive behaviour in animals, are linked to the photoperiod so that they can be synchronised with the seasons.

The need for light by plants affects the structure of communities. A clear example of this is seen in the different layers of vegetation in a woodland, where shade-tolerant species are found underneath the tree canopy. In aquatic habitats, the plants are confined to the surface of the water or to the shallow water at the margins. Even in shallow water, the penetration of light for photosynthesis may be affected by the turbidity due to suspended particles.

Wind

Air movements, or wind, are of significance in the distribution of organisms as they interact with other physical factors. This is particularly apparent in exposed coastal or upland areas where the prevailing wind may affect the growth of trees and shrubs. Air movements accelerate the rate of evaporation of water and hence affect transpiration. Increased wind speed increases the liklihood of freezing so that the buds on the more exposed side of a tree are more likely to suffer frost damage. Some seed and spore dispersal mechanisms rely on air movements and strong winds can affect the migration of birds.

Water availability

Water is essential for all forms of life and may be a limiting factor in some terrestrial habitats. There is a continuous cycling of water, called the **hydrological cycle**, illustrated in Fig.4.3. This summarises the events which determine the availability of water in terrestrial habitats.

Plants vary widely in their ability to tolerate a shortage of water. **Hydrophytes** are adapted to living in waterlogged or submerged conditions, whereas **xerophytes** show adaptations which reduce water loss and can survive in conditions where water is scarce. **Mesophytes** are those plants which thrive best in conditions where there is an adequate supply of water, so that water loss through transpiration can usually be replaced by water uptake from the soil. In temperate regions, these plants do show some adaptations to seasonal variations in water availability: woody plants may shed their leaves in autumn

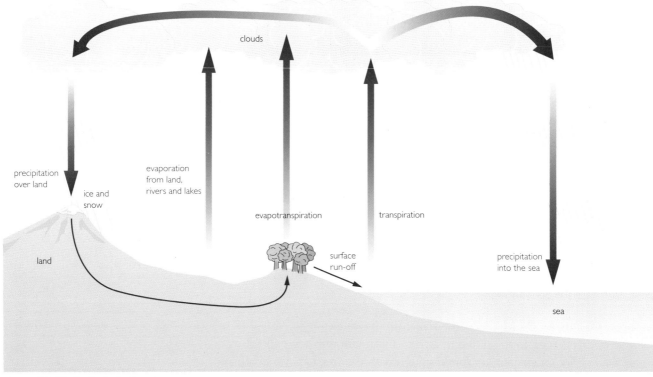

Figure 4.3 The hydrological cycle

and the aerial parts of herbaceous perennials may die down, thus reducing water loss during the unfavourable season.

Many terrestrial animals show adaptations which enable them to regulate water loss and thus survive in conditions of water shortage. Emphasis is on the conservation of water as in areas where water is in short supply, the rate of evaporation is also high. Aquatic animals also need to be able to maintain a constant internal environment, so must regulate water uptake and water loss. Those animals living in fresh water, where the concentration of body fluids is often higher than the concentration of ions in their environment, have a tendency to take up water by osmosis. They show adaptations which prevent or reduce water uptake and also possess a means of getting rid of the excess water. In marine habitats, the concentration of ions in sea water is higher than that in fresh water, so animals may show adaptations to counteract the tendency to lose water.

Salinity

Plants which can tolerate high levels of salt are referred to as **halophytes**, and are typically found growing in estuaries and saltmarshes, where their roots may be immersed in sea water. The degree of salinity to which they are exposed can vary, depending on their location and the tides. Salinity levels can be lower than sea water in a tidal estuary, but higher in a saltmarsh due to the evaporation of water from the soil at low tides. Adaptations to these changing conditions often involve the maintenance of high salt concentrations in the tissues or the development of special tissues in which water can be stored.

Some animals can adapt to changes in salinity, but as many animals are motile they will generally move to an area where they are best suited to the prevailing conditions.

Water movements

Water movements include currents and the ebb and flow of tides, the latter possibly interacting with air movements to bring about wave action. Any movement of water will have an eroding action on soil and rocks, as well as moving living organisms from one place to another, unless they are rooted or attached to a substratum. The churning action of water movements results in aeration of the water and it may also contribute to the turbidity. Organisms in aquatic habitats show a wide range of adaptations to life in water.

Rate of flow in streams or rivers is an important parameter because of its influence on the organisms inhabiting the water. As the current increases, organisms which are unable to swim against it, or to take hold, are likely to be washed away. Faster flowing water is likely to be better oxygenated than sluggish or still water, because of the mixing effect.

Oxygen concentration

Oxygen is of vital importance to aerobic organisms in any habitat. The oxygen concentration of the atmosphere is more or less constant at around 21 per cent, but that of the soil atmosphere is slightly lower due to respiration of soil organisms. The oxygen concentration in aquatic habitats can vary greatly. It may be very low in still, undisturbed water, with anaerobic conditions in the mud at the bottom. Any disturbance of the water will bring about aeration, as will the presence of actively photosynthesising plants. An increase in temperature will reduce the amount of dissolved oxygen in water and could result in a reduction in the numbers of organisms present.

Most living organisms are unable to tolerate anaerobic conditions and will die quickly if the level of oxygen falls. One of the effects of organic pollution in freshwater is a reduction in dissolved oxygen levels as a consequence of the activity of aerobic bacteria. Some invertebrates are more tolerant to low levels of oxygen than others and can survive in poorly oxygenated water. These tolerant organisms include *Tubifex* worms and the larvae of certain midges, notably *Chironomus*. These organisms both contain haemoglobin which helps them to obtain oxygen. Larvae of the stone fly, for example *Perla* spp., will only live in well-oxygenated water, which is characteristic of fast-flowing streams. These organisms are illustrated in Figure 4.4.

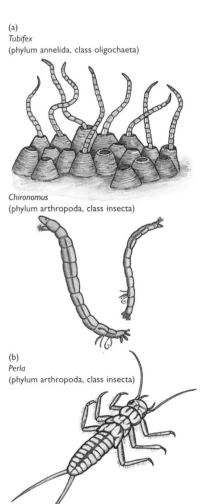

(a)
Tubifex
(phylum annelida, class oligochaeta)

Chironomus
(phylum arthropoda, class insecta)

(b)
Perla
(phylum arthropoda, class insecta)

Figure 4.4 Organisms typically found in (a) poorly oxygenated and (b) well-oxygenated freshwater

Table 4.4 *Table of mineral particle sizes in soils*

Particle type	Diameter/mm
clay	<0.002
silt	0.002–0.02
fine sand	0.02–0.2
coarse sand	0.2–2.0
gravel (small stones)	>2.0

ADAPTATION TO THE ENVIRONMENT

Soil

Soil covers a large part of the land surface of the Earth and consists of:

- rock, or mineral particles, derived from the breaking up or weathering of rocks (46–60 per cent)
- organic material called humus (about 10 per cent)
- water (25–35 per cent)
- air (15–25 per cent)
- living organisms, such as bacteria, fungi, protozoa, insect larvae, earthworms and moles in variable numbers depending on the soil type and location.

The mineral particles vary in size from gravel, with particles greater than 2 mm in diameter to clay, with particles less than 0.002 mm in diameter. Table 4.4 categorises soil particles according to size. The proportions of the different sized particles determine the texture and properties of a soil. These proportions can easily be determined by shaking up a sample of soil with water in a measuring cylinder, allowing it to settle and then estimating the volume of different sized particles as they appear in the sample. The larger, heavier particles sediment out more quickly than the lighter, smaller ones. Clay particles may remain in suspension but can be caused to settle out by the addition of calcium hydroxide, which flocculates the particles. Any organic material usually floats on the top. The results of this technique are illustrated in Figure 4.5.

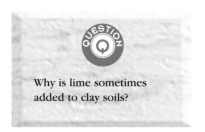

Why is lime sometimes added to clay soils?

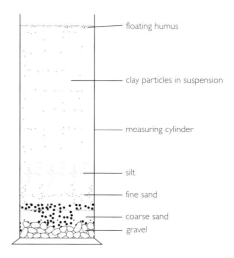

Figure 4.5 Sedimentation to show soil composition

Water drains more easily through a soil with a high proportion of larger sized particles, such as a sandy soil, than it does through a clay soil, where the smaller sized particles predominate. Clay soils can become waterlogged if rainfall is high, although they will retain water in drier conditions. Sandy soils do not retain much water and the rapid drainage results in the loss of mineral ions through leaching.

The presence of organic matter in the soil is an important source of food for the decomposers and detritivores. These organisms bring about the breakdown of the dead remains of plants and animals, releasing mineral ions into the soil, which are then available for uptake by plants.

Soil type can have an effect on the habitat, because it will determine the type of vegetation that is present, which in turn will determine the number and species of animals. Soil type can be determined by digging a **soil profile** to expose the different layers. Figure 4.6 illustrates two different soil profiles. A soil profile usually shows the following characteristic layers or **horizons**:

- **A horizon**: the topsoil, from which minerals tend to be removed by leaching. This layer is often subdivided into A_0, the litter layer, A_1, the humus layer and A_2, the leached layer
- **B horizon**: the subsoil, which tends to collect the minerals leached out of the A horizon
- **C horizon** consisting of weathered parent material
- **D horizon:** the parent rock, or bedrock.

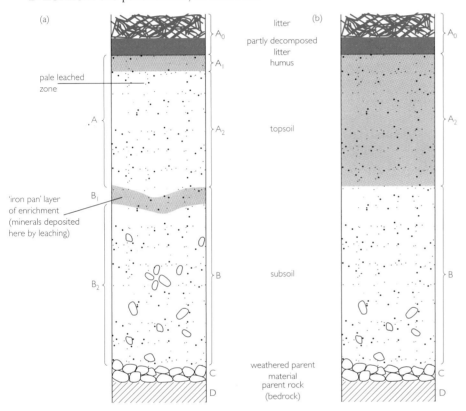

Figure 4.6 Soil profiles of (a) a podzol and (b) a brown earth

Brown earths or brown forest soils are associated with temperate deciduous woodland at low altitudes. The profile is a simple one, with a relatively thin litter layer and A and B horizons of a similar brown colour. These soils tend to have a pH range of 4.5 to 8.0 and support large numbers of soil organisms which break down the litter into humus. There are usually earthworms present which pull litter down into the soil as well as mixing the soil layers in the course of their burrowing activities.

Podzols are typically associated with coniferous woodland and heathland at higher altitudes in temperate regions. The profile is more complicated than that of a brown earth, with a much deeper litter layer, an ash-grey A_2 horizon and an 'iron pan' in the B horizon. The litter consists of leaves from the pine trees, which take a long time to break down. The pH range of these soils is in the range of 3.0 to 6.5, so there are fewer soil microorganisms and earthworms are scarce. Most of the breakdown of the litter is achieved by fungal action. The soils are sandy and occur in regions of high rainfall, so a great deal of leaching occurs, leaving the A_2 horizon an ash-grey colour. The minerals, including iron, and humus accumulate in the B horizon and give rise to a dark layer called the 'iron pan', which is hard and impenetrable to plant roots.

pH

The pH value is a measure of the hydrogen ion concentration of an aqueous solution and indicates the level of acidity or alkalinity. The pH value of a soil has an effect on the availability of mineral ions and, to some extent, can determine the type of vegetation which will grow. Certain plants, such as heather (*Erica* sp.), grow well in the acid conditions of heathland, while others, such as dog's mercury (*Mercurials perennis*), will only grow in more alkaline conditions. The type of vegetation in an area will have an influence on the animal populations. In Britain, soil pH values range from pH 4.0, which is acid, to pH 8.0, which is slightly alkaline. A pH of 7.0 is neutral. Many plants appear to be sensitive to the presence or absence of calcium ions in the soil. Plants which can thrive on calcareous soils, where calcium ions are abundant, are termed **calcicoles**, or **calcicolous**. Plants which cannot tolerate calcium are termed **calcifuges** and their presence indicates the absence of calcium in the soil. When these species are grown on their own in the 'wrong' type of soil, they still thrive, so it is likely that some other factor, such as the availability of other mineral ions, climatic conditions or competition, is involved.

The pH of aquatic environments can cause variations in the distribution of organisms. The range of pH in lakes and ponds is from 4.7 to 8.5, and although some aquatic species, such as *Gammarus* (the freshwater shrimp), can tolerate a wide range, others are only found in either acid or alkaline conditions. The presence or absence of calcium ions also influences the distribution of members of the Mollusca and Crustacea. Both these groups need calcium ions, the molluscs for their shells and the crustaceans for their integuments, so both are restricted to situations where calcium is present.

Qualitative and quantitative field techniques

Before undertaking an ecological survey, it is necessary to consider the factors which we want to measure. Within ecology, studies are of two broad types, **autecology** and **synecology**. Autecology considers the ecology of a single species, such as the distribution of limpets on a rocky shore, whereas synecology focuses on organisms and their environment. In this case, the whole of the rocky shore community would be investigated.

In an ecological investigation, careful planning is required to identify the sorts of biotic and abiotic factors which are important in a particular ecosystem. Measurement of these factors will, of course, depend on the resources available, but useful data can be obtained with simple equipment. It is the design of the investigation which is important, rather than the amount of elaborate equipment which is used. Figure 4.7 illustrates a variety of simple equipment used in ecological studies.

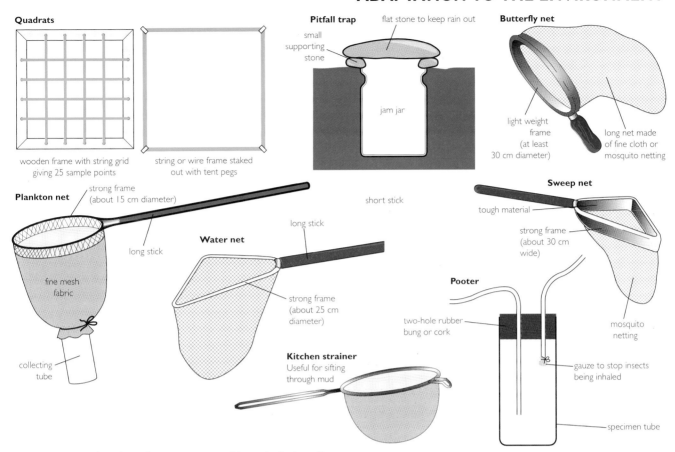

Figure 4.7 Examples of simple equipment used in ecological studies

Random sampling

Suppose we want to determine the number of dandelions (*Taraxacum officinale*) present in a field, or the number of limpets (*Patella vulgata*) on a rocky shore. It would obviously be impractical to count every individual, so instead we need to use a sampling technique. In a study of this sort, random sampling with a quadrat frame is used (see Figure 4.7). A quadrat frame is usually made of wood or metal and is used to take a sample of the area under investigation. It is assumed that the area within the quadrat frame is representative of the entire area. There are three questions which need to be considered when using a quadrat is this way:

- What size quadrat should be used?
- How many samples should be taken?
- How should the quadrat be positioned?

Quadrat size

Quadrats of various sizes are available, for example 0.25 m² and 1 m². In general, smaller quadrats are more reliable than a large quadrat, but there are practical considerations which need to be taken into account. The shape of the quadrat is another factor you could consider. Quadrats are normally square, but does the shape of the quadrat have any effect on estimations of species density? This could be investigated by using quadrats of different shapes, for example, square, rectangular and circular, but having the same area in each case.

ADAPTATION TO THE ENVIRONMENT

Sample size

The distribution of many species is aggregated, that is they are not uniformly distributed. One sample, therefore, may be completely unrepresentative of the entire population and a calculation of the density based on that sample would be inaccurate. To make the results statistically significant, a large number of samples should be taken, but identifying and counting all the species present in a very large sample can be too time-consuming. We therefore need a method to investigate the effect of sample size on the estimate of density. This can be done by working out the running mean of the density, with each sample taken. To illustrate this method, suppose we were finding the density of specimens of common periwinkles (*Littorina littorea*) on a rocky shore, using a 0.5 m^2 quadrat. The results might look something like Table 4.5.

Table 4.5 *Density of* Littorina littorea *on a rocky shore*

Sample No.	Specimens per quadrat	Density /m^{-2}	Cumulative mean density /m^{-2}
1	7	14	
2	2	4	$(14 + 4) \div 2 = 9$
3	12	24	$(14 + 4 + 24) \div 3 = 14$
4	8	16	$(14 + 4 + 24 + 16) \div 4 = 14.5$
5, etc.	15	30	$(14 + 4 + 24 + 16 + 30) \div 5 = 17.6$

Initially, the estimated population density fluctuates greatly, indicating that there are too few samples for a reliable estimation of the density. Eventually, the fluctuations decrease and the estimated density will remain more or less the same with each successive sample. The actual number of samples required will depend on the distribution of organisms, but for practical purposes, a minimum of 20 samples should usually be taken. The effect of sample size on population density estimates could form the basis of an individual study.

Positioning the quadrat

We adopt a scientific approach to obtaining a number of random samples, by using pairs of random coordinates. Random coordinates are used to position the quadrat using a fixed reference point, for example one corner of a field being studied. Two tape measures can be set out at right angles to each other to set up a pair of axes. Pairs of random numbers, for example 5 and 13, are then used to indicate the directions in the x and y axes, in this case it would be 5 m along in one direction, then 13 m along perpendicular to that axis.

Random numbers can be obtained in several ways. Numbers representing the distances in metres can be written on separate pieces of paper, placed in a small bag or box, shaken well, then one number is drawn out (the 'numbers out of a hat' method). Each time a number is drawn out, it should be noted, replaced, and the procedure repeated. Tables of random numbers are available in books of statistical tables or ecological techniques. Two adjacent columns can be used as the x and y coordinates. A computer (and some calculators) can be used to generate random numbers.

A belt transect

A transect is a form of systematic sampling, where the samples are arranged in a linear sequence. Transects are particularly useful for recording changes in populations of plants where some sort of transition exists, for example, from the edge of an open field into an adjacent woodland, or across a salt marsh from the high water mark to the low water mark.

A belt transect is a strip usually 0.5 m in width, located across a study area in such a way that the transitions in plant populations are highlighted. A tape measure is laid across the sampling area and a 0.25 m² quadrat frame laid down at 0.5 m intervals alongside the tape measure. Animals and plants within each quadrat are counted, identified and recorded. With a transect over about 15 m in length, this procedure becomes too time consuming and it is more usual to carry out quadrat sampling at 1 m intervals.

Temperature

Temperature is an important parameter as it can affect the metabolic activities of many organisms and, in turn their growth or reproductive rates. The solubility of oxygen in water is inversely related to the temperature, so this is an important factor to consider in aquatic ecosystems. The ordinary glass thermometer is not really suitable for field work as it is too fragile. There are a number of electronic thermometers available, which have several advantages. For example, if they have a temperature probe, it is possible to measure the temperature of water at different depths, or the probe can be pushed into soil to measure soil temperature. Some temperature probes are ideally suited to interfacing with datalogging equipment which makes it possible to make continuous recordings over extended periods of time.

Figure 4.8 A temperature probe being used to take measurements of soil temperature

Light

Light is a particularly important parameter in most ecosystems, as light influences the primary productivity and therefore, in turn, the rest of the ecosystem. However, measurement of light intensity is not particularly easy. Some light meters will give an indication of the relative light intensity and, unless the instrument is calibrated, will give readings in arbitrary units. This is adequate if you are recording, for example, the change in light intensity as you move into woodland from an area of open ground. When taking readings, it is important that the solar cell is always pointing in the same direction, usually towards the Sun, and that readings are taken at the same height above ground. To make comparisons meaningful, it is important that the light source remains constant while you are taking the readings. If the Sun becomes hidden by a cloud, it will obviously change the reading in one place. Recordings should, therefore, be made as quickly as possible. Light sensors are also available which make continuous recording possible over a period of time.

Soil type

As previously discussed, soil type can be investigated by sedimentation and the digging of soil profiles. Different soils can be compared and contrasted by these methods.

Soil water

Soil water is divided into several categories. After heavy rain, before water has had a chance to drain away, the soil is said to be at **saturation**. Much of this water (referred to as the **gravitational water**) will move down through the spaces between soil particles, but some will remain adhering to soil particles. This is known as **capillary water**. Soil that contains all the capillary water it can hold against gravity is said to be at **field capacity**. This can be determined by placing a soil sample in a funnel, adding water and leaving it until water stops dripping out. The percentage of water in the soil sample can then be determined using the method described on page 77.

Measurement of pH

The pH of soil is an important factor which influences the organisms which it will support. Most crop plants grow well at a pH of 6.5 and this pH will support a wide range of soil organisms such as earthworms. Soil pH can vary widely from about 3 to over 8. Low pH soils are relatively infertile, with a low calcium content, and support plants such as heathers which are adapted to these conditions. At the other extreme, alkaline soils have a high calcium carbonate content and support plants such as *Clematis*.

Measurement of salinity in aquatic habits

Salinity is a measure of the salt content of sea water. The salts in the sea are mainly sodium and chloride, but smaller concentrations of other ions are also present. These include potassium, magnesium, calcium and sulphate ions. The salt concentration is usually given the symbol ‰ (parts per thousand). Salinity can be determined by titrating a sample of sea water against silver nitrate solution, using the method described on page 78.

Measuring wind and water movements

Several hand-held meters are available, which can be used in the field. Examples are wind speed indicators, and flow meters, including the stream flow meter which has the advantage of being able to be used at specific depths. Flow meters are generally expensive and as an alternative, adequate measurements of flow rate can be made using the timed float method described on page 79.

Measurement of dissolved oxygen

The solubility of oxygen in water is relatively low and varies inversely with the temperature, as shown in Table 4.6. The amount of oxygen dissolved in water is usually measured either in mg dm^{-3} or as a percentage of air saturation (the amount of oxygen present expressed as a percentage of the amount dissolved in water at equilibrium with air at the same temperature).

There are two main approaches to the measurement of dissolved oxygen, either using a dissolved oxygen meter and suitable electrode, or chemical determination. Dissolved oxygen meters are expensive but convenient to use in field work. Chemical analysis is simple and relatively inexpensive. A method is given on page 79.

Table 4.6 *Oxygen content of water saturated with air at normal pressure*

Temperature/°C	Dissolved oxygen/mg dm^{-3}
0	14.66
5	12.37
10	10.92
15	9.76
20	8.84
25	8.11

Succession

The development of a stable community, or **climax community**, takes place in a number of stages and long-term changes in the composition of a community are known as **succession.** Succession occurs because the activities of living organisms over a period of time have a modifying effect on the nature of the environment. Very few large plants can grow where there is no organic matter in the soil, such as on the bare rock of newly-erupted islands, sand dunes or glaciated surfaces. Such areas can be colonised by algae and lichens, forming a **pioneer community**. Their activities result in the accumulation of organic debris which, together with the weathering of the rock particles, leads to the formation of a soil in which other organisms can live. Once a soil has been formed, mosses, ferns and small herbaceous plants can grow, replacing the pioneer organisms. Eventually, these smaller plants may be replaced by larger plants until the climax community, a woodland or forest, is formed. Such a succession from bare rock to forest is called a **sere** and each different community in the succession is called a **seral stage**, or **seral community**. This type of succession, from bare rock to woodland, is known as a **primary succession**. Where a succession occurs on an area of burnt heathland or cleared woodland which has previously had vegetation on it, then the succession is called a **secondary succession** (illustrated in Figure 4.9). In both types of succession, the plant and animal species of neighbouring areas will have an influence on the composition of the flora and fauna. Spores and seeds can be carried by air movements and animals can move independently, both over quite large distances. Larger animals such as mammals and birds are amongst the later residents of communities, although they may visit an area undergoing colonisation to seek food. It is unlikely that the conditions will be suitable for them to build nests or find shelter until shrubs and trees have colonised the area. It may take hundreds of years for a primary succession to reach the climax community, but less time for a secondary succession as the soil has already been formed.

Some general points can be made about the changes taking place during succession.

ADAPTATION TO THE ENVIRONMENT

- The types of animals and plants differ in their characteristics from one seral stage to another. In the early stages, the plants will be annuals, completing their life cycles in one growing season, succeeded later by herbaceous perennials and later still by woody perennials.
- The succession is always associated with increase in biomass. As the soil becomes deeper, it can support a greater amount of vegetation and hence larger numbers of animals.
- The diversity of plant and animal species tends to increase. The greatest number of animal species will be present when the climax has been reached. The number of plant species may be greater before the climax vegetation is reached. This is particularly noticeable where the climax is a beech wood; the dense canopy reduces the number of ground species due to the availability of light.

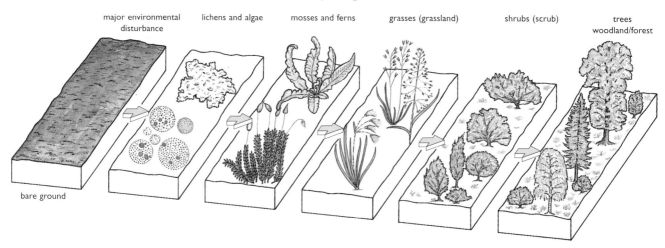

major environmental disturbance   lichens and algae   mosses and ferns   grasses (grassland)   shrubs (scrub)   trees woodland/forest

bare ground

Figure 4.9 Diagram showing secondary succession from bare ground to woodland

When conditions are dry, the initial colonisers must be adapted to survive where water is scarce, so tend to be xerophytes. Such a succession is known as a **xerosere**. As succession proceeds, the conditions change and with the development of an organic soil, more water retention is possible enabling mesophytes to grow. If the succession starts in shallow water, then a **hydrosere** is formed. In this case, the accumulation of organic debris results in the conditions becoming gradually dryer and more favourable to mesophytes. These types of succession are illustrated in Figure 4.10.

A climax community is stable and in equilibrium with the climatic conditions. For these reasons it is known as a **climatic climax**. In most of Britain, where the climate is cool temperate, the climatic climax is deciduous (broad-leaved) woodland. Humans and other animals can interfere with this climatic climax and a different equilibrium may be reached, known as a **biotic climax**. Any form of management of the land or aquatic environment can produce a **biotic climax**. One of the most familiar effects is that of intensive grazing by sheep or rabbits. The growth of shrubs and trees is prevented because the grazing animals feed on the young shoots and the plants never have a chance of becoming established. In this case the biotic climax is grassland. It is seen extensively in the North and South Downs, where sheep are farmed and where

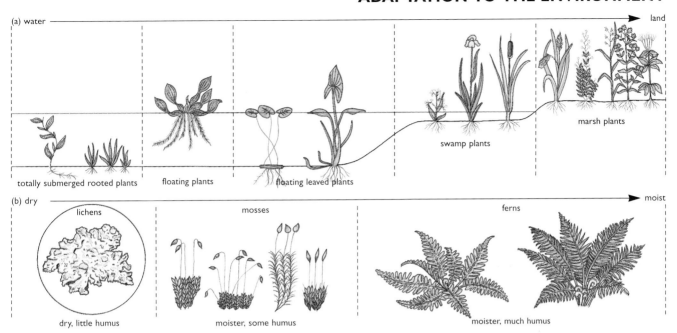

(a) water → land

totally submerged rooted plants floating plants floating leaved plants swamp plants marsh plants

(b) dry → moist

lichens mosses ferns

dry, little humus moister, some humus moister, much humus

Figure 4.10 (a) Hydrosere and (b) xerosere succession

rabbits are plentiful. When most of the rabbit population was wiped out due to myxomatosis, large areas of chalk grassland quickly developed into scrub, with hawthorn and other shrubs. Conservationists became concerned as the distinctive chalk flora was in danger of being lost in some areas, so the scrub was cut back in order to preserve the diversity of species.

The type of biotic climax illustrated by chalk grassland is sometimes called a **deflected climax**. Other examples of deflected climaxes are brought about by seasonal mowing, burning and weed spraying. Most of the land with which we are familiar comes into this category, from our lawns, playing fields and parks, which are mown to keep them as grassland, to grouse moors, where burning is carried out to keep the heather in the right condition for the game birds, and roadside verges, where until fairly recently both mowing and spraying to get rid of weeds has been carried out.

Sometimes a deflected climax results from a combination of climatic conditions. When a hydrosere develops in an area of high rainfall and low temperatures, acid conditions prevail and the plant remains do not break down completely but accumulate, forming peat. The resulting climax community is moorland or bog instead of deciduous woodland, with large amounts of *Sphagnum*, a moss characteristic of wet, acid conditions.

Zonation

Plant succession can be observed in the zonation seen in sand dune systems, where it is possible to see all the seral stages of a primary succession. Nearest the water's edge are the newly-formed **young dunes**, characterised by plants such as sea holly (*Eryngium sp.*) and sea spurge (*Euphorbia*), which are highly adapted to living in dry conditions. Behind the young dunes are **yellow dunes**, reaching heights of up to 20 m, mainly colonised by marram grass (*Ammophila*), but still showing large areas of bare sand. The colonisation is not complete and these dunes are often referred to as **partially fixed**. Behind these are **grey**

Figure 4.11 Sand dunes in North Wales showing Ammophila *colonising sand near the sea (in the foreground) and dunes building up and stabilising (in the background). The fences have been placed across sandy areas as a conservation measure to help maintain the dune system.*

ADAPTATION TO THE ENVIRONMENT

dunes, where the sand is more stable and there is a greater diversity of plants. The ridges of dunes are separated from each other by **slacks**, which are low-lying damper areas with their own characteristic plants. Further back, grassland or heathland may become established. In some situations, a scrub of brambles (*Rubus*), elder (*Sambucus*), hawthorn (*Crataegus*) and dwarf willow develops, which may eventually give rise to woodland. This is illustrated in Figure 4.12.

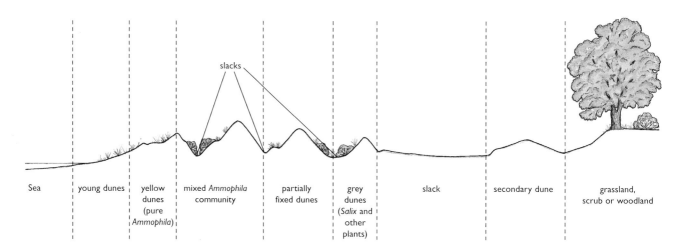

Figure 4.12 Zonation in a sand dune system

Vertical zonation can be seen on mountains, where variations in the abiotic environment due to altitude have an effect on the spatial distribution of species. For each 100 m rise in altitude, the temperature drops by 0.5 °C and this has an effect on the plants and animals. Depending on the altitude, there is usually little soil on the top of a mountain so few plants, apart from lichens and some mosses, can grow. Lower down there is often grassland, scrub and eventually woodland at the bottom. In addition to the changes in temperature, there may be differences in rainfall, with one side of the mountain drier than the other. This type of zonation, illustrated in Figure 4.13, reflects the differences in the climatic conditions: a climax community develops within each zone.

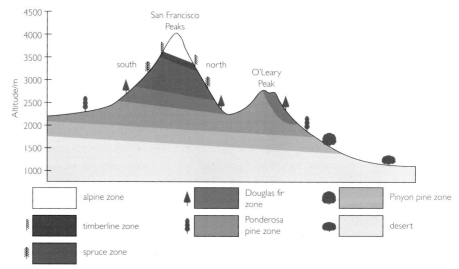

Figure 4.13 Zonation of vegetation on San Francisco Peaks, Arizona, as viewed from the south-east

Another example of zonation can be seen on a rocky shore such as that depicted in Figure 4.15, where seaweeds and marine animal species occupy specific bands or **zones** between the low tide and high tide levels. The distribution of organisms depends on their degree of tolerance to exposure and is affected by variations in the tidal height, which in turn cause variations in temperature and degree of desiccation. The tides range from the upper limit of **EHWS** (extreme high water of spring tides) to **ELWS** (extreme low water of spring tides). This range means that some organisms living high up on the shore will only be covered with sea water twice a month, whereas those living at the lower limit will only be uncovered twice a month. Between these limits, organisms will be exposed to the air for varying periods of time, most of them twice a day.

Figure 4.14 A rocky shoreline in Pembrokeshire, Wales

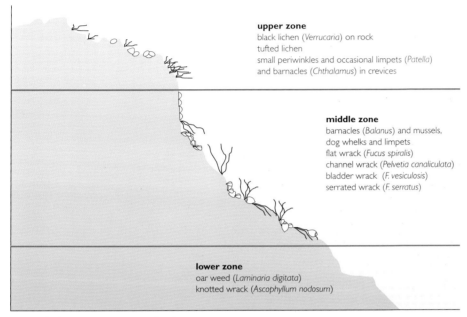

upper zone
black lichen (*Verrucaria*) on rock
tufted lichen
small periwinkles and occasional limpets (*Patella*)
and barnacles (*Chthalamus*) in crevices

middle zone
barnacles (*Balanus*) and mussels,
dog whelks and limpets
flat wrack (*Fucus spiralis*)
channel wrack (*Pelvetia canaliculata*)
bladder wrack (*F. vesiculosis*)
serrated wrack (*F. serratus*)

lower zone
oar weed (*Laminaria digitata*)
knotted wrack (*Ascophyllum nodosum*)

Figure 4.15 Zonation on a rocky shore

When the organisms are covered by the water, the temperature is more or less uniform, but when they are exposed to the air the temperature is much more variable. Organisms which can tolerate the greater variability of the air temperature tend to be found higher up the shore. Such organisms are also adapted to withstand drying out, many of them producing mucus coatings (e.g. *Actinia*– sea anemones) or attaching themselves firmly to the rocks (e.g. *Patella* – limpets).

The effects of wave action will depend on the slope of the shore and which way it is facing. Seaweed species growing on sheltered shores tend to be more abundant and larger than those on exposed shores, where wave action is greater. The following account of an investigation illustrates these differences.

A sample of ten specimens of the seaweed (*Fucus vesiculosus*) was collected at random from a sheltered shore, P, with little wave action, and a further sample of ten specimens was collected from a nearby exposed shore, Q, with considerable wave action.

ADAPTATION TO THE ENVIRONMENT

The length of the frond and diameter of the holdfast were measured for each specimen collected. The results of this investigation are given below.

Table 4.7 *Differences in length of seaweeds on exposed and sheltered shores*

Sample from sheltered shore P		Sample from exposed shore Q	
Length of frond /cm	Diameter of holdfast /cm	Length of frond /cm	Diameter of holdfast /cm
61.0	0.7	40.0	1.0
53.0	0.5	24.0	1.0
46.0	0.7	27.0	0.8
62.0	0.5	29.0	0.8
72.0	1.0	50.0	0.7
89.0	0.2	37.0	0.5
64.0	0.6	52.0	1.0
49.0	0.7	28.0	0.5
50.0	0.4	33.0	0.6
44.0	0.4	47.0	0.7

Neither the zonation on the rocky shore nor that on the mountain show true succession in the same way as the sand dune system. The rocky shore and the mountain illustrate spatial distribution due to variations in the physical factors of the environment, whereas zonation in sand dunes enables us to observe a succession of seral stages which have taken place over a period of time.

Determining water content of a soil sample

Materials

- Soil sample
- Evaporating dish or crucible
- Accurate balance
- Thermostatically controlled oven
- Desiccator
- Tongs

Method

1 Weigh the evaporating dish and record the mass.
2 Add about 10 g of soil to the dish and reweigh.
3 Subtract the mass of the dish to obtain the fresh mass of soil.
4 Place the dish plus soil in a thermostatically controlled oven at 110 °C. Leave for 24 hours.
5 Remove the dish and allow to cool in a desiccator. Record the mass.
6 Replace in the oven for 24 hours and repeat the previous stage until a constant mass is obtained.
7 Calculate the loss in mass of the soil sample. This represents the mass of water in the soil which can be expressed as a percentage of the fresh mass:

Percentage of water in the soil = (loss in mass ÷ fresh mass) × 100

8 Use this method to compare the percentage of water in different soil samples.

Determining the pH of a soil sample

Materials

- Soil sample
- Test tube
- Barium sulphate
- Distilled water
- Universal indicator solution
- Colour chart

Method

1 Place about 1 cm of soil and 1 cm of barium sulphate in a test tube. Barium sulphate causes clay particles to settle leaving a clear solution.
2 Add 10 cm^3 of distilled water and shake the tube thoroughly.
3 Add a few drops of universal indicator solution and compare the colour with the colour chart against a white background.
4 Record the pH value.

This method has the advantage of being suitable for use in the field. More accurate results may be obtained by using a pH electrode and meter to find the pH of the soil solution.

ADAPTATION TO THE ENVIRONMENT

Determining the salinity of a sample of sea water

Materials

- Sample of sea water
- Silver nitrate solution, 27.25 g dm^{-3}
- Burette and stand
- Small conical flask
- Graduated pipette
- Potassium chromate indicator solution, 10% w/v

VERY TOXIC
potassium chromate
indicator solution

Method

1 Pipette 10 cm^3 of sea water into a small conical flask.
2 Add a few drops of potassium chromate indicator solution.
3 Add silver nitrate solution slowly from a burette, swirling the flask constantly to mix.
4 The end point occurs when there is a distinct change in colour of the indicator from bright yellow to deep orange.
5 The volume of silver nitrate used (cm^3) is equivalent to the salinity of the sea water (parts per thousand, ‰), subject to a minor correction shown in Table 4.9.

WEAR EYE
PROTECTION

Table 4.9 *Table of corrections for salinity*

Salinity found / ‰	Correction
40	−0.15
38	−0.08
36	−0.03
34	+0.03
32	+0.07
30	+0.11
28	+0.15
26	+0.17
24	+0.20
22	+0.22
20	+0.23
18	+0.23
16	+0.23
14	+0.20
12	+0.19
10	+0.16

Timed float method to determine current velocity

Materials

- Tape measure
- Stop watch
- Suitable float. An orange is recommended because it is conspicuous and will float mainly below the surface

Method

1 Choose a suitable straight stretch of water. Measure the distance with the tape measure.
2 Accurately time the float over the measured distance. Repeat three times to obtain a mean.
3 Divide the mean time by the coefficient 0.85. This will give a more accurate velocity for the stream because the water at the surface flows faster than that beneath.
4 Calculate the velocity using the formula: velocity = distance ÷ time

As an example:

Mean time of floats	= 14 seconds
Mean time divided by 0.85	= 16.47 seconds
Distance	= 10 metres
Velocity	= 10 ÷ 16.47
	= 0.61 m sec^{-1}

Chemical determination of dissolved oxygen

Materials

- Solution A: 45 g of manganese (II) chloride in 100 cm^3 of distilled water
- Solution B: 70 g of potassium hydroxide and 15 g of potassium iodide in 100 cm^3 of distilled water
- Solution C: 50% v/v sulphuric acid
- Solution D: 1.55 g of sodium thiosulphate (IV) dissolved in distilled water and made up to 1.0 dm^3
- Solution E: 0.25% w/v starch in saturated sodium chloride solution

CORROSIVE
potassium hydroxide
sulphuric acid

Method in the field

1 Obtain carefully and without any air bubbles a 250 cm^3 sample of water.
2 Add 1.0 cm^3 of Solution A, which sinks to the bottom of the bottle.
3 Immediately add 1.0 cm^3 of Solution B and close the bottle without introducing air bubbles.
4 Mix the contents for 1 minute, then allow the precipitate to settle.

The dissolved oxygen in the water sample will now be 'fixed' since the manganese (II) salt is oxidised to manganese (III). The sample can be taken back to the laboratory.

HARMFUL
manganese(II)
chloride

Method in the laboratory

1 Using a safety bulb pipette, carefully add 2 cm^3 of solution C down the inside wall of the sample bottle.
2 Close the bottle and mix the contents.
3 Titrate 100 cm^3 of the sample against solution D until pale yellow.
4 Add 0.5 cm^3 of Solution E and continue adding solution D drop by drop until the blue colour disappears.
5 Calculate the dissolved oxygen content using the following formula:

$$\text{Dissolved oxygen content in mg dm}^{-3} = \frac{1000 \times V_2}{V_1 \quad 10}$$

Where V_1 = sample volume
V_2 = volume of 0.0125 mol dm^{-3} sodium thiosulphate solution used.

WEAR EYE PROTECTION
(in laboratory AND in field)

Ecosystems

Components of ecosystems

Figure 5.1 A deciduous woodland in April

Ecosystems are complex associations of plants, animals and microorganisms which interact with each other and with their non-living environment. The flow of energy and flow of nutrients within ecosystems are regulated by the organisms living there. Through photosynthesis, green plants incorporate about 1 per cènt of the sun's energy falling on them into organic compounds. The plants may then be eaten by herbivores (primary consumers), which in turn may be eaten by carnivores (secondary consumers). This type of sequence forms a **food chain** and the different levels within it are called **trophic levels.**

An **ecosystem** was first defined in 1935 by Sir Arthur Tansley as 'the living world and its habitat'. This definition includes both the organisms within the ecosystem (the **biotic** component) and the physical and chemical factors (the **abiotic** component) which influence them. Different ecosystems may have more or less clearly defined boundaries, but they also merge into one another, so it may not always be easy to see the boundaries. For example, there is a gradual transition between a freshwater ecosystem, such as a pond, through the surrounding marshy ground to a field.

An ecosystem will contain a number of different **habitats**. A habitat is the place where an organism lives within an ecosystem. The pond shown in Figure 5.2 is a habitat for a whole range of different species, including pond snails and phytoplankton. Many organisms will only occupy one particular part of the total habitat, for example the pea mussels living in the mud at the bottom of the pond. This location within the overall habitat is sometimes referred to as a **microhabitat**.

An ecosystem thus comprises several habitats with their communities of organisms, and includes both biotic and abiotic components. A useful way of thinking of an ecosystem is as a network of habitats interlinked by the flow of

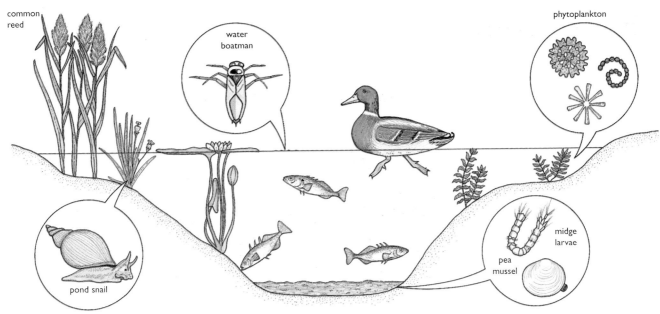

Figure 5.2 A freshwater ecosystem. Organisms in the inserts are shown enlarged, but not to the same scale.

energy and nutrients. Abiotic components are the non-living factors which will influence the distribution of organisms within an ecosystem. These components include physical factors (such as light intensity, temperature, rainfall, wind and water currents) and chemical factors (including all the elements and compounds in the ecosystem). Biotic components also affect the populations of organisms in an ecosystem; these include the availability of food, competition and predation. We will discuss these factors in detail later in this chapter.

Population and community

A **population** is defined as a group of individuals of one species found in the same habitat. For example, the woodland habitat shown in Figure 5.1 includes a population of bluebells (*Endymion non-scriptus*) and the leaf litter contains a population of brown-lipped snails (*Cepaea nemoralis*).

The term **community** is used to refer to either all the organisms present in a particular habitat, or to one group within the habitat, such as the plant community. The plant community in a woodland habitat will contain all species of trees, shrubs and herbaceous plants. The mollusc community in a pond might contain *Lymnaea stagnalis*, *L. peregra* and *Planorbis crista*.

Population dynamics
In this section, we will consider the growth of populations and some of the factors which are responsible for influencing their growth rate. Populations in ecosystems interact and, as a result, the **population density** (that is, the number of individuals per unit area or per unit volume) changes with time.

Factors affecting population size
One of the fundamental characteristics of living organisms is that they reproduce and so populations tend to increase in number. A single bacterial cell, for example, may divide into two cells every twenty minutes under favourable

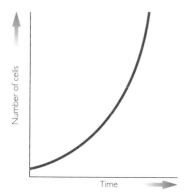

Figure 5.3 Exponential growth

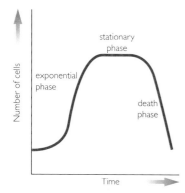

Figure 5.4 Growth curve for a bacterial population

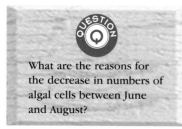

What are the reasons for the decrease in numbers of algal cells between June and August?

conditions. This means that the number of cells in a culture will double every twenty minutes. A graph of the number of cells plotted against time (referred to as a growth curve) shows **exponential growth** (Figure 5.3).

This growth cannot continue indefinitely – the food supply may be limited, or the organisms may produce toxic by-products of metabolism. Within a relatively short period of time, the growth rate will decrease and the cells will stop dividing. This is referred to as the **stationary phase.** Eventually, cells may begin to die, so that the number of viable cells will decrease. The culture has then entered the **death** (or **decline**) **phase** (Figure 5.4).

Similar changes can occur in natural populations. For example, rapid growth of algal cells can occur in ponds or lakes as a result of enrichment with mineral ions such as nitrate. Figure 5.5 shows the growth and subsequent decline in numbers of the brown alga *Dinobryon divergens*.

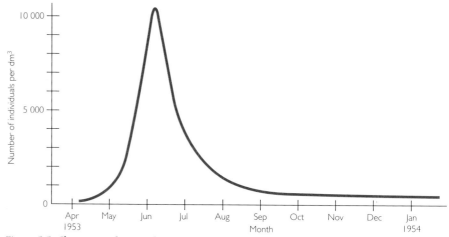

Figure 5.5 Changes in the population of the alga Dinobryon divergens *between April 1953 and January 1954*

Figure 5.6 shows another example of changes in a population, that of the numbers of sheep in Tasmania following their introduction to the island in 1814. The fine line shows the year-to-year variations in numbers, the thick line shows the overall population trend. Notice that the numbers increase up to a maximum and then the population levels out and remains approximately constant.

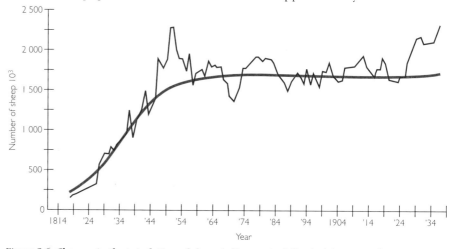

Figure 5.6 Changes in the population of sheep in Tasmania following their introduction in 1814

Carrying capacity and environmental resistance

Although this population of sheep could, theoretically, continue to increase, it does not and levels out at a population of about 1.5 million. The maximum size of a population that a particular environment can support is called that environment's **carrying capacity.** Various environmental factors, referred to as **environmental resistance**, reduce the growth rate of a population. These factors include disease, competition, predation and unfavourable climatic conditions.

The number of individuals in a population changes as a result of fluctuations in the birth and death rate. To illustrate this, we will consider the results of a study on the numbers of great tits (*Parus major*) in Sweden. Figure 5.7 shows the changes in both the population of great tits and the availability of beech seeds upon which they feed. Dramatic increases in the numbers of these birds followed large crops of beech seeds. This extra available food meant that more juvenile birds were able to survive the winter and their numbers increased the size of the population the following spring.

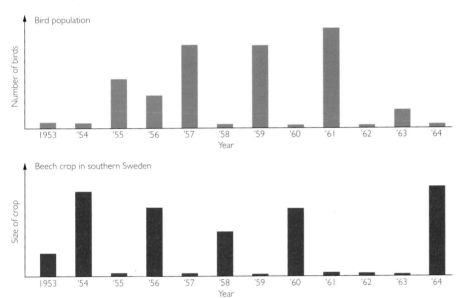

*Figure 5.7 Changes in the breeding numbers of great tits (*Parus major*) and the size of the beech crop in southern Sweden*

As we have seen, populations do not keep increasing indefinitely. Population size may be limited by either increases in mortality (deaths) or by decreases in natality (births). Mortality and natality are often expressed as rates, for example, a mortality rate of 2/100 (or 2 per cent) represents two deaths in a population of 100.

Density-dependent and density-independent mortality rates

Death rates that change with population density are described as **density-dependent** mortality rates. Density-dependent mortality is an important factor in limiting the growth of a population when it reaches a high density.

To illustrate this factor, we will consider the results of a detailed study of the population of tawny owls in Wytham Wood, near Oxford. This showed that the mortality rate of young owls was much greater in years when the wood already had a high owl population. In other years, when the overall population was low, many more young owls survived and established territories of their own. The mortality rate of young owls is therefore density dependent and this has an important regulating effect on the population density. Figure 5.8 compares density-dependent and density-independent mortality rates.

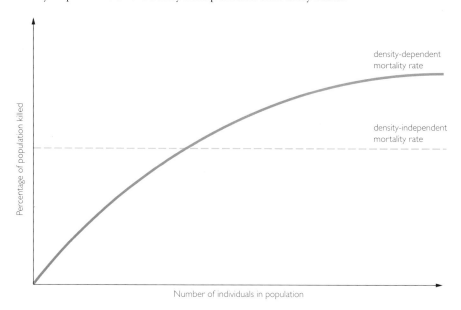

Figure 5.8 Density-dependent and density-independent mortality rates compared

The solid line shows an example of a density-dependent mortality rate, where the mortality rate is proportional to the population density. The dotted line indicates a density-independent mortality rate, showing that a fixed percentage of the population is likely to die irrespective of the population density. In the case of tawny owls, this could be a result of some of the birds being killed by the winter cold.

Density-dependent mortality factors are always biotic (such as competition for food or space), whereas density-independent mortality factors may be either abiotic (such as winter cold) or biotic.

Intraspecific and interspecific competition

In an ecosystem, organisms which are using a limited resource, such as food, space or light, will compete with each other for that resource. As a result of this competition, some individuals may have a reduced growth rate, or increased risk of mortality. Competition therefore reduces the rate of population growth.

Intraspecific competition

This occurs between organisms of the same species. These are likely to have similar resource requirements, so they will compete for any that are in limited supply. The more competitors there are, the greater the effect of intraspecific competition is likely to be. In other words, intraspecific competition is density dependent. Competition between tawny owls, described previously, is one example of intraspecific competition. As another example, we will consider the effect of intraspecific competition on the limpet *Patella cochlear*. This is a marine mollusc which feeds by grazing on algae which grow on the limpets' rocky habitat. As the density of limpets increases, competition for food also increases. The effects of population density on the maximum length and biomass of *Patella cochlear* are shown in Figure 5.9. As the population density rises, intraspecific competition increases, resulting in a reduction in length and maintaining an approximately constant population biomass.

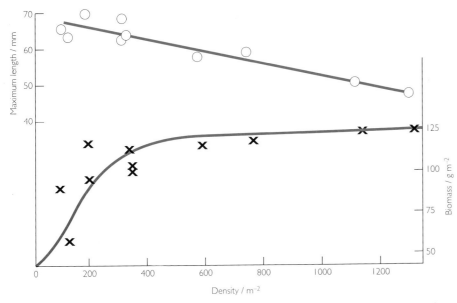

Figure 5.9 Effect of intraspecific competition on length and biomass of the limpet Patella cochlear

Interspecific competition

In a complex ecosystem, with many different populations competing for resources such as space, food or light, it is inevitable that competition will occur between different species. This is known as **interspecific competition.**

Over 50 years ago, the Russian ecologist G.F. Gause proposed the **principle of competitive exclusion.** This principle states that if two species are competing with each other for the same limited resource, then one species will be able to use that resource more efficiently than the other, so the former species will eventually eliminate the latter. Gause carried out a series of experiments using two species of *Paramecium: P. caudatum* and *P. aurelia* (kingdom Protoctista, phylum Ciliophora). The results of some of these experiments are shown in Figure 5.10.

These experiments show that, when these two species are grown together in a mixed culture, *P. aurelia* drives *P. caudatum* to extinction, in other words, *P. caudatum* has been competitively excluded. This is because *P. aurelia* uses the available food resources more efficiently, and reproduces more quickly than *P. caudatum* under these conditions.

Subsequent experiments with both plants and animals, by other workers, have confirmed the principle of competitive exclusion, but the results are not always easily predictable and can vary according to the actual conditions. For example, experiments on two species of the flour beetle, *Tribolium,* carried out by T. Park in 1954, showed that the species which survived varied according to the climatic conditions in which they were cultured. *T. confusum* won under cool, dry conditions, while *T. castaneum* usually won under hot, moist conditions. The results of these mixed culture experiments are shown in Table 5.1.

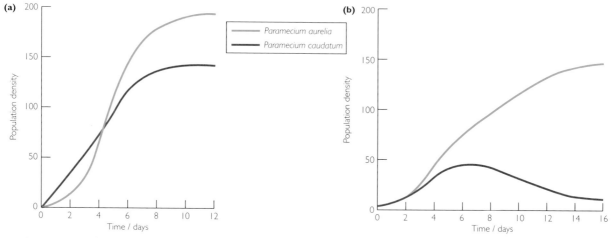

Figure 5.10 Interspecific competition between two species of Paramecium (a) cultured separately; (b) cultured together

Further experiments have shown competitive exclusion in populations of plants, for example two species of clover, *Trifolium repens* and *T. fragiferum*. These two species were sown at various densities, but *T. fragiferum*, although initially slower growing, eventually overshadowed *T. repens*, competing more effectively for light and excluding the other species.

Competitive exclusion has been observed in natural populations. One of the species is not, however, usually driven to extinction by the other, but remains rare and may increase in numbers if the population of its competitor should decrease.

Predator–prey relationships

A **predator** is an organism which feeds on all or parts of animals, referred to as the **prey**. Predation is an important biotic factor which influences the abundance of organisms in an ecosystem.

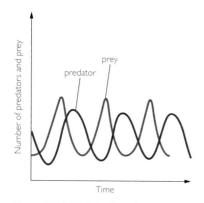

Figure 5.11 Model of predator–prey interaction

The size of a population of a predator will depend on the availability of its food source, so the larger the prey population, the larger the population of predators it can support. However, increased numbers of predators will decrease the numbers of prey organisms. The populations of each would be expected to show a series of regular cycles, where the population of predators is out of phase with the population of prey. This is illustrated in Figure 5.11, which shows a hypothetical model of predator–prey numbers.

Table 5.1 *Interspecific competition between two species of* Tribolium

Climate	Percentage wins	
	T. confusum	*T. castaneum*
hot – moist	0	100
temperate – moist	14	86
cold – moist	71	29
hot – dry	90	10
temperate – dry	87	13
cold – dry	100	0

What are the biotic and abiotic factors influencing the populations of *Tribolium* in this experiment?

This model has been supported by experimental work involving two species of mites. *Typhlodromus occidentalis*, a predatory mite, was added to cultures of its prey, *Eotetranychus sexmaculatus*, a herbivorous mite, feeding on oranges in a tray. Figure 5.12 shows the resulting fluctuations in the populations of each species, where the numbers of each can be seen to oscillate and the cycles are slightly out of phase with each other.

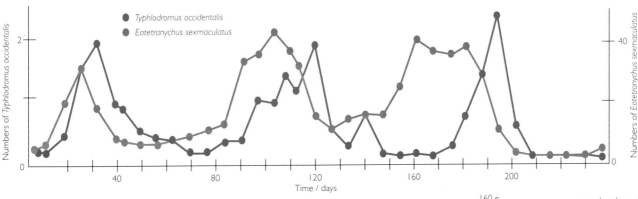

Figure 5.12 Predator–prey interaction between two species of mite

Similar fluctuations in populations can be seen in, for example, the numbers of glasshouse whiteflies (*Trialeurodes vaporariorum*) and the predatory insect *Encarsia formosa*. This predator is used as a means of biological control of the whitefly, which is discussed in Chapter 7. There is also evidence that predator–prey interactions are responsible for the fluctuating numbers of snowshoe hares (*Lepus americanus*) in northern Canada. One of its main predators, the lynx (*Lynx canadensis*) shows similar, but out of phase, fluctuations in population density (Figure 5.13).

Flow of energy through an ecosystem

All organisms in an ecosystem depend on an adequate supply of energy for their survival. Energy from the sun, trapped by photosynthesis, provides the source of energy for all living organisms. Carbohydrates, which are produced in photosynthesis, are used both as building blocks for the growth of green plants and as an energy source for the plant. Carbohydrate is respired in plant cells to provide ATP for endergonic reactions, such as protein synthesis.

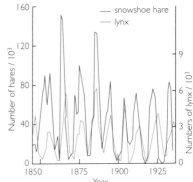

Figure 5.13 Predator–prey interaction between lynx and snowshoe

Heterotrophs cannot fix carbon dioxide by photosynthesis and therefore depend on obtaining ready-made organic compounds from their food. In heterotrophs, these organic compounds serve the same functions as they do in autotrophs, that is, they provide building blocks for the synthesis of new cell compounds for growth, and provide substrates for respiration and the production of ATP.

What are the advantages and disadvantages of exploiting predator–prey relationships in biological control of pests?

Producers and photosynthesis, consumers and decomposers

We have already stated that energy from the sun enters an ecosystem through organic compounds produced, in photosynthesis, by green plants. Green plants are described, in this context, as **primary producers**. The Earth's surface receives between 0 and 5 joules of solar energy per square metre every minute. However, only a small percentage of this energy is captured by chlorophyll and converted into chemical energy of newly synthesised organic compounds. Some of the light striking a leaf will be reflected, or will pass straight through, and about half is made up of wavelengths which are not used in photosynthesis. There is also some wastage of energy through biochemical inefficiency of the reactions of photosynthesis.

The bodies of all the living organisms within a unit area constitute the standing crop of **biomass**. Biomass is defined as the mass of organisms per unit area of ground (or water) and is usually expressed in either units of energy (for example $J\ m^{-2}$) or as units of dry organic matter (tonnes ha^{-1}). The primary productivity of an ecosystem is the rate at which biomass is produced per unit area by green plants. It is expressed either in units of energy (such as $kJ\ m^{-2}\ yr^{-1}$) or as the mass of dry organic matter produced (such as $kg\ ha^{-1}\ yr^{-1}$). The total fixation of energy by photosynthesis is referred to as **gross primary production** (**GPP**); the most commonly used units for GPP are $kJ\ m^{-2}\ yr^{-1}$. Some of the GPP will be used by the plant for respiration and will ultimately be lost as heat energy. The rate at which organic compounds are used in this way is referred to as **plant respiration** (**R**) and is also measured in $kJ\ m^{-2}\ yr^{-1}$. The difference between GPP and R is known as the **net primary production** (**NPP**). This represents the actual rate of production of new biomass that is available for consumption by heterotrophic organisms.

We can summarise the relationship between GPP, NPP and R by the equation:

$$GPP = NPP + R$$

This relationship and the fates of solar radiation falling on a leaf are shown in Figure 5.14.

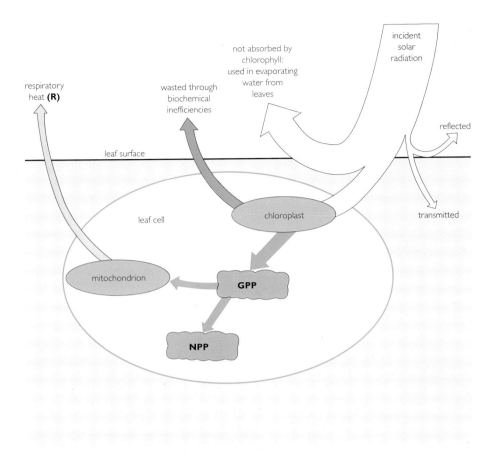

Figure 5.14 Fates of solar energy falling on a leaf

It has been estimated that the global terrestrial net primary production is 110 to 120×10^9 tonnes dry mass per year, and in the sea, 50 to 60×10^9 tonnes per year. Since only the net primary production is available for consumers to eat, NPP values are often used to compare the **productivity** of ecosystems. Productivity varies widely in different ecosystems, as shown in Table 5.2.

Table 5.2 *Mean values for NPP in a range of ecosystems*

Ecosystem	Mean NPP/kJ m^{-2} yr^{-1}
extreme desert	260
open ocean	4700
temperate grasslands	15 000
temperate deciduous forest	26 000
intensive agriculture	30 000
tropical forest	40 000

ECOSYSTEMS

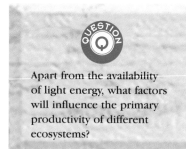

Apart from the availability of light energy, what factors will influence the primary productivity of different ecosystems?

What happens to the net primary production in an ecosystem? Some NPP will remain stored within plants and increase plant biomass, and some may be eaten by herbivores. This energy input to herbivores is referred to as **herbivore consumption**. In a grass field, about 40 per cent of the annual production may be eaten by cows, but in a forest only about 2 per cent of the NPP passes to herbivores. Herbivores may themselves be preyed upon by predators, the first carnivores. The energy input to these animals is termed **carnivore consumption**. A final part of the NPP, contained in, for example dead leaves and flowers, reaches the ground where it forms litter or **detritus**. This provides a source of food for a variety of organisms. Some of these are earthworms, termed **detritivores**, others are soil fungi and bacteria. These soil microorganisms are collectively known as **decomposers** and are the ultimate consumers of all dead organic matter in an ecosystem.

Food chains, food webs and trophic levels

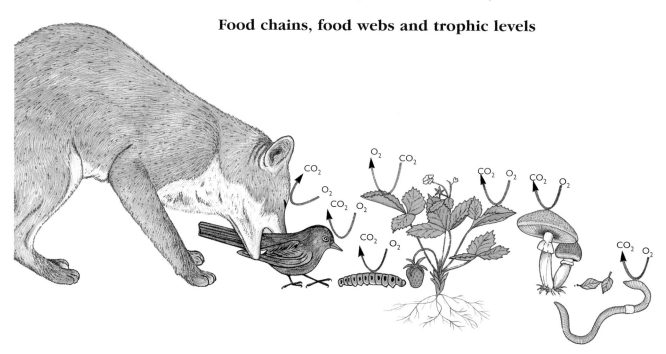

Figure 5.15   Heterotrophs depend on autotrophs

Figure 5.15 shows an example of a **food chain**, that is, a sequence of organisms eating other organisms. One sequence shown is:

plant $\rightarrow$ caterpillar $\rightarrow$ bird $\rightarrow$ fox

Each organism within the chain occupies a particular **trophic level**. The food chain 'plant to fox' shown above has four trophic levels and can be written in general terms as:

primary producer $\rightarrow$ herbivore $\rightarrow$ first carnivore $\rightarrow$ second carnivore

There are very large numbers of different food chains in the range of ecosystems on Earth, but it is rare to find chains with more than five trophic levels. Green plants fix only about 1 per cent of the sun's energy falling on their leaves, and successive members of a food chain incorporate into their own biomass about 10 per cent of the energy available in the organisms they consume. The loss of energy at each trophic level is so great that very little of the original energy remains in the chain after it has been incorporated successively into the biomass of organisms at four trophic levels.

If you look carefully at Figure 5.15, you will see other relationships between the organisms. As examples: dead plant leaves provide a food source for earthworms; fungi are decomposers and will also obtain nutrients from dead leaves; worms and caterpillars are eaten by both birds and foxes. It is clear that in this diagram there are several food chains, and one organism may feed at more than one trophic level – foxes can be first or second carnivores, or even herbivores, as they are partial to fruit. There is a complex series of feeding relationships between organisms in an ecosystem. This set of relationships is referred to as a **food web**. Figure 5.16 shows a generalised food web based on Figure 5.15.

Constructing a food web involves detailed field and laboratory work. The main approaches include observation of predator–prey relationships, laboratory food preference experiments, analysis of gut contents and the use of radioactive isotopes such as ^{32}phosphorus (^{32}P). Plants can be labelled with ^{32}P and the passage of this isotope is then followed in nearby animal populations.

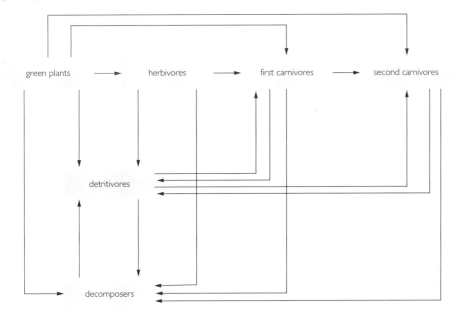

Figure 5.16 A generalised food web. The arrows show the direction of energy flow.

An example of a food web in a salt marsh is illustrated in Figure 5.17, showing the complex relationships between organisms. (Meiofauna are small invertebrate organisms which are adapted for living in the spaces between grains of sand and mud in marine deposits.)

Pyramids of number, biomass and energy
In an ecosystem, there are usually far more organisms at lower trophic levels than at higher trophic levels. Similarly, the biomass of primary producers present in an ecosystem is much greater than that of the herbivores, and successive trophic levels have a progressively smaller biomass.

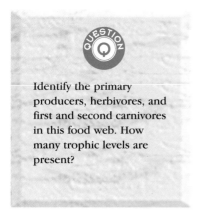

Identify the primary producers, herbivores, and first and second carnivores in this food web. How many trophic levels are present?

Figure 5.17 A food web on a salt marsh

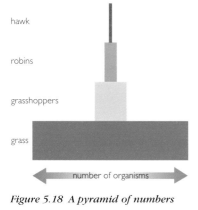

Figure 5.18 A pyramid of numbers

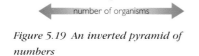

Figure 5.19 An inverted pyramid of numbers

We can represent these relationships as diagrams, known as **pyramids**. An example of a **pyramid of numbers** is shown in Figure 5.18. The width of each block is proportional to the numbers of organisms present at each trophic level.

One disadvantage of representing a food chain in this way is that it does not take into account the biomass of each trophic level. For example, one rose bush might support a very large population of aphids, so if this was represented diagrammatically, we would have an inverted pyramid of numbers (such as that illustrated in Figure 5.19).

A **pyramid of biomass** shows the mass of material at each trophic level. This usually results in an upright pyramid – the biomass of a rose bush would be much greater than the biomass of the aphids it supports. However, there are situations where an inverted pyramid of biomass can be obtained. Pyramids of biomass show the standing crop, that is, the biomass at one particular time, and do not take into account the fact that the biomass at each trophic level may vary over a period of time. For example, in January, the biomass of zooplankton in the English Channel is greater than that of the phytoplankton (the primary producers). However, over the whole year, the total biomass of primary producers far exceeds that of the consumers in this ecosystem.

A **pyramid of energy** gives a more accurate representation of the transfer of material from one trophic level to the next. Rather than plotting numbers, or biomass, of organisms at each trophic level, we plot the productivity for each level in the ecosystem. Productivity is a measure of the energy content of each level and can be obtained by converting the mass of new organic material produced per unit area per year into an equivalent energy value. This energy value for each trophic level is expressed in units of kJ m^{-2} yr^{-1}. A pyramid of energy can be seen in Figure 5.20.

Productivity and energy loss

We have already stated that successive members of a food chain incorporate into their biomass only about 10 per cent of the energy available in the food they eat. What happens to the remaining 90 per cent?

Let us consider the possible fates of the energy available in grass eaten by a herbivore, such as a sheep. Some of the grass will remain undigested and will be egested as faeces, so this represents one source of energy loss. The grass which is digested will be absorbed from the gut and assimilated by the herbivore. There are three possible fates of the energy contained in this assimilated grass:

1. It can be used in respiration and will be lost as heat.
2. It can contribute to an increase in the biomass of the herbivore, termed energy of production.
3. A small amount will be lost in urine.

These possible fates are illustrated in Figure 5.21.

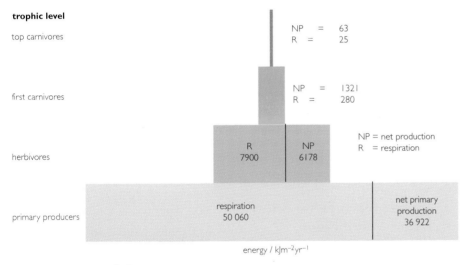

trophic level

top carnivores — NP = 63, R = 25

first carnivores — NP = 1321, R = 280

NP = net production
R = respiration

herbivores — R 7900, NP 6178

primary producers — respiration 50 060, net primary production 36 922

energy / kJm^{-2}yr^{-1}

Figure 5.20 A pyramid of energy

We can write an equation which summarises the fate of all the energy consumed by the herbivore:

$$E_C = E_P + E_F + E_U + E_R$$

This equation accounts for all of the energy entering the herbivore. The energy of production, E_p, will be available as energy of consumption for a carnivore, and the sequence is repeated at the next trophic level. Only about 10 per cent of the energy entering one trophic level (the energy of consumption) is available for consumption by the next trophic level.

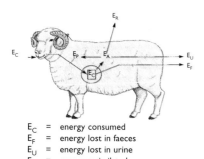

E_C = energy consumed
E_F = energy lost in faeces
E_U = energy lost in urine
E_A = energy assimilated
E_R = energy lost as heat in respiration
E_P = energy of production

Figure 5.21 Fates of energy in grass consumed by a herbivore

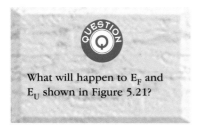

What will happen to E_F and E_U shown in Figure 5.21?

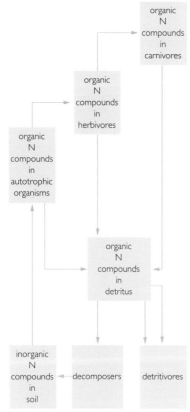

Figure 5.22 The biological part of the nitrogen cycle

Biogeochemical cycles

We have seen that a continuous supply of energy is essential for all living organisms. Organisms also require a range of chemical substances, including water, mineral ions and organic compounds. The original supply of energy is the sun, but there are only fixed amounts of chemical substances available on Earth. The chemicals needed to build the tissues of living organisms are used and re-used repeatedly. These substances move in cycles from the soil, water or the atmosphere, into plants and animals and back again. Cycles that operate in this way are known as **biogeochemical cycles** and include the carbon, sulphur and nitrogen cycles.

The nitrogen cycle

Nitrogen is required by all living organisms because it is a component of many substances, including nucleic acids and proteins. Plants and animals obtain their nitrogen in different ways. The source of nitrogen for green plants is in the form of inorganic ions, either nitrate (NO_3^-) or the ammonium ion (NH_4^+) present in soil or water. Heterotrophic organisms obtain their nitrogen only from organic substances.

In a typical green plant, nitrate ions taken up by the roots are reduced to ammonium ions in the cells and are then incorporated into organic compounds to form, for example, amino acids. These organic compounds then provide the source of nitrogen for detritivores, herbivores and, subsequently, carnivores.

Organic nitrogen-containing compounds in dead tissues, faeces and urine can be converted back into inorganic nitrate by the action of saprophytic bacteria and fungi, referred to as **decomposers**. We can illustrate the biological part of the nitrogen cycle in a simplified form, as in Figure 5.22.

Bacterial and fungal decomposers break down organic nitrogen compounds in detritus to release ammonium ions (NH_4^+). If no oxygen is available, or if the soil is waterlogged, cold or acidic, the process stops here and only ammonium ions are available to plants. In 'good' soil conditions, ammonium ions are oxidised by other soil bacteria (such as *Nitrosomonas*) into nitrites (NO_2^-). Another group of soil bacteria (such as *Nitrobacter*) further oxidise nitrite ions into nitrate ions (NO_3^-). The overall conversion of ammonium ions to nitrate ions is termed **nitrification**.

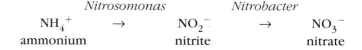

$$NH_4^+ \xrightarrow{\textit{Nitrosomonas}} NO_2^- \xrightarrow{\textit{Nitrobacter}} NO_3^-$$
ammonium → nitrite → nitrate

Nitrate ions produced in this way are available for uptake by plant roots. However, nitrate ions are very soluble in water and, unlike other ions present in the soil, do not bind tightly to soil particles. As a consequence, nitrate ions are easily washed out of soil after heavy rain, in a process called **leaching**. Rainwater draining through soil carries nitrate ions into ponds, lakes, rivers and the sea. Leached nitrate (and phosphate) ions enrich bodies of water with minerals, a process called **eutrophication**. This can result in the excessive growth of algae in ponds and rivers. The environmental consequences of eutrophication are described in Chapter 7.

Another way in which nitrate ions can be lost from soil is via the process of **denitrification**. Under anaerobic conditions, such as when soils are waterlogged, denitrifying bacteria convert nitrate to nitrite and then to nitrogen gas which escapes into the atmosphere. Denitrifying bacteria include *Pseudomonas denitrificans* and *Thiobacillus denitrificans*.

There are several ways in which nitrates can be added to the soil. They can be added as nitrogen-containing fertilisers in managed ecosystems. Small amounts of inorganic nitrogen compounds are also formed by the action of lightning in the atmosphere. This produces oxides of nitrogen which combine with rainwater to form nitrate ions.

Nitrogen fixing organisms, known as diazotrophs, which live in soil, are able to reduce nitrogen gas to ammonia. This is a biological version of the Haber–Bosch process. Unlike this chemical nitrogen fixation, which requires temperatures of 300 to 500 °C, high pressures and an iron catalyst, biological nitrogen fixation is much more efficient, occurring at low temperatures and atmospheric pressure. Biological nitrogen fixation is catalysed by nitrogenase, a complex enzyme containing iron and molybdenum.

$$N_2 \quad + \quad 3H_2 \quad \xrightarrow{\text{nitrogenase}} \quad 2NH_3$$
nitrogen hydrogen ammonia

Only certain bacteria and cyanobacteria can fix nitrogen gas in this way. Some of these organisms, such as *Azotobacter vinelandii*, are free-living in the soil. One genus of nitrogen fixing bacteria, *Rhizobium*, forms a symbiotic relationship with legumes (plants such as peas, beans and clover). These plants develop swellings on their roots, called **root nodules**, containing *Rhizobium* (Figure 5.23). It is a symbiotic relationship because the bacteria receive carbohydrates and ATP from the plant, which obtains fixed nitrogen, in the form of ammonia, in return. Ammonia is then incorporated into organic compounds to synthesise amino acids.

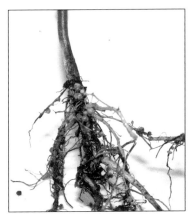

Figure 5.23 Nodules on the roots of the runner bean, Phaseolus multiflorus

One of the long-term aims of gene technology is to incorporate genes for nitrogen fixation into non-leguminous plants. This could have very considerable environmental and economic benefits, avoiding the need for expensive inorganic fertilisers, only about half of which are actually taken up by plants. The complete nitrogen cycle, incorporating both biological and chemical elements, is illustrated in Figure 5.24.

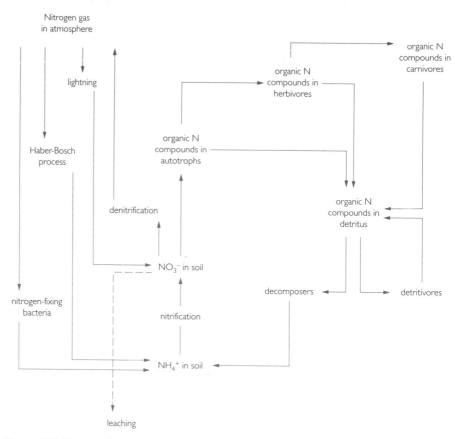

Figure 5.24 The complete nitrogen cycle

Estimation of pyramids of numbers and biomass

Introduction

The purpose of this practical is to obtain data which can be used to construct pyramids of numbers or of fresh biomass. The ecological community chosen will, of course, depend on accessibility, but the exercise could be carried out on a rocky shore, in woodland or in open grassland.

Materials

- 1 m² or 0.5 m² quadrat
- Trowel
- Scissors
- Large white sorting tray
- Hand lens
- Beakers
- Forceps
- Pooter
- Identification key
- Balance

Method

1 Select an area and place your quadrat carefully. On a rocky shore you should ensure that the entire quadrat is occupied by only one type of community.
2 Collect the leaf litter, or cut plants at the base, and place in the white tray. If appropriate, record the number of individual plants.
3 Search carefully and remove all the animals present. Smaller animals may be removed with a pooter (Figure 5.25), larger animals should be handled with forceps, or fingers. Place in suitable containers, such as plastic beakers.
4 Weigh the plant material.
5 Sort the animals into two groups: primary consumers (herbivores) and secondary consumers (carnivores).
6 Weigh the groups of animals separately and record the total number of animals in each group.
7 Return the animals to their habitat.

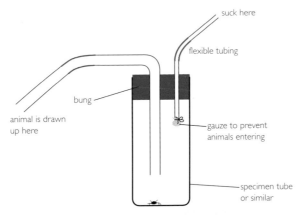

Figure 5.25 A pooter

Results and discussion

1 Construct pyramids of numbers and of biomass. A horizontal scale is chosen to represent either the numbers of organisms present per m^2 or the biomass per m^2. The lower block represents the producers (plants), the middle block the primary consumers (herbivores) and the upper block the secondary consumers (carnivores).

2 Consider the advantages and disadvantages of using pyramids of numbers and biomass to represent an ecosystem.

Further work

1 You could use this method to compare two communities, such as different areas of leaf litter. It is important that comparable samples are used in each case.

2 How could you adapt the method to construct a pyramid of numbers and of biomass in a pond?

Land use, biodiversity and conservation

Human activities and their impact on ecosystems

Early humans had relatively little impact on their surrounding environment. They gathered food plants, hunted animals, took small quantities of materials to build shelters and used wood as fuel to provide warmth and energy for cooking. As settled agriculture developed, the sphere of influence widened and humans began to change the environment in a more permanent way. Even so, as long as their activities remained on a small scale, disturbance of ecosystems was minimal and an equilibrium was more or less maintained.

By the late 20th century, the scale of activities has escalated to a stage where the influence of humans on their environment has become overwhelming. There is considerable concern that some of these activities will lead to irreversible changes to our planet Earth as we know it today. Pressures on the environment arise from increased population and the demands associated with an increasingly complex and technological way of life. The greatest changes have taken place during the last 150 years, coinciding with the period of the Industrial Revolution, particularly in Europe, and of colonial expansion by European nations. Towards the end of the twentieth century, the human population continues to increase and industrialisation is still expanding in the developing nations (Figure 6.1).

Over much of the globe, natural ecosystems have been converted to agricultural land, designated for producing food. Agricultural land often approaches a monoculture and therefore has low species diversity. Further land is absorbed into living space as villages, towns and large sprawling cities. There are industrial areas and a supporting infrastructure, including roads, railways, airports and

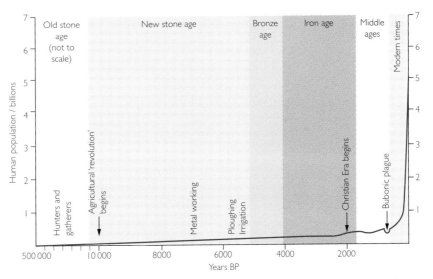

Figure 6.1 Increases in the human population over the last half-million years. Note the very rapid rise up to the present day (BP = before present)

seaports. Exploitation of natural resources, such as mining for minerals, has created further disturbance. Demands for energy have led to exploitation of wood for fuel, coal from mines and both oil and natural gas have been piped away from its source. In this chapter we look at some of the consequences of these human activities, and how they have affected natural ecosystems.

Figure 6.2 Use of land for agriculture leads to loss of natural ecosystems – fields and hedges in England (left); terraced hillside in Nepal (right)

Figure 6.3 Loss of natural ecosystems through urbanisation – buildings, roads and car parks

Changes in land use

Any change in land use is likely to alter the ecosystem that would naturally be present in the area. Here we look at **deforestation** and **desertification**, two specific examples of how natural ecosystems have been altered or destroyed through the activities of humans.

Deforestation

For thousands of years humans have invaded forests, cut down the trees and converted the area into agricultural land, to be used for growing crops or grazing animals. Often the debris has been burnt as part of the clearing process. In Britain, without human influence, much of the landscape would be dominated by woodland, whereas now only about 10 per cent of the land remains wooded. A similar situation is evident worldwide. Agricultural expansion during the second half of the nineteenth century led to large-scale deforestation in temperate regions, but by the late twentieth century there are particular concerns over the loss of tropical forest. Estimates suggest that an area approaching 12 million hectares (about the size of England) is disappearing from tropical forests each year. Amazonia, mainly in Brazil, remains the largest area of rainforest. As the rate of deforestation accelerates, recovery of natural forest may be impossible, and some of the changes may have long-term effects on global climate.

On a small scale, loss of forest to traditional systems of shifting cultivation (often cleared by slash and burn methods) may be short-lived. After a few seasons the people who cleared the forest move on to a new area. Provided the human population is of low density, the forest regenerates, and in a few years is likely to recover. If forest is managed successfully for its productivity (usually for timber), it remains reasonably stable. Controlled **reafforestation** replenishes the stock of trees, thereby maintaining the ecosystem.

Table 6.1 *Population pressure and deforestation in areas with tropical moist forest*

Country	Area / million km^2	Forest cover / thousand km^2			Estimated population in millions		
		Original	1989	Primary forest in 1989	1950	1989	2020
Latin America							
Bolivia	1.1	90	70	45	3	7	15
Brazil	8.5	2860	2200	1800	53	147	234
Central America	0.5	500	90	55	9	28	54
Colombia	1.1	700	278	180	12	31	49
Ecuador	0.3	132	76	44	3	11	20
Guianas	0.5	500	410	370	1	2	3
Mexico	2.0	400	166	110	27	87	142
Peru	1.3	700	515	420	8	21	35
Venezuela	0.9	420	350	300	5	19	23
Asia							
Burma	0.7	500	245	80	18	41	69
India	3.3	1600	165	70	362	835	1374
Indonesia	1.9	1220	860	530	77	185	287
Malaysia	0.3	305	157	84	6	17	27
Papua New G.	0.5	425	360	180	2	4	8
Philippines	0.3	250	50	8	20	65	131
Thailand	0.5	435	74	22	20	56	82
Vietnam	0.3	260	60	14	24	67	121
Africa							
Cameroon	0.5	220	164	60	5	11	24
Congo	0.3	100	90	80	1	2	6
Côte d'Ivoire	0.3	160	16	4	3	12	35
Madagascar	0.6	62	24	10	5	12	30
Nigeria	0.9	72	28	10	41	115	274
Zaire	2.3	1245	1000	700	14	35	89

Large-scale damage to forest ecosystems is occurring as a result of mechanisation, leading to massive clearance on an unprecedented scale. Pressures on forests, well illustrated by the Amazonian rainforest, arise from world-wide commercial demands as well as smaller but significant demands from indigenous people. The forests are exploited for their timber and fuelwood, mined for minerals (such as manganese, tin and iron), used as a source for extraction of medicines (by large pharmaceutical companies as well as local people) and to provide food. Roads have been built to regions in the forest that were previously inaccessible, except to the indigenous people. Rivers have been utilised to generate electricity through construction of hydroelectric schemes. In central America, areas of forest have been destroyed to provide grassland for 'ranching' of cattle and overpopulation has encouraged semi-permanent settlements of small-holders to encroach upon forest land.

An immediate effect of deforestation is on the soil. Trees and associated vegetation tend to act as a sponge, retaining water and releasing it slowly to the soil and streams, and as water vapour to the air. With removal of the forest this stops abruptly; the soil surface loses its roughness and surface run-off of water increases. Large amounts of soil particles and nutrients are washed into streams and rivers. In tropical rainforests, the soil itself is relatively poor in nutrients, though considerable quantities are locked away in the biomass of the forest plants and become available to future plant life through natural recycling. Removal of forest products as timber, takes large quantities of nutrients away from the area. Disturbance to the soil and removal of surface organic litter upsets the microbial population and the natural nutrient cycling mechanisms. The remaining soil is thus relatively poor in quality for any future agricultural use. Sudden exposure of the soil means that the surface is heated up; it may become desiccated and is also subjected to the direct effects of rain and wind. This leaves the soil in the area more liable to erosion, which becomes particularly acute in sloping areas. Erosion can also affect more distant locations through deposition of sediment in rivers, which may lead to flooding. The change from a forest ecosystem to agricultural crops or grassland for grazing may alter the utilisation of carbon dioxide in photosynthesis. If less carbon dioxide is used by the replacement crop, there will be a net increase in carbon dioxide in the atmosphere. The possible contribution to the greenhouse effect is discussed in Chapter 7.

Probably the greatest resource of tropical forests lies in the biological diversity. Any attempt to estimate numbers of species is bound to be conservative, as many have never been recognised or described. Many of these unknown species are likely to be invertebrates, especially insects. A few examples from the Amazonian forest will give a hint as to the diversity. There are exotic birds, mammals and insects, often hunted for various trade outlets. Among the birds are toucans, parakeets and trogons; the monkeys include capuchins, howler and squirrel monkeys. There are anacondas, perhaps 10 m in length, boa constrictors and many other snakes, lizards, turtles, crocodiles, toads and frogs. The mammals include tapirs, anteaters, jaguars and capybara. A range of products from tropical forests have direct value to humans as food sources. Many vegetables and fruits eaten around the world originated in tropical forests. These include rice, maize and potatoes, pineapples, avocados and bananas, coffee and cocoa, palm oil and Brazil nuts. Plants with medicinal value are used by the indigenous people, and some have been used in modern medicine. Examples are quinine for malaria, and curare as a muscle relaxant in surgical operations. Curare is obtained from the bark of trees and for centuries was used by South American Indians as an arrow poison.

It is difficult to predict the long-term effects of the loss in biological diversity as forests are destroyed. Inevitably species are lost. Some species may adapt to life elsewhere, but others depend critically on factors in the forest ecosystem, say a food plant or the shelter provided. The giant panda exists in only a few locations in China, largely dependent on bamboo for food, but its survival is seriously threatened by the destruction of forests in these crucial areas. Endangered species from other forests include apes, lemurs, elephants, leopards, tigers, some parrots and crocodiles.

Figure 6.4 Deforestation in Sichuan province, western China (left); natural forest in Sichuan province in western China, one of the few remaining areas still inhabited by the giant panda. The natural habitat is being eroded because of deforestation, encroaching agriculture and spread of industrialisation.

The impact of humans is not a new problem. In prehistoric Britain there were large mammals: lions, leopards and hyenas, similar to species currently living in Africa, and more recently there were bears and wild swine. The demise of these large mammals was probably due to the combination of hunting and loss of the protective cover of native woodland as it was cut down. The last known aurochsen (ancestors of modern domesticated cattle) were observed in forests in Poland in 1627. Reduction of woodland to the point where surviving animals could be hunted and caught almost certainly contributed to their extinction.

The genetic reserves in both animal and plant species in the rich forest ecosystem are potentially highly valuable resources. The gene pools in commercial varieties of domesticated animals and crop plants used in agriculture have become dangerously narrow, with the risk that whole populations could be destroyed by disease or by pests. Wild populations may, therefore, provide sources of fresh genetic material, with the potential of introducing favourable characteristics, such as flavour, improved disease resistance, or allowing the crop plants to be grown in different climatic regions.

Desertification

Natural deserts occur in both hot and cold regions. They are characterised by low and intermittent rainfall, usually less than 250 mm of precipitation per year. Sometimes total drought persists from one year to the next. Because of the low water availability, desert areas support only limited vegetative cover. The term **desertification** has been defined as 'land degradation in arid, semi-arid and dry sub-humid areas arising mainly from adverse human impact'. This definition recognises that human activities are the main cause of contemporary desertification and associates its progress with increased population and consequent pressure on the natural resources of the area. Concern arises from the loss of productivity from land which could be used for agriculture, leading to worsening poverty or even famine among the people the area has traditionally supported.

Global change is not new and some of the concerns today are echoed in this passage written by Plato, 2300 years ago.

There are mountains in Attica which can now keep nothing but bees, but which were clothed not very long ago, with fine trees producing timber suitable for roofing the largest buildings ... . There were also many lofty trees, while the country produced boundless pastures for cattle. The annual supply of rainfall was not lost as it is at present, through being allowed to flow over the denuded surface to the sea, but was received by the country ... where she stored it ... and so was able to discharge the drainage of the heights into the hollows in the form of springs and rivers with an abundant volume and a wide territorial distribution ...

How far are these words, written by Plato about Greece, still true today? What effects does deforestation also have on species diversity?

LAND USE, BIODIVERSITY AND CONSERVATION

Pressures on the land come mainly from grazing flocks (often sheep, cattle or camels) and from gathering fuelwood. Some arid areas are just able to support the growing of crops, probably helped by irrigation, but the situation is usually already fragile. A dry season can be disastrous and result in crop failure. Any additional pressure, from increased human population or worsening of the drought conditions, means that grazing of animals and fuel gathering spreads over a yet wider area. Demands by the herds then exceed the carrying capacity of the land and the vegetation cannot support them, nor can it recover unless the pressure is removed.

Lack of vegetation leads to a downward spiral in terms of deterioration of the land. The bare soils, exposed to direct sunlight, suffer further desiccation and, combined with exposure to winds, are more likely to suffer from **erosion**. Deposition of soil particles elsewhere, perhaps over other marginal crops or grazing land, causes additional losses. High rates of evaporation in the absence of plant cover are likely to alter the water–salt balance. Salts drawn up from the ground water are left behind in the surface layer of the soil. This leads to **salinisation** which contributes to the deterioration of the soil as it makes it unsuitable for plant growth. Salinisation may also result from poorly designed irrigation systems, where the existing drainage is unable to handle an increased water supply. As the groundwater level rises, dissolved salts are brought to the surface, but the area effectively becomes waterlogged so the salts remain in the water. With high rates of evaporation, the salts then accumulate at the soil surface. The situation may be exacerbated if chemicals have been used on the land, either as fertilisers or for pest or weed control. Lack of vegetation alters the amount of moisture in the atmosphere which may in turn affect the climate pattern in the area, particularly rainfall.

Two examples will emphasise how human pressures on the land have brought about desertification in areas where an equilibrium had previously existed. Disaster struck the Sahel region of Sudan in the early 1970s. This semi-arid zone had had a higher than average rainfall for about twenty years. This encouraged greater cultivation in the region, including cash-crops such as peanuts. The pastoral people with their grazing herds were pushed further north into the fringe of the Sahara desert, crowded into land with already poor carrying capacity. When the drought returned, thousands of people and millions of animals died because they had no food, inadequate water and nowhere to go.

In the second example, in northern China, extensive areas of semi-arid grassland are undergoing desertification, seen as deterioration of the land with more sand blowouts and shifting dunes. Increased human population in the area is linked to political movements of people. Sedentary agriculture is replacing traditional practices which depended on seasonal movement of grazing animals. Since 1949, in an area of Inner Mongolia, the number of livestock has increased while the land available for grazing has decreased (see Table 6.2). There has also been introduction of breeds of cattle, such as the European Friesian or Simmental, with potentially higher productivity, but without the necessary adjustment in herd size to take account of their higher feed intake. Collection of fuelwood to support the increased population has worsened the situation.

Figure 6.6 Areas of the world at risk of desertification. It is the arid and semi-arid lands that are at risk, whereas the hyper-arid zones are already natural deserts

hyper–arid

arid

semi–arid

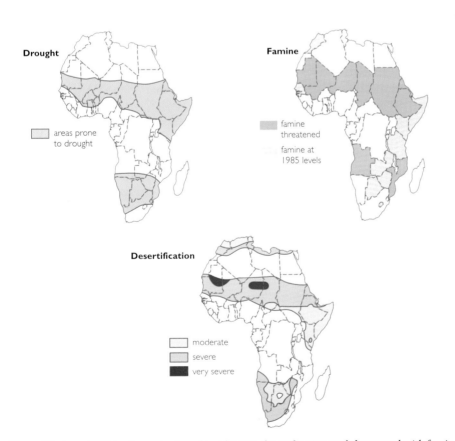

Drought

areas prone to drought

Famine

famine threatened

famine at 1985 levels

Desertification

moderate

severe

very severe

Figure 6.7 Areas in Africa prone to drought, subject to desertification and threatened with famine

Figure 6.5 Desert areas in Afghanistan: sparse vegetation (top); grazing pressure (second); fuel gathering (third); erosion in area denuded of vegetation (bottom)

Table 6.2 *Grazing pressures and desertification – Xilingole League, China (1949 to 1980)*

Year	Number of livestock	Grazing area / km²	Density / number km⁻²
1949	1 740 000	193 000	9.01
1958	4 750 000	193 000	24.63
1964	8 210 000	179 000	46.00
1970	5 780 000	123 000	47.04
1980	5 270 000	142 000	37.02

Use of biomass for fuels

Biomass is defined as the total mass of living material in an area. Growing plants utilise solar energy in the process of photosynthesis. Solar energy is thus incorporated into biomass which has the potential for use as a fuel by humans. The **biomass value**, usually expressed as dry mass in kilograms per square metre (or tonnes per hectare), gives a measure of the available energy in an ecosystem. Biomass includes plants, animals and microorganisms, but in practice, biomass values usually refer to the plant mass above ground level. Ecosystems in warm, damp environments, where plant growth is rapid and often continuous, have higher biomass values than those in cooler or drier environments, where growth is slower and usually interrupted at certain seasons (see Chapter 5).

Fossil fuels

Fossil fuels (coal, oil and natural gas) are derived from biomass that was living millions of years ago. Coal is formed from former vegetation which accumulated in peat beds in waterlogged, anaerobic conditions. Later it became buried under sediments, followed by slow processes of being compacted and compressed, which eventually turned it into hard rock (lithification). Oil (petroleum) also developed from anaerobic decomposition of former organic material. It is found as a liquid, trapped in reservoirs within sedimentary rocks, usually of marine origin. Natural gas probably arose in the same way and is found as a gas in underground reservoirs in rocks, often associated with oil deposits. The main constituent of natural gas is methane (about 85 per cent), and it also contains ethane, propane and small quantities of other hydrocarbons. On a global scale, these three fossil fuels together account for most of the energy used by humans, though compared with industrialised countries, less developed countries depend to a greater extent on recently harvested biomass. This is usually burnt as wood but often converted to charcoal.

Reserves of fossil fuels are limited because the rate at which they are formed is far slower than the rate at which they are being used. On a global scale, since 1950, production and use of these fossil fuels has more than doubled and in the case of natural gas it has increased ten-fold. Despite concerns that sources of fossil fuels will become depleted, known reserves of coal, oil and natural gas should be sufficient to supply increasing energy requirements into the 21st century.

Fossil fuels are relatively cheap and easy to obtain, transport and store. The chief disadvantage of burning fossil fuels lies in their impact on the environment, because of the pollution caused by various emissions from industry and from motor vehicles. The effects of acid rain, and the build-up of carbon dioxide with possible long-term effects on global climate, are largely attributed to the use of fossil fuels (see Chapter 7). Coal is considered to be the dirtiest of the three, causing most environmental damage, whereas natural gas is the cleanest.

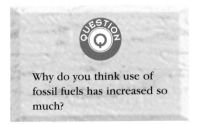

Why do you think use of fossil fuels has increased so much?

There is pressure for development of alternative energy sources, which are both renewable and minimise damage to the environment. In some countries, government legislation requires electricity generating companies to incorporate some non-fossil fuel sources in their generation of electricity. In England, this is called the non-fossil fuel obligation (NFFO) and has stimulated considerable research into development of viable alternative energy systems. The emphasis now is to explore their use economically, on a commercial scale, so that they become realistic alternatives to meet global energy demands.

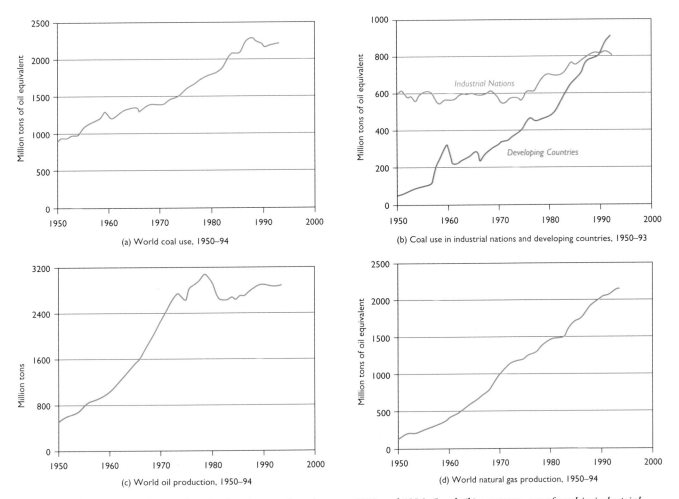

Figure 6.9 Increases in world use of coal, oil and natural gas between 1950 and 1994. Graph (b) compares use of coal in industrial nations with that in developing countries. Values for coal and natural gas are given in millions of tons of oil equivalent

Alternative energy systems include generating power from wind and movement of water, conversion of solar energy into electric power through photovoltaic cells. Recent technology is exploring the use of geothermal energy, extracted from deep in the Earth.

Why do you think there is pressure for development of alternative energy sources?

Figure 6.10 Traditional coppicing – these ash stools may be hundreds of years old

Figure 6.11 Short-rotation coppicing – a way of exploiting fast-growing woody species as a renewable energy source. This photograph shows young willows

Summarise some of the benefits of growing short-rotation coppice.

Energy from renewable biomass

Traditionally, wood is the main form in which biomass has been used as an energy source, though other materials such as animal dung are also used. The energy in the biomass is usually extracted by direct burning with a supply of air, in a stove to provide heat, or in a boiler to produce steam to drive turbines for generating electricity. In Britain, most of the commercial production of wood from the growth of trees in the forestry industry, is directed into timber. When a conifer plantation is felled, there is considerable waste in the form of small branches, needles and low-grade timber and this can be diverted for use as fuel.

Wood can be exploited as a renewable energy source by using land for plantations of fast-growing woody species. One system is known as **short-rotation coppicing** and areas of land used in this way are described as 'energy plantations'. The practice of coppicing has been used for several hundreds of years as a technique for management of woodland, particularly in Europe. Traditional coppicing is described in detail on page 114.

Short-rotation coppicing uses the same principle as traditional coppicing but with a reduced cutting cycle. In Britain, the species most commonly used are willow and poplar, because they are easy to propagate and fast-growing. Varieties are selected to give high yields. Cuttings are planted closely, then at the end of the first year they are cut off (coppiced) near the ground. Both willow and poplar then produce several side shoots. The crop of wood from these shoots is harvested after 2 to 4 years growth. The coppice is harvested by machines which cut and chop the material. Wood can probably be harvested from these stools for about 30 years, then the land would be returned to other agricultural crops. Some farmers use the wood chips produced on their own farms for running electricity generators; others sell the harvested crop to electricity generating companies as a fuel source.

The NFFO in Britain gives encouragement to the development of short-rotation coppice both on a large scale and in small plantations integrated within an existing farming system. Short-rotation coppicing is an attractive option to some farmers as an energy crop. It may be an appropriate way to exploit surplus land which, because of government controls, has been taken out of cultivation for food crops, such as cereals. Creation of coppiced areas with willow and poplar has benefits for wildlife, particularly in an arable farming area, because they provide an attractive habitat for birds and an increased diversity of plant species appears along the paths and within the coppice. Other fast-growing species could be used as energy crops. One which shows potential is *Miscanthus*, an annual bamboo-like grass.

Straw left over after harvesting cereal crops is another material which has potential as a fuel source. In Britain, about 12.4 million tonnes of straw are produced annually. Some is used as bedding or for feeding livestock, but there is an unwanted annual surplus of about 4.5 million tonnes. Until recently, surplus straw was usually burnt along with the stubble in the fields at the time of harvest, leaving the field clear for planting the next crop, but burning of stubble has been banned since 1992 so farmers have had to find other ways of disposing of the straw. Burners that can take straw bales have been developed and are used in some countries, notably Denmark and the USA, and a straw-fired electricity power station is planned in Britain. Compared with coal and oil, straw has very low levels of sulphur, but the main difficulties with utilising straw arise from the cost of collecting, storing and transporting a rather bulky material.

Biomass to gasohol

Fermentation of sugar by yeasts (*Saccharomyces* spp.), can convert the energy in biomass into ethanol which can be used as a fuel. An example is the fuel called **gasohol** which consists of 80 to 90 per cent of unleaded petroleum spirit with 10 to 20 per cent ethanol, and is used in motor vehicles. Most of the ethanol produced for gasohol uses sugar crops as its source material, though other plant species (such as maize and manioc), and waste materials (including wood and animal products), are also used.

Sugar is obtained from two different crops: sugar cane (*Saccharum officinum*) and sugar beet (*Beta vulgaris*). Sugar cane is grown in tropical and semi-tropical countries, whereas sugar beet is grown in more temperate regions. In the cane sugar industry, the cane is cut then processed. The initial milling yields two products: cane juice and bagasse (a residue of fibrous material). The juice can be further treated to the stage when sugar crystallises out, leaving a viscous sugary liquid known as molasses. Cane juice, bagasse and molasses can all be fermented to produce ethanol. Cane juice must be processed soon after harvest because it cannot be stored, whereas the molasses can be stored and fermented at a later date. Production of ethanol tends to be seasonal, linked to the time of harvest for the crop. Usually, small distilleries are set up at sites close to the crop. Bagasse is mainly cellulose, hemicellulose and lignin, and is often used to fuel the boilers used in the distillation process.

There is considerable potential for large scale industrial production of ethanol for use as gasohol, but the decisions regarding its development are economic and political as well as environmental. Sugar cane and sugar beet are already grown in a wide range of climates and countries. Production of sugar crops could be expanded but at risk of competing with other food crops. Ethanol has the advantage of being a relatively clean fuel, producing less pollution than the petrol it would replace if used as gasohol in motor vehicles. However, production of ethanol for gasohol is uneconomic compared with conventional petrol and would thus require both political and financial backing if it is to make a major contribution to the fuel used. Sugar cane growers may prefer to sell their crop to be refined as sugar rather than to be processed as ethanol. The most extensive gasohol programme is that in Brazil, initiated during the 1980s. Yields of biomass from sugar cane are high in Brazil, and because of lack of available funds at that time there were difficulties in purchasing oil from overseas. However, a shift in oil prices or other political changes could alter the balance away from ethanol production and its development as a renewable fuel source.

Biomass to biogas

Fermentation by bacteria can convert the energy in biomass into **biogas**, a gaseous fuel which consists mainly of methane. This process exploits the metabolic activities of different groups of bacteria which digest organic matter under anaerobic conditions. A typical fermentation would produce biogas with

Why do you think burning of stubble has been banned? What is the main carbon compound in straw that gives off energy when burnt? What benefits to the soil would there be from burning the straw on the fields? In what other ways could surplus straw be used?

Figure 6.12 Bales of straw after harvest of cereal crops – surplus straw can be used as fuel

*Figure 6.13 Other crops have potential as alternative energy sources – bright yellow oilseed rape (*Brassica napus*), grown for the harvest of oil from its seeds, can be used to produce a fuel known as biodiesel*

a composition of about 65 per cent methane, 35 per cent carbon dioxide and traces of ammonia, hydrogen sulphide and water vapour. The methane in biogas burns with a clear flame to produce carbon dioxide and water without any hazardous air pollutants. The process is used mainly with dung or slurries from animals so has the added benefit of turning waste material into a useful product. After digestion, the residue has value as a fertiliser.

There are three stages to the rather complex digestion process. The material to be digested is likely to contain mainly carbohydrate with some protein and lipid. Initially, aerobic bacteria utilise these substrates and convert them by **hydrolysis** to simple sugars, amino acids and glycerol and fatty acids. As the available oxygen is used up, acetogenic bacteria convert the sugars and other substrates to short-chain fatty acids, mainly acetic acid, with some carbon dioxide and hydrogen. This stage is described as **acetogenesis**. The final stage is **methanogenesis**, which is carried out only in anaerobic conditions by methanogenic bacteria and involves conversion of the acids to methane. Methanogenic bacteria are **obligate anaerobes**, which means they are active only when there is no oxygen present. It is essential that conditions are anaerobic for the digestion to produce methane. For successful operation, temperatures are usually maintained between 30 and 40 °C. The methanogenic bacteria are sensitive to temperature changes and if fluctuations of more than 5 °C occur, the material goes sour due to build up of undigested volatile acids.

The digestion process is carried out in an enclosed tank, called a **digester**. The design of the digester may depend on locally available construction materials, but essential features are that it is strong enough to hold a large volume of the material to be digested and withstand the build-up of pressure inside. It must be gas-tight and allow the anaerobic conditions to be maintained. It should have an accessible inlet for loading the material, an outlet for the gas and a means of recovering the residue when digestion is completed. A plan of a simple but effective domed model, used widely in China, is shown in Figure 6.14. Sinking the digester in the ground helps to provide both support and insulation with respect to temperature. When digestion is completed, access to the digester pit is through the slurry reservoir. This allows the residue to be taken out and the reservoir cleaned. Often several digesters are used together to ensure continuous supplies of gas.

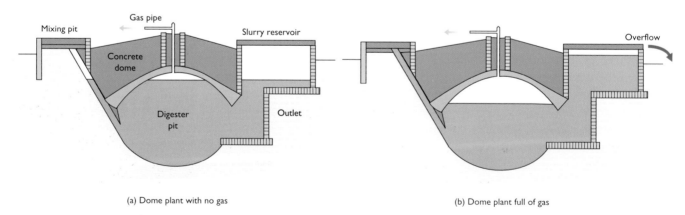

(a) Dome plant with no gas

(b) Dome plant full of gas

Figure 6.14 Fixed dome biogas plant, commonly used in China. Gas given off collects in the dome and slurry is displaced into the reservoir. As the gas is used, the slurry flows back from the reservoir into the digester pit

Biogas has been produced on a small scale in China for more than 50 years. Other countries such as Nepal, India and developing countries in Africa and South America also find biogas useful for small-scale production of fuel, particularly in rural areas, where it also provides a way of disposing of animal wastes and human excreta. In Britain, biogas digesters are being used increasingly as a means of disposing of the large quantities of animal wastes derived from intensive farming methods. Ideally the digester is located close to the source of biomass (slurry from cattle or pigs, or chicken litter) to avoid transport costs.

Figure 6.15 Electricity from poultry litter: fuel hoppers containing poultry litter at a power station in Suffolk

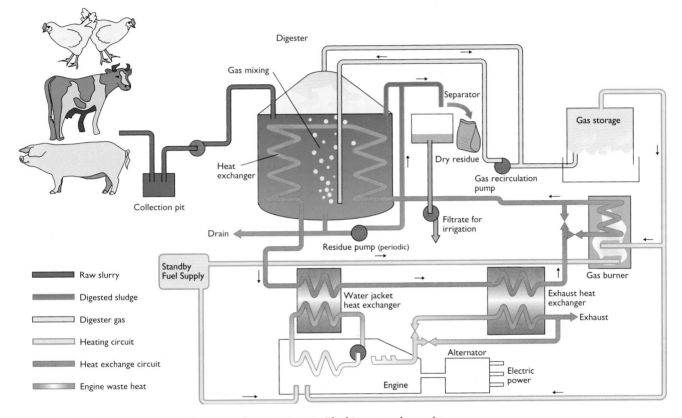

Figure 6.16 A biogas plant designed for use on farms in Britain. The biogas can be used to power an engine which generates electricity. In the digester, a suitable temperature is maintained by internal heating coils and heat exchangers, with biogas as the source of heat. Waste heat from the engine cooling system and exhaust can also be utilised, with a standby fuel supply for emergency. The contents of the digester are mixed by recirculation of the gas produced

Rubbish or energy?

Disposal of domestic and industrial waste has become a very large problem. In the UK, approximately 46 million tonnes of waste are produced each year, equivalent to 300 kg of domestic waste per person per year. These mountains of waste give rise to a number of environmental concerns. Dumping the waste occupies considerable areas of land and there is danger of leakage of toxic substances and other pollutants from the dumps. However, the waste material represents a source of energy which could be converted to fuel. In Britain, nearly 90 per cent of waste is disposed of in landfill sites (Figure 6.17). Biodegradable

Figure 6.17 Mountains of domestic and industrial waste – most is dumped in landfill sites

Figure 6.18 Separation of domestic waste – the brown bin is for biodegradable materials which can be converted into compost

In some areas in the UK, domestic households are requested to separate their rubbish into different bins. Materials that can be made into compost go into a separate bin from other waste.

What are the environmental benefits of separation? What can be recovered from the compostable materials and what about the rest?

materials in the deposited waste (mainly paper, garden waste and foods) start to decay and soon use up available oxygen. Conditions thus become anaerobic which leads to the production of methane. This process has been described earlier in the production of biogas (page 110). The gas is usually known as **landfill gas** when produced in the landfill site. The gas may seep out from the layers of compacted rubbish with the potential danger of igniting or causing explosions. A number of landfill sites are now being constructed so that landfill gas can be collected and the methane used as a fuel. A landfill site begins to produce gas about a year after landfill is completed and may be viable for up to 15 years. Sometimes rubbish is incinerated (burnt) and part of the energy from the biomass in the rubbish could be recovered as useful heat.

Conservation

There is inevitable conflict between the pressures to support the global demands of human society which exploit ecosystems and the less visible and more elusive benefits of conservation. Conservation is defined as the protection and preservation of natural resources and of the environment. But conservation is dynamic rather than static, achieved through active intervention and management rather than passive preservation. In this section we will look at ways of maintaining a range of habitats and a diversity of species within them and consider particular examples to see how conservation is achieved.

Succession and species diversity

Any community of plants and animals is strongly influenced by the physical factors of the surrounding environment. Over a period of time, changes may take place in the community that lead to the development of a stable or **climax** community. The progression from one stage to the next is known as a **succession**.

A typical area of mown grass, such as a playing field, lawn or in a park, can eventually develop into woodland, though it may take up to 30 years for the woodland to become established (Fig 6.19). Changes in vegetation occur if there is no mowing or grazing which would keep the area as grassland. At each of the stages of this succession, there would be populations of invertebrates associated with the vegetaion. As cover becomes more dense, small mammals and an increasing number of birds would become evident.

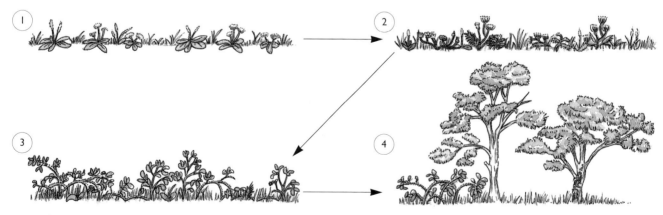

Figure 6.19 Example of a succession: development from grassland to woodland (1) mown frequently in growing season; (2) mown once or twice a year; (3) grassy area not mown; (4) woodland established

Similarly, from the edge of a freshwater pond or lake into the water, there is a range of plants, each associated with the different depths of water. Animal life amongst the plants depends on the plants and also requires water or wet marshy conditions. If the open stretch of water is left alone for a period of time, plant material, mud and other debris accumulate. Slowly this material builds up from the bottom of the pond or lake, and eventually may rise above the original level of the water. As the habitat becomes less watery, the nature of the plant communities changes, as it goes through the stages of a succession from open water to dry land and eventually to woodland. Those plants and animals associated with open water would gradually disappear. Both in the grassland and in the water habitat, if we wish to maintain the range of habitats representing different stages of the succession, some active intervention is required.

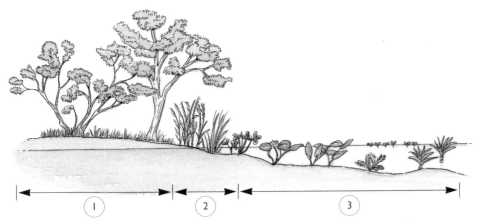

Figure 6.20 Profile from open water to edge of pond or lake, representing stages of succession: (1) Dry land - can support climax community; (2) Bank zone - sometimes flooded; (3) Open water

In the open water, some plants are totally submerged, others have leaves on the surface; some float freely, others are rooted on the bottom. In this zone, plants include duckweed (free-floating), water milfoil (submerged), water crowfoot and water lilies. In shallower water, towards the bank, plants include reeds, sedges and grasses, horsetails and perhaps marsh marigold or bogbean. Woody species associated with the banks of wetlands include willows and alder.

Conservation in two nature reserves – Bradfield Woods and Wicken Fen

Bradfield Woods

Bradfield Woods is a National Nature Reserve, situated in East Anglia. It occupies an area of about 65 hectares and is surrounded by arable farmland. Woodland has probably existed continuously on the site from the end of the last glaciation some 12 000 years ago. Over the last few hundred years the wood has dwindled to its present size from a much more extensive area. Historical records from the Abbey of Bury St Edmunds indicate that the area has been managed by humans as woodland for at least 750 years.

In the early 1970s, the remaining woodland was on the verge of being destroyed and converted to arable land, but was saved by a small group of local people who saw its potential as a conservation area. Instead of chainsaws and bulldozers moving in, the area has gradually and sympathetically been restored to a working coppice woodland. In the history of Bradfield Woods, the harvested wood has been used for the manufacture of various products, including wooden rakes, handles for scythes, thatching pegs, hazel for daub and wattle (used in local timber-frame buildings), fencing materials and hurdles as well as firewood. Even today, certain wood products are sold, either from the wood itself or (until recently) through a nearby factory. The scientific value of Bradfield Woods is seen in the richness of its plant and animal life. Its aesthetic

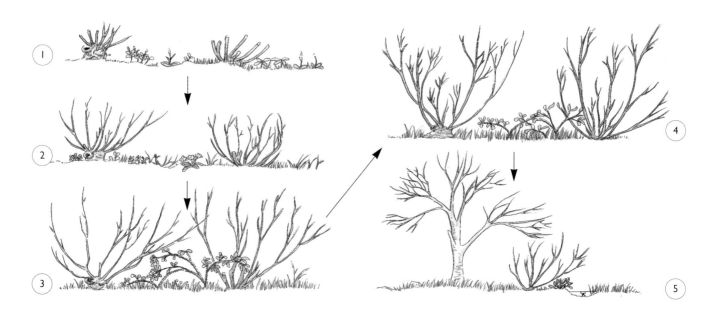

Figure 6.21 Stage 1: first year after coppicing. Ground species include dog's mercury and water avens. Herb paris and early purple orchid may be seen, both indicators of ancient woodland.

Stage 2: second and third years after coppicing. Richest ground flora seen at this stage, in spring. Ground species include sweet violet, bugle, wood anemone, primrose, dog's mercury, patches of oxlip (now rare), bluebell and wood spurge.

Stage 3: fourth and fifth years after coppicing. More grasses, brambles and other species invade.

Stage 4: eight and nine years after coppicing. Ground flora more sparse, and has lost its colour and diversity (though the cycle starts again when the area is freshly coppiced).

Stage 5: climax woodland stage.

value is enjoyed by many visitors, and it is used as an educational resource for adults and children. Volunteers contribute to the conservation programme, and other visitors may be interested in the bird or insect life or just enjoy the spring flowers. In 1994 Bradfield Woods was declared a National Nature Reserve, giving recognition to its value as a conservation site.

A visitor entering Bradfield Woods walks down wide grassy tracks, known as rides, usually with a shallow ditch either side. The wooded areas are laid out in blocks (or fells) and the coppicing or cutting of the woody species takes place in these blocks. Coppicing is a method of woodland management that has been practised for many hundreds of years. At intervals, certain tree species (here mostly alder, ash, birch, hazel and sallow) are cut down close to ground level. New side shoots then grow out from the stumps (also known as **stools**) and are allowed to grow for say 10 to 15 years, then are harvested as poles. Different blocks are coppiced in successive years, giving a crop of wood every year. Among the coppiced stools, a few trees (usually oak and silver birch) are allowed to grow to their full size as mature trees. These are known as **standards**. When felled, these trees supply timber, used in building houses. Trees regenerate naturally from seedlings. Some of the ash stools reach a diameter of up to 6 m, with an estimated age of at least one thousand years. They have probably yielded crops of wood continuously over this period. Figure 6.10 shows coppiced stools in Bradfield Woods. The series of coppiced areas representing the cutting intervals within the 10 to 15 year cycle show successive stages of regrowth of the shoots from the stools. Comparison of the ground flora in these areas reflects stages in a succession, from open ground through to dense shading by the tree cover (see Chapter 4).

The open rides offer sites for other flowers and along the edges are shrubs and taller vegetation, providing attractive habitats for insects including butterflies. The rides are relatively dry because of the drainage ditches along the sides.

These ditches introduce yet another habitat for a different range of species. As well as the plant life, there are plenty of birds and the abundant birdsong includes that of willow warblers, blackcaps and nightingales. Other animal life includes adders, grass snakes, frogs and toads and a range of typical small woodland mammals. All these are rare or absent in the surrounding farmland.

The different stages of the coppicing cycle create a range of habitats and hence a diversity of species. At least 42 native trees and shrubs have been described from Bradfield Woods, about two-thirds of the total in the whole of Britain. Over 350 species of flowering plants are known from the woods, including a number of rarities. However, if left to grow without any management, the diversity of species within the wood would gradually diminish. Hazel is an example of a woody species that does not compete well in dense, mature woodland. The rides need to be mown to keep them open and stop invasion of woody species. The shrubby growth at the edges of the rides must be cut back, preferably in stages to ensure continuity of habitats suitable for the insect life.

Wicken Fen

Wicken Fen is a nature reserve situated in a low-lying region in eastern Britain. Today its 300 hectares represent a small relic of what was an extensive area of fenland, spreading over more than 3000 km². The original East Anglian fen developed because drainage was poor and the land was liable to become flooded. In these waterlogged conditions, dead vegetation accumulates and becomes peat instead of decaying. As the peat builds up, the natural succession tends towards woodland (known as fen carr). For centuries, people have exploited the fenland and harvested its products. Around Wicken Fen, people dug peat to provide fuel. They cut reeds for thatching and sedges for thatching and covering floors. They gathered mixed vegetation for animal fodder and bedding, and from the flooded areas they caught fish and wildfowl for food. The activities of digging and cutting helped to create and maintain a range of habitats, from open water through wet reed beds to drier meadows. The cutting also helped to prevent progression to woodland. The range of habitats supported a diversity of wetland species, both plant and animal.

Figure 6.22 Management practices in Wicken Fen

Peat is dug in a few places to create fresh hollows or pools of water. Ditches and the main water channels (lodes) are kept clear. In the past windmills have been used to pump water up into the fen. There has been a small-scale revival of the harvesting of crops by cutting sedge. Meadows and paths are mown to ensure this stage of the succession is retained. Different cutting regimes within the year, or on a two or three year rotation, encourage greater diversity of the vegetation. A high proportion of the area has developed into woodland (carr), but parts are cleared to prevent the woodland stage taking over.

LAND USE, BIODIVERSITY AND CONSERVATION

Figure 6.23 Management practices at Wicken Fen help maintain diversity – different stages of succession from open water (ditch in lower photograph) through to woodland. Intermediate stages are maintained by mowing and cutting, and the waterways must be cleared from time to time

Figure 6.24 Conservation in a churchyard – a Suffolk churchyard in May, where mowing is delayed until after the flowering period as a means of encouraging diversity (top); Conservation in a garden – at the edge of a lawn, a strip is left unmown until autumn, allowing spring flowering of lady's smock (Cardamine pratensis) (bottom)

The big change to the East Anglian fen region came in the seventeenth century, when extensive drainage systems were built so that the fenland could be converted into agricultural land. Wicken Fen was one of the relatively few areas that was not drained and continued to be used in the traditional manner. However, by the end of the nineteenth century, it had become uneconomic to continue these activities as the demand for peat, sedge and reeds had largely disappeared. Very little cutting was done in the early part of the twentieth century and gradually the area became invaded by shrubs and trees. Today Wicken Fen is managed as a nature reserve, under the ownership of the National Trust. The aims of the management strategy are to encourage and maintain the diversity of species associated with wet fenland communities. Some management practices can be seen in Figure 6.23.

The main threat to the fenland is the lowering of the water table. The surrounding farmland now lies several metres below the level of the fen, because the land shrank lower as it was drained and became drier. Water must therefore be pumped up into the fen, to help maintain the wet conditions and because the water gradually seeps out into the farmland below, parts of the perimeter of the fen have been sealed to try to keep the water in.

The descriptions of Bradfield Woods and Wicken Fen illustrate some of the reasons for conservation and the principles behind conservation strategies. These are highlighted below:

- change of land use may lead to loss of diversity in terms of habitats and species (contrast Bradfield Woods and Wicken Fen with surrounding arable farmland)
- historically, human activities have already influenced these two sites (people have interacted with the wood or fen exploiting their productivity)
- management for conservation requires active interference to maintain diversity by creating a range of habitats (coppicing, mowing, maintaining ditches)
- scientific interest lies in the range of communities, rare plant and animal species
- nature reserves provide opportunity for academic research, educational study
- leisure benefits (medieval rides, present-day visitors)
- conservation aims to protect existing ecosystems, maintain genetic resources.

Managing grassland and maintaining diversity

Agricultural land frequently includes grassland. A successful farmer must maintain productivity at an economic level and this may be seen as the crop removed from the area or through the animals grazing on the land. Elsewhere, grassland is found in parks, golf courses, racecourses, playing fields and gardens, as well as in areas designated for conservation. Maintenance of grassland is achieved through mowing, grazing and sometimes burning, but the profile of plant and animal species present and the diversity achieved depend on the management strategy adopted.

Mowing

Mowing is an unselective way of controlling plant growth. When mowing is done by machine, the sward is cut to an even height and any unevenness, such as molehills or tussocks, tends to be levelled off. Traditional mowing with a scythe is more sympathetic and retains some unevenness which can provide useful habitats for invertebrates or ground-nesting birds. The cut vegetation may be removed for use as a hay or silage crop. Removal of the crop means loss of nutrients from the area. In traditional farming systems, nutrients are replaced by dung from animals allowed to graze after mowing, but in modern farming practice, artificial fertilisers are often applied. High levels of nutrient encourage growth of vigorous grasses which may exclude other less competitive species. If mowings are left on the ground, they may smother certain plant species or prevent seeds reaching the surface and germinating. Accumulated mowings tend to discourage ground-dwelling invertebrates and also create pockets of high nutrient. Mowing tends to lead to uniformity of species.

Grass grown for silage normally consists of high-yielding varieties and chemical fertiliser is added to maximise the crop. Mowing for silage crops generally starts in late April or early May, followed by two more cuts at six-week intervals. The emphasis is on high yields, uniformity of growth and mowing at the stage which produces good quality silage. More diversity is achieved when grass is cut for hay, usually in June or early July. The later cutting date allows more species to flower and produce seeds.

Increased diversity of species can be achieved by varying the season and frequency of mowing and the height of the cutter. Some plants can survive frequent cutting or trampling but many need to flower and set seed if they are to survive in the area. Table 6.4 gives the flowering period of some grassland species typically found in churchyards. For spring flowering species, the mowing would be done towards the end of June, but if there is a predominance of species which flower in June and July, the cut should be delayed until late July or August.

Generally the even, close cut of frequently mown grassland is inhospitable to insects and other invertebrates, whereas they are encouraged by rough tussocky grass, which has pockets of bare ground or patches of uncut vegetation amongst mown grass. Sometimes an area can be left unmown for 2 to 3 years or even longer. Patches of long grass are important as cover for birds. Reduction in numbers of some birds (such as the corncrake) is probably associated with unsympathetic cutting of grassland. It also helps if fields are cut from the centre towards the outside, so that any birds, particularly those with young chicks, have a chance to keep under cover and escape from the cutter.

Grazing

Grazing animals are selective in their choice of species. In Britain the main domesticated grazing animals are sheep, cattle and horses, though wild animals, particularly rabbits, have considerable influence on the nature of the grassland and in prevention of the succession to scrub. Sheep bite vegetation close to the ground and thereby maintain a short, even sward. This habitat is favoured by birds such as stone curlews, woodlarks and wheatears. During the day sheep drop their dung across the area but at night dung is deposited in specific areas, which then become enriched with nutrients which is deleterious to the flora. Trampling by sheep has relatively little effect, except where the soil is loose or on steep hillsides. Cattle wrap their large tongue around the vegetation and, compared with sheep, can consume relatively tall and coarse plants. The resulting sward is more uneven than with sheep because cattle tend

Table 6.4 *The flowering periods of some grassland species typically found in churchyards*

Species	Mar	Apr	May	Jun	Jul	Aug	Sep	Oct
lesser celandine	- -	——	——	- -				
primrose	——	——	- -					
cowslip		- -	——					
meadow saxifrage		- -	——	——				
greater stitchwort		- -	——	——				
germander speedwell	- -	- -	——	——	- -			
bulbous buttercup	- -	- -	——	——				
bugle		- -	——	- -				
meadow vetchling			- -	——	——	- -		
bush vetch			- -	——	——	- -		
sorrel		- -	——	- -				
hoary plantain			- -	——	——	- -		
ox-eye daisy				- -	——	——		
hedge bedstraw				- -	——	——		- -
lesser stitchwort			- -	——	——	- -		
lady's bedstraw					- -	——		
common knapweed				- -	——	——	- -	
dark mullein				- -	——	——	- -	- -
burnet saxifrage					——	——		
common toadflax					- -	——	- -	- -

to select certain patches. They also trample the ground leaving hoof marks, bare soil and muddy areas if wet. Their dung, in the form of cowpats, results in local enrichment of nutrients. Around the cowpat the area is unpalatable, but the cowpat itself provides the opportunity for another community to become established, at least for a short time. Trampling by cattle along the edge of ditches can increase marshy and muddy areas, but if the trampling pressure is too heavy it is harmful. Horses are much more selective than sheep or cattle. They may almost eliminate some species from an area yet ignore others, producing rather patchy vegetation.

Table 6.5 *Grassland butterflies show different preferences for sward height. Some survive the winter on their foodplant as eggs or larvae, so heavy winter grazing or cutting would be detrimental (ø). Some prefer very short turf (*), others (†) prefer bare patches around the foodplant.*

Species	Sward height / cm
small skipper ø	15+
Essex skipper ø	20+
Lulworth skipper ø	30+
silver-spotted skipper *	1–4
large skipper ø	8–20
dingy skipper ø	2–5
grizzled haristreak	4–10
small copper brood 1	10–30
brood 2/3	1–9
small blue	4–15
silver-studded blue ø*	2–5
common blue	4–10
chalkhill blue ø	2–6
Adonis blue †	0.5–2.5
marsh fritillary	4–15
grayling *	2–6

Wet grasslands

Wet grasslands are part of a traditional farming system. In summer, these meadows are cut for hay then used for grazing; in winter they are often flooded. These expanses of open water attract a range of birds, including waders (such as lapwing, snipe and redshank), and ducks, geese and swans, though prolonged flooding may have an adverse effect on both the vegetation and on the invertebrates in the soil. The area is more attractive to birdlife if there is some variation in the surface features. It is particularly important to have plenty of 'edge' habitat and pool margins to provide suitable conditions for feeding waders and their chicks (see Figure 6.25).

In many places, 'improvement' of wet grassland has been undertaken to enhance its agricultural potential but with consequent loss of this type of habitat and the species associated with it. The changes have been effected by

LAND USE, BIODIVERSITY AND CONSERVATION

Figure 6.26 Retaining wetland habitats is important to maintain diversity of species.

So-called improvement is not always negative for wildlife. Geese, for example, vary in their food preferences: Greenland white-fronted geese prefer rough, tussocky grassland whereas improved grassland, though floristically poorer, makes good feeding for pink-footed, brent and barnacle geese

Redshank

require surface water
feed at pools and edges of creeks at coastal sites
prefer short swards (<15cm) and tussocks to build nests
best management by cattle grazing

Snipe

probe for soil invertebrates in soft, damp soil
need tussocks and taller vegetation (>25 cm) for concealment
move only short distances for feeding
use edges of pools and ditches
best management by cattle grazing

Lapwing

least dependent on wet or damp conditions (breed on wide range of farmland types)
short vegetation required only during breeding season
pick up invertebrates from muddy edges of shallow surface pools
best management by cattle or sheep grazing

Figure 6.25 Redshank, Snipe and Lapwing are three bird species associated with wet grasslands. They have specialised feeding requirements, so grassland can be managed to encourage these and other species

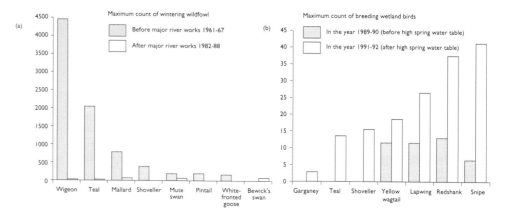

Figure 6.27 Effect of drainage schemes on bird populations at Pulborough Brooks, Sussex. (a) shows a marked decline in bird population, whereas (b) shows the changes after management of the site by the RSPB. Changes were made to the water control structures to allow shallow winter floods to be retained and the water table has been kept high in spring and early summer. Similar successes have been reported at other sites where water levels have been controlled to keep the water table high at certain times of year

drainage, mowing and enrichment with fertilisers. Sometimes weedkillers have been used to select the required grasses in the sward, or the land may have been ploughed up and then re-seeded with more vigorous grasses. Drainage of the wet grasslands lowers the water table, making the land drier, but feeding is then difficult for waders. If there is no winter flooding, the wildfowl do not come to the area. Suitable management can lead to restoration of wet grasslands and encourage the birdlife and other associated species. Water levels in surrounding ditches or water channels are controlled to allow deliberate flooding at certain times of year and it may be appropriate to re-introduce grazing by cattle during the summer.

Managing wet habitats and maintaining diversity

Rivers, canals and ditches, together with their banks and the habitats along their edges, can be managed so that they enhance the diversity of species within farmland or other areas. Open water needs to be kept clear, at least in parts, because of mud and vegetation which would gradually accumulate. Clearance can be carried out by small-scale digging, preferably in stages. When larger scale dredging is required, a sympathetic approach, section by section over a period of time, helps to retain a range of habitats which reflect different stages in the succession from open water to the drier habitats. In some places it may be possible to maintain or create shallow ledges at the edge of the water, particularly if the banks need to be supported to prevent collapse. A vertical edge is unfriendly and uninhabitable for many plant and animal species. The vegetation on the banks is often cut to prevent encroachment into the water. This is best done in rotation, avoiding sudden loss of one of the stages in a succession. The timing of the cutting may be important to allow some species to flower.

Table 6.6 *Changes in farming practice that have resulted in loss of habitat and of species diversity*

Changes in farming practice	Effect on habitat and species
uncropped land converted to cropped land	• reduced vegetation diversity • reduced invertebrates, small mammals, birds
large woods fragmented to small woods	• decline of woodland species
hedges removed (increased size of fields)	• decline in small mammals, owls, overwintering predatory invertebrates • removal of berry and nut food source for winter thrushes and other birds
ponds filled in	• reduced aquatic vegetation • decline in newts, frogs, aquatic • invertebrates (e.g. dragonflies)
wet meadows drained	• change in vegetation • reduced invertebrate diversity • decline in breeding waders
saltmarshes drained	• loss of plant species • reduction in geese, waders and wildfowl
permanent pasture converted to temporary ley (grassland)	• reduced diversity of vegetation • reduced earthworms and other invertebrates • reduced birds such as skylarks, lapwings, thrushes, gulls, golden plovers
mixed crops replaced by arable monoculture	• reduced invertebrate diversity • decline in birds such as lapwings, skylarks, stone curlews
crop rotation replaced by same crop each year	• reduced diversity of invertebrates including butterflies • reduced diversity of birds
undersowing with grass or clover replaced by no undersowing	• reduced summer invertebrates • nitrogen inputs needed on land
application of herbicides, insecticides, fungicides	• direct toxicity • reduced invertebrate abundance and diversity, effects through food chains on food supplies

Integrating conservation and production

In this chapter, we looked first at deforestation and desertification as two extreme examples of how ecosystems can be destroyed. In the section on conservation, we looked closely at some British habitats and saw how, through careful management, a diversity of species can be maintained. Specially protected areas, like Bradfield Woods, Wicken Fen and many other reserves, are very valuable in conservation terms but they occupy only a tiny fraction of the land (probably less than 2%). There is, however, considerable scope for integration of conservation with current agricultural practices in ways that have benefits in terms of encouraging species diversity.

In Britain, demand for increased food production over the last 50 years has meant that traditional small-scale farming systems have largely been replaced by intensive large-scale systems, which rely heavily on machinery, fertilisers, pesticides and herbicides. In addition, more land has been brought into cultivation by clearing woodland, draining wetlands and ploughing up meadows. Table 6.6 lists a range of farming practices that have resulted in loss of 'natural' habitat with an associated reduction in species diversity.

Even on intensively farmed land, there are ways of improving the diversity of wildlife, often with minimal economic loss in terms of crop yield. Areas with semi-natural habitats can be increased and diversity can be created within other habitats. The margins of cropped areas can be valuable provided the area is large enough for species to become established. Any such areas must also be managed in a way that encourages species diversity. Some farm management practices and how these can be modified to benefit conservation are summarised in Table 6.7.

Farmers, like any other business managers, must make a profit out of their activities. They may also need advice as to how to implement conservation and encourage diversity of species on their land. A number of schemes are in operation which provide the opportunity for creation of wildlife habitats, or direct funding often with advice to farmers. These include ESAs (Environmentally Sensitive Areas), set aside scheme, farm woodland scheme, FWAG (Farming and Wildlife Advisory Group), RSPB (The Royal Society for the Protection of Birds).

As an oversimplification, we summarise below the choices in terms of land use, its management and the maintenance of plant and animal species diversity in the interests of conservation:

- progression or succession to natural climax community
  → less species diversity
- intervention in a succession and management for conservation
  → more species diversity
- intensive agriculture → less species diversity
- integration of conservation with farming practice
  → more species diversity.

Some questions about conservation

This section has looked at conservation in practice and decribed some specific examples in a limited number of sites, all in Britain. The principles that would be applied in many other habitats are similar, where the aim is to maintain a range of habitats and with it a diversity of species.

How might a farmer manage strips around an arable field to encourage invertebrates? How do beetle banks and short-rotation cropping contribute to conservation?

List some of the benefits of maintaining species diversity and give some definite examples. Think about the importance of maintaining genetic resources as well as the aesthetic and scientific value of management for conservation.

Table 6.7 *Some farming practices and ways that management for conservation can be integrated within the farm system*

Habitat	Farm use and management practices	Management for conservation	Benefits
grassland (pasture)	• used for grazing, hay, silage • managment involves weed control, fertiliser and lime, farmyard manure, reseeding, adjustment of stocking density	• use minimum amounts of fertiliser and herbicide • retain damp areas and corners • avoid re-seeding • prevent colonisation of scrub and avoid planting trees where flora is rich	important for plants, butterflies and other insects, birds including winter thrushes, waders
woods, copses	• used for timber, firewood, game conservation • management involves felling and replanting, coppicing and thinning in rotation; maintenance of rides and glades	• plant or retain native species • manage part at a time to allow recolonisation • use coppice management on rotation where suitable • maintain wide rides, cutting alternate sides each year	important for plants, butterflies, birds, mammals
hedges	• used as stock-proof fences, for shelter and game conservation • managed by laying, cutting, coppicing, annual trimming	• maintain hedges to be thick at the base and of reasonable height • keep some trees in the hedge • laying or coppicing at suitable intervals	important for plants, butterflies, birds, mammals
water courses	• useful as barriers for stock and for drainage and irrigation • need regular cleaning, involving mechanical and chemical weed control	• manage part at a time to allow recolonisation • keep gentle profiles • ensure water is present throughout the year	important for plants, dragonflies, fish, birds, mammals
ponds	• used for drinking by stock, for irrigation, angling, shooting	• manage part at a time to allow recolonisation • make sure water remains unpolluted throughout the year • avoid too much shading by surrounding trees	important for plants, dragonflies, fish, amphibia, birds, mammals
lanes, roadside verges	• needed for access to fields and buildings • managed by mowing, herbicides	• cut late, after flowering and remove cuttings if possible • avoid herbicides	important for plants, insects including butterflies

Bradfield Woods and Wicken Fen are two nature reserves, described in this section. Compare these two reserves, or compare one of them with another reserve that may be familiar to you. Try to answer these questions for each reserve, and draw up a list of common features.

• What human activities over the last 500 years or more have influenced the vegetation in the area?
• What would be the natural climax vegetation in the area?
• In what ways has the productivity of the area been exploited and harvested by local people?

• What sort of land surrounds the reserve today? How does the reserve differ from the surrounding land in terms of diversity?
• What management strategies are being used to help maintain the diversity within the nature reserve?
• Suppose you were involved with a group of people interested in creating a small nature reserve out of a piece of waste land. Draw up a list of principles that you would wish to apply to make it an attractive conservation area. Then give examples of some of the practical things you would do to create and maintain the reserve.

A comparison of two closely related communities

Introduction

The aim of this investigation is to compare the distribution of species in two closely related communities and to consider the factors which may influence the distribution of organisms. The choice of sites to compare will, of course, depend upon access, but valuable data can be obtained using, for example, a footpath across a lawn or school field to investigate the effect of trampling. Other possible sites include:

• managed and unmanaged woodland, with the same dominant species
• north-facing and south-facing slopes
• grazed and ungrazed grassland
• football pitch and margins of the pitch
• well-drained and marshy land.

In each case, a **belt transect** is used to investigate changes in the communities (see Chapter 4). Ideally, the quadrats should be placed immediately after each other, although it may be more appropriate to sample at, for example, 1 metre intervals.

Materials

• Tape measure
• Marking pegs
• A quadrat frame, e.g. 0.5 m × 0.5 m
• Identification keys
• Recording sheets

Method

1 Set out the tape measure across the area you are going to study.
2 Place the quadrat at suitable intervals and, each time, record the presence of species, estimate their percentage cover, note the growth form of the plants and measure their maximum height.
3 If possible, obtain soil samples at each site to determine, for example, pH, moisture and humus content.
4 Repeat this procedure using a parallel transect.

Results and discussion

1 Draw a simple map of your study area and include relevant background information.
2 Record your results carefully and present using appropriate graphs.
3 Which abiotic factors show changes along your transect?
4 Consider the extent to which changes in plant communities can be explained by human activities and related biotic factors.

Further work

It is possible to make a quantitative comparison of contrasting communities by calculating a **diversity index**. In general, a complex community (consisting of a large number of different species but relatively few individuals of each species), is more stable than a community containing relatively few species. Polluted freshwater, for example, may contain very few species of invertebrates, but large populations of each.

A community can be evaluated by counting the numbers of individuals of each species. One way of expressing this relationship is **Simpson's Diversity Index (DI)**, which is based on the probability of randomly collecting a pair of organisms of the same species from a population. The higher the value of the calculated index, the greater the species diversity of the community. Simpson's Diversity Index is calculated using the formula:

$$DI = \frac{N(N-1)}{\Sigma n(n-1)}$$

where DI = diversity index, N = total number of individuals of all species and n = number of individuals of a species.

Factors affecting the distribution of *Pleurococcus*

Introduction

Pleurococcus is a green alga which is commonly found growing on the surface of walls or on the bark of trees, where it forms a powdery encrustation. Observation will indicate that the distribution is not uniform as it is more abundant on some aspects that others. This distribution could be influenced by light intensity, water availability, or aspect. Aspect will not only influence light intensity, but also factors such as prevailing wind and rain.

It is difficult to use a quantitative method to measure the actual density of *Pleurococcus*, but we can adopt an approach to judge how much of the alga is present by estimating the intensity of the green colour, in other words, the darker the green colour, the higher the density of *Pleurococcus*. In order to compare the densities, you could use a series of comparisons, as indicated in Table 6.8.

Table 6.3 *Colour comparison for estimation of population density of* Pleurococcus

Density 1 (minimum)	Plant apparently absent, or green tint extremely faint
Density 2 (intermediate)	Much greener than 1, but underlying surface still visible
Density 3 (maximum)	Intensely green, cells so close together that underlying surface is hidden

LAND USE, BIODIVERSITY AND CONSERVATION

In this practical, we use a sampling technique to estimate the density of *Pleurococcus* on a tree.

Materials

- Flexible plastic quadrat
- String
- Drawing pins
- Chalk

Method

1 Fix a piece of string around the tree to be studied. Hold in position with drawing pins.
2 Choose grid lines within the quadrat enclosing an area of suitable size.
3 Starting at one end of the string make a record of the density of *Pleurococcus* within the quadrat frame.
4 Slide the quadrat along the string with the same edge in contact with it and repeat the procedure for successive areas. Before moving the quadrat along, make a mark with chalk to show its position.
5 Make a careful note of the kind of surface, the slope, the height above ground, the size of the sampling area, the place and the date.

Results and discussion

1 Plot your results as a column graph to show the densities of *Pleurococcus* in each successive position of the quadrat. The position of the quadrat should be shown on the *x* axis and the density present at each position on the *y* axis.
2 Repeat this experiment, if possible, in a different type of locality.

Further work

1 Consider the factors which could account for the differences in the density of *Pleurococcus*.
2 You could measure differences in light intensity around the tree, and use a compass to determine the aspects of the tree. Is the density greater on the south or the north aspect of the tree?
3 As a long term study, you could investigate the influence of run-off on the distribution of *Pleurococcus*. Water drains down the surface of the tree after rain and snow. Four collecting vessels per tree should be adequate, two for low density areas and two for high. Measure the volume of water collected during a known period of time. Some methods for collecting water flowing down a tree are shown in Figure 6.28.

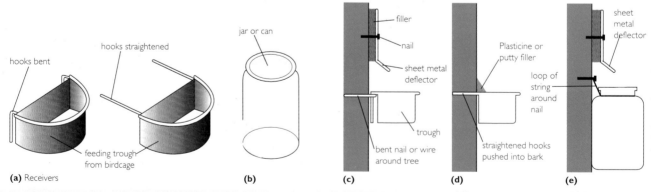

Figure 6.28 Methods for the collection of water flowing from a tree

Pests and pollution

Human activities and their effects on the environment

In this chapter we look further at demands made on the environment by the increasing human population. The quest for higher yields from food crops to feed the world population has to some extent been satisfied by the use of chemicals. These chemicals come in the form of fertilisers to raise the nutrient levels in the growing medium, or as pesticides to reduce losses from disease and herbicides to minimise competition from weeds. We see how the effects of these chemicals often extend beyond their immediate purpose. **Pollution** is a familiar word in our everyday vocabulary and we are repeatedly reminded of how a range of human (anthropogenic) activities produce harmful substances which contaminate the air, the water and the land which together make up the environment for living organisms. Pollutants are derived mainly from the waste materials produced in industrial activities (including agriculture), and from motor vehicles, with some coming from domestic sources. The consequences of pollution are observed in the effects on living organisms and, on a different scale, in the possible long-term effects on climate.

Control of insect populations

Insects are are usually present in considerable numbers in any habitat, such as grassland, roadside verges, hedges, woodland and cultivated land. They occur in the soil, among decaying litter on the ground and feeding on plants or other animals in the area. Estimates suggest there can be as many as 25 million insects in a hectare of soil, and perhaps 25 000 in flight over a hectare of land at any one time. A sugar beet crop infested with black.bean aphids could even carry up to 200 million aphids per hectare.

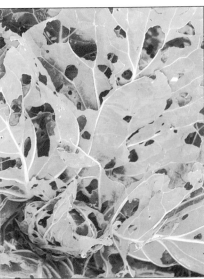  

Figure 7.1 Damage to plants caused by insect pests. (left) direct damage to cabbage by large white butterfly caterpillars; (centre) North American lupin aphids (Macrosyphum albifrons) *feeding on lupin stem. (The large female on the left is giving birth to a live nymph.) Aphids cause direct damage by piercing the phloem and sucking out sap, and they also cause indirect damage by transmitting viruses which cause disease; (right) viral damage to leaf of the cowpea plant.*

As with any populations of organisms, numbers of insects within a population are never static and, in any defined place, are liable to fluctuate over a period of time. Actual numbers in a population depend on the number of individuals of the species that are being born or dying, and those migrating into or out of the area. Changes in population numbers are influenced by season, weather conditions and competition between organisms. A discussion of competition between organisms is given in Chapter 5. Remember also that insects are part of food webs so that any alteration in numbers of one species is likely to affect other interrelated organisms.

Insects are described as pests when they cause damage or nuisance, directly or indirectly, to crop plants, domesticated animals or humans. The discussion here will focus on pests in relation to crop plants in which there may be losses in yield or quality of the produce as a result of insect pests. Considerable expenditure is devoted to the control of insect pests of agricultural and horticultural crops. If there were no insect pests, on a global scale food production could probably be increased by 30 per cent. A successful insect pest is often one that can establish itself and spread rapidly in the crop. A high rate of reproduction is a feature which may contribute to success. The monocultures associated with modern intensive agricultural systems often present an ideal situation for the insect pest by providing easy access to an abundant food supply, with few if any natural predators in the area.

The Colorado beetle (*Leptinotarsa decemlineata*) illustrates how friend can turn into foe. In 1824 the Colorado beetle was described by a collector from the Rocky Mountains in north America. This attractive, striped beetle was considered a rarity. It fed on the weed Buffalo-bur, a member of the potato family. Thirty years later, settlers to the region brought potatoes as one of their crops. The potatoes provided a nutritious and plentiful new food source for the beetles and their effect on the crops was devastating. Populations of the Colorado beetle rapidly increased and spread eastwards across the US in the second half of the nineteenth century, causing famine as the crops failed to produce enough food. Finally a decision was taken to spray the potato crops with Paris Green, an insecticide containing arsenic, as a means of controlling the beetle. The Colorado beetle is now treated as a pest on an international scale, with strict regulations governing procedures to be taken if any are found.

Using chemicals to control insect pests

An **insecticide** is a chemical substance which can kill or repel insects. Naturally occurring insecticidal chemicals which have been used for hundreds of years include pyrethrum from the flower of *Pyrethrum cinereafolium* (in the daisy family) and nicotine from *Nicotiana tabacum* (the tobacco plant). Many artificial chemical substances have now been synthesised for use as insecticides. These fall into four main groups, the **organochlorines**, **organophosphates**, **carbamates** and **pyrethroids**, summarised in Table 7.1. These insecticides are complex organic substances. You will see, for example, that the full name of DDT is 1,1,1-trichloro-2,2 *bis*(p-chorophenyl)ethane.

Table 7.1 *the main groups of chemical insecticides*

Insecticide group	Examples	Mode of action	Behaviour in plant, relative toxicity and other comments
organochlorines e.g. DDT	DDT, HCH, aldrin, dieldrin, endosulfan	DDT inhibits enzyme cytochrome oxidase, destabilises nervous system by interfering with permeability of nerve axon membrane	• low toxicity to humans, so relatively safe when being applied • slow breakdown, so persistent in soil • broad spectrum **contact** insecticide
organophosphates e.g. parathion	parathion, malathion, dimethoate, metasystox, schradan	combine with cholinesterase, thus inhibiting hydrolysis of acetylcholine produced at nerve endings, so interfere with transmission of nerve impulses across synapses	• high toxicity to humans • easily broken down so less persistent in soil • considerable flexibility in the group, most show **systemic** action; malathion is a **fumigant** insecticide
carbamates e.g. carbaryl	carbaryl, carbofuran, aldicarb, methomyl, pirimicarb	interfere with the nervous system by acting as competitors with cholinesterase, thus inhibiting hydrolysis of acetylcholine	• some very toxic to humans (carbofuran, aldicarb) • persistence lies between organochlorines and organophosphates • pirimicarb has low mammalian toxicity, biochemical selectivity for aphids and some flies • methomyl has good **contact** action, is also a **fumigant** and to some extent has **systemic** action
pyrethroids e.g. resmethrin	(*natural*) pyrethrum, nicotine, rotenone (*synthetic*) allethrin, cypermethrin, resmethrin	similar to DDT	• high toxicity to insects, low toxicity to mammals • similar mechanism to DDT, so some cross resistance • **contact** insecticide, so damaging to natural populations of insects

Insecticides are usually dissolved in water or oil and applied by spraying in fine droplets. In some cases they may be dispersed on an inert solid carrier. In closed spaces, such as glasshouses, the insecticide may be burned to give off smoke (**fumigant** insecticides). Spraying may be done from small hand-operated tanks or by means of large-scale machinery using tractors or aircraft. It is important that the sprayed insecticide reaches the target (either the insect itself or the plant) at a suitable time and in the required concentration, without drifting or causing damage elsewhere, and that the insecticide remains on or in its target long enough to exert its effect on the insect pest. In practice, often only a small fraction of the insecticide actually lands on the required target.

The insecticides get into the insects in different ways. Some insecticides have their effect by direct **contact** with the insect and penetrate the cuticle. **Fumigants** are inhaled by the insects. Others act as **stomach poisons** and are taken in when the insect eats the leaf. **Systemic** insecticides are taken into the plant by absorption through the leaves or roots, or by the seed when it germinates, then circulate through the plant. Systemic insecticides then enter the insect when they feed on the sap of the plant and are particularly useful for controlling sap-sucking insects such as aphids.

Figure 7.2 Spraying crops with insecticide to minimise future damage by pests

PESTS AND POLLUTION

In most cases, the toxicity of the insecticide is due to its interference with the nervous or respiratory systems. Often an insecticide is also toxic to other species, including harmless insects, humans or other mammals. Some of these species may even be beneficial due to their predatory action on the pests (hoverflies and ladybirds on aphids, for example) or because they carry out pollination (bees). Systemic insecticides have an advantage in that they only affect insects which feed on the plant. They are thus useful for the control of aphids sucking sap but do not affect predators or other harmless insects walking over the leaves. Pirimicarb is an example of an insecticide (systemic and fumigant) that has a biochemical selectivity for aphids and some flies, but has a very low toxicity to mammals. When applying insecticides, strict precautions must be taken to avoid unwanted harmful effects on humans and other organisms.

Insecticides differ with respect to the length of time they persist in the plant, or in the soil, before breaking down. The organochlorines are relatively stable and resistant to breakdown, whereas most organophosphates are biodegradable and rapidly broken down by metabolic reactions in animals and excreted as harmless substances. The persistence associated with organochlorines may at first sight appear to be an advantage in terms of controlling the pest insect, but, over a period of time, can result in undesirable toxic side-effects. This is well illustrated by **DDT**. In the 1940s and 1950s, DDT was welcomed as a solution to many pest problems, for control of pests on crops and to minimise the spread of malaria by killing the mosquito (a vector of the pathogen which causes malaria). However, since the 1960s, there has been increasing concern over the longer term effects of using DDT and other persistent insecticides.

The concern arises from the effects that insecticides may have on other harmless or beneficial organisms, by being passed on through food chains. Toxic residues in the soil may seep into water courses and contaminate them, or they may persist on the crop plants and then be found in human food. In the 1960s, residues of organochlorines were detected in a very wide range of organisms, including humans and many food sources. Figure 7.3 summarises the results of studies (carried out in the 1960s) of British birds and their eggs, which revealed detectable quantities of organochlorine insecticide residues. The accumulation appeared to increase at higher trophic levels, being greatest in the top carnivores.

Over a period of time, the effectiveness of some insecticides has become less because of the development of resistance within the insect populations. The situation provides an example of genetic change in a population in response to selection pressure. Within the gene pool of the population, there are likely to be some individuals which are not killed by the insecticide, or have a natural resistance to its harmful effects. If these insects survive and reproduce, their offspring may inherit the characters which produced the resistance. In subsequent generations an increasing proportion of the population is likely to show resistance to the insecticide.

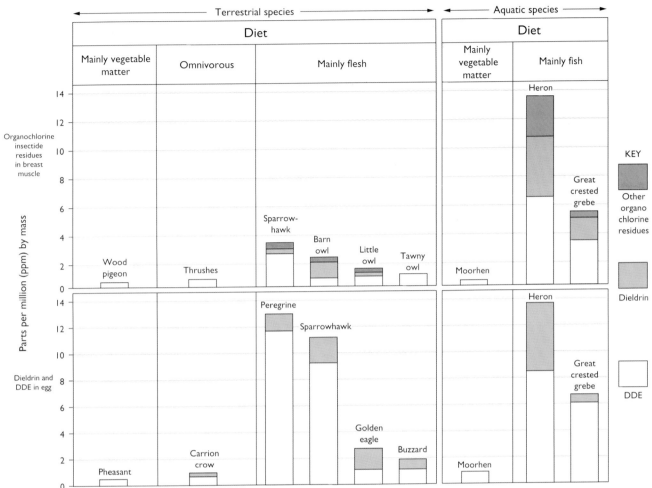

Figure 7.3 Persistent pesticides and how they pass along the food chain. Residues of organochlorine insecticides in British birds (DDE is derived from DDT by metabolism. DDE is more stable and persistant than DDT.)

One further problem associated with the use of insecticides is that of **resurgence**, illustrated in Figure 7.4. When no pesticide is applied, population numbers of a pest and its natural predators are likely to fluctuate around an equilibrium level. However, if a broad spectrum pesticide is used, natural predators are likely to be killed as well as the pest. With lower numbers of predators, any surviving pests may increase very rapidly leading to a population explosion.

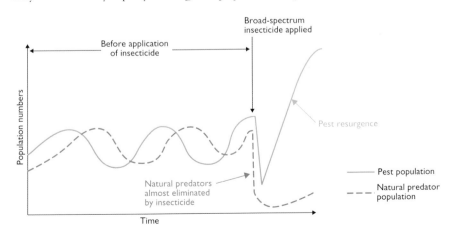

Figure 7.4 Resurgence of a pest after application of a broad-spectrum insecticide. Before application of this insecticide, the pest population numbers in a given area (density) are controlled by natural predators – numbers of the pest increase, then the natural predators increase after a short time lag. After application of the insecticide, the natural predators are also killed so they can no longer control the numbers of the pest. Surviving pests may then show a population explosion

Figure 7.5 Ladybirds are natural predators of aphids and other insects and can be used as a means of biological control

Biological control of insect pests

Biological control exploits interactions between organisms. The term usually implies the deliberate introduction of one species to control another species (the pest), either through a predator–prey relationship or as a parasite which weakens or causes disease in the pest. The controlling organism may already be present in the area as a natural enemy (but probably in low numbers) or may be an alien, introduced to the area by humans. Biological control has been used for centuries: cats are used to prey on rats in grain storage barns and ducks are used in China to control various pests on crops. In 1762, the Mynah bird was brought to Mauritius from India to help control locusts. Perhaps the beginning of the use of biological control in the modern sense came in the 1880s, when ladybirds were successfully used to control a scale insect attacking citrus crops in California.

The biological control agent may be another insect, or control may be through viruses, bacteria or fungi. Some examples of biological control agents used against insect pests are given in Table 7.2. The success of a control agent depends on the nature of the relationship between the two species and on the population dynamics. If the control organism is introduced in low numbers, it may take time to build up sufficient numbers to harm the pest; meanwhile the pest may have done considerable damage to the crop. Generally the aim is to keep the damage from the pest at a level acceptable to the grower rather than to eliminate the pest completely. If the control organism is present in large numbers or is very effective at eliminating the pest, the control organism is then deprived of its food source. Numbers of both the pest and the control organism are likely to fluctuate, since one depends upon the other. In some cases the biological control organisms are effective only in a confined space and where conditions can be controlled, such as inside a glasshouse.

Table 7.2 *Examples of biological control of insect pests*

Biological control agent		Insect pest	Example of use and other comments
viruses	baculoviruses	larval stages of butterflies and moths; ants, bees and wasps; flies, gnats, midges, beetles, caddis flies	• different strains of virus attack different species of insect • used in USA to protect cotton and fir trees, in France for vegetables, in Brazil for soya bean
bacteria	*Bacillus thuringiensis*	larval stages of butterflies and moths; beetles, flies	• different subspecies produce toxins active against different insects • used widely to protect edible and ornamental flowering plants
fungi	*Verticillium lecanii*	aphids and whitefly	• used on cucumber, eggplant, chrysanthemums
nematodes	*Heterorhabditis megidis*	vine weevil larvae	• used to protect bedding plants, e.g. primulas
insects	*Encarsia formosa*	aphids and whitefly	• parasitic wasp, used widely for control of whitefly in glasshouses
	Phytoseiulus persimilis	mites	• predatory mite, used for control of red spider mite in glasshouses

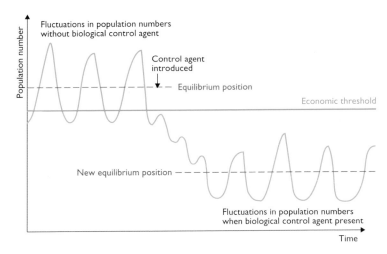

Figure 7.6 Biological control, showing how the numbers of a pest insect in a given area (density) can be reduced from a high equilibrium level to a new lower equilibrium level. The aim is to bring the numbers of the insect pest below the level which causes unacceptable economic damage

An understanding of the ecology and behaviour of both organisms is needed to ensure the control organism is introduced at appropriate times and in suitable numbers to exert effective control. Some examples will help to illustrate this. *Phytoseiulus persimilis* is a predatory mite which feeds only on the red spider mite (*Panonychus ulmi*) and can be used successfully as a biological control agent. Red spider mites are pests of a number of glasshouse crops, including cucumbers and tomatoes. In Britain, the main attack by red spider mite starts when overwintering females emerge from the soil and walls in April or May. Even though the predatory mite can survive at lower temperatures (down to 5 °C) it would be inappropriate and uneconomic to introduce the predatory mite before the red spider mite has emerged. Whitefly is another pest of glasshouse crops. Control of whitefly by the parasitic wasp *Encarsia formosa* is more critical in terms of temperature requirements. The wasp does not overwinter in Britain in unheated glasshouses because it does not survive below 10 °C, so fresh supplies must be introduced at an appropriate time. The wasp requires a temperature above about 18 °C to keep pace with increases in numbers of whitefly. However, at temperatures above 26 °C the wasp reproduces much more rapidly than the whitefly and is likely to eliminate the pest. This means the wasp population also dies out and needs to be replenished if further control of the pest is required.

Encouraging natural predators

Natural predators of insect pests are likely to be present in hedges or uncultivated margins around the edges of fields. These natural predators can be encouraged by maintaining suitable habitats close to or amongst the crops. One way is to create strips of uncultivated land within fields. The strips (sometimes known as 'beetle banks'; Figure 7.7) are sown with suitable vegetation which allows natural predators to overwinter close to the site of the crop. The spacing of the strips is such that predators (such as hoverflies and lacewings) can penetrate into most of the crop when foraging for food. Even though this practice means a small reduction in land available for growing the crop, this is more than compensated for by the reduction in pest damage. Such habitat areas give further benefit by creating diversity within stretches of land dominated by monoculture.

Figure 7.7 Beetle banks encourage overwintering of natural predators adjacent to the crop

Figure 7.8 Blue flowers of Phacelia tanacetifolia *can be used to attract natural pests. Strips of* Phacelia *grown along the margins of cereal crops attract hoverflies, which lay their eggs within the crop. Hoverfly larvae are voracious predators of aphids*

Figure 7.9 Intercropping – potatoes intercropped with maize, in western China

Intercropping provides the opportunity for predators to live on one crop and attack the prey on the adjacent crop. As examples, undersowing with ryegrass provides suitable conditions for ladybirds to control aphids on cereals; cabbage intercropped with red and white clover helps to regulate cabbage aphids, probably because of increase in ground beetles. Interruption of one crop by another may also reduce the spread of insect pests through the affected crop. Similarly, mulches and even weeds can be beneficial by providing a suitable habitat for natural predators. Other cultivation techniques can be used to attract or maintain populations of natural predators (Figure 7.9).

Integrated pest management

A review of the ways used to control insect pests on crops shows that each approach has both advantages and disadvantages. Some of the features are summarised below.

Use of toxic chemicals (insecticides) to kill or repel pests:
- most are general rather than specific in their effects
- lose effectiveness when resistant strains of pest insect increase in the population
- concern over effects of toxicity on other animal species which might naturally help to control pests
- persistence of some insecticides means their effects last over a period of time and spread through food webs to other organisms
- some insecticides taint or spoil the flavour of food crops
- high costs involved (research development, manufacture, machinery to apply the insecticide).

Deliberate biological control by introducing or increasing the numbers of predators or parasites of the pest insect:
- can target a specific pest species or group of pests
- acts relatively slowly and may be unpredictable
- usually does not eliminate the pest completely
- population can be self-perpetuating (but not if the control species eliminates the pest)
- requires careful control with regard to timing of release
- generally no harmful side effects
- limits use of pesticides on other pests on the same crop, which may not be controlled by the biological agent
- relatively inexpensive, both in development costs and in application.

Relying on natural predators, encouraged by maintenance of suitable habitats in the vicinity of the crops (hedges, beetle banks, intercropping, rotation of crops)
- keeps the pest population at low level rather than eliminating it
- level of control unpredictable; populations liable to fluctuate
- natural; no damage to environment
- little or no cost.

In practice, the best control of populations of insect pests is often achieved by a combination of approaches. The following example illustrates how an integrated approach has helped overcome problems.

The spotted alfalfa aphid was first seen in California in 1954. Organophosphates were used to control the aphids but by the end of the 1950s resistance to the insecticide had developed and losses of crops (such as lucerne) had become critical. The usual response would be to increase the application of pesticide, but instead a *reduced* dosage of insecticide (dimethoate) was applied. Even though this meant fewer aphids were killed by insecticide, it allowed some of the natural enemies to survive and these helped to control the aphids. Two further measures were then taken which helped to control the pest. Firstly, the local natural enemies were reinforced by introducing additional parasitic species. Secondly, the lucerne crop was harvested in strips. This allowed some aphids to remain on the newly cut strips with their natural enemies, while at the same time older strips of lucerne were treated by insecticides.

Atmospheric pollution

The term **atmospheric pollution** implies change in the constitution of the atmosphere brought about by human (anthropogenic) activities, causing harm to humans or to other living organisms in the environment. In the late twentieth century, atmospheric pollution affects all nations of the world. The increase in pollution over the last 150 years is attributed to the increasing human population and the rapid growth of urban and industrial societies. However, the effects are not confined to urban or industrial areas, because the pollutants travel in air currents and spread across international boundaries.

Damage caused by atmospheric pollution affects people, their crops, buildings and wildlife as well as the global climate. There is considerable concern that some of the effects are irreversible because the atmosphere has limited ability for recovery. Increasing awareness of the effects of pollution has led to some changes in the relevant human activities, but a realistic reduction can only be brought about by efforts involving individual, local, national and international controls. The following sections look at the causes and consequences of three aspects of atmospheric pollution: ozone depletion, the enhanced greenhouse effect and increasing acid rain.

Ozone and ozone depletion

Ozone is a gas, made up of three atoms of oxygen (O_3). It is formed from a reaction between an oxygen molecule (O_2) and an oxygen atom, in the presence of ultraviolet radiation. The oxygen atoms are derived from oxygen (O_2), which dissociates into two atoms, using energy from UV radiation. Ozone can be destroyed by certain atoms or 'radicals' (parts of molecules in a short-lived reactive state), such as oxygen, chlorine and oxides of nitrogen (collectively represented as NO_x). These reactions are illustrated in Figure 7.10.

Ozone is normally present in air at a concentration of about 0.01 parts per million. It is generally believed that most of the ozone is produced in the regions over the equator and is then distributed to other parts of the globe by high air currents. We will consider three different ways in which ozone is linked with pollution: firstly high in the atmosphere where its presence forms the **ozone layer**, secondly as a pollutant near ground level and thirdly as a **greenhouse gas**. The contribution of ozone to greenhouse gases is described on page 138.

How does knowledge of the biology and ecology of a pest help in deciding how to manage the control methods used? Think about timing of application (of insecticide or of biological control agent). What cultivation techniques would you suggest to someone who did not wish to use chemical insecticides? What features would you look for if you were asked to develop the 'perfect pesticide'?

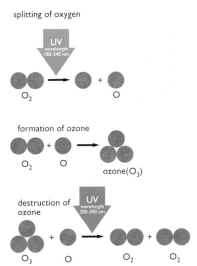

Figure 7.10 Formation and destruction of ozone

PESTS AND POLLUTION

Moving away from the surface of the Earth up into the atmosphere, concentrations of most gases decrease. Ozone is unusual because it is found in higher concentrations at heights between 10 and 50 km above the surface, in the region known as the **stratosphere**. Within the stratosphere, the greatest concentration of ozone lies at heights between 23 and 30 km, and this zone is referred to as the **ozone layer**.

At these heights, the layer of ozone is useful to living organisms in the Earth below. Ultraviolet (UV) radiation is absorbed by the reactions which form ozone and by those which break it down to oxygen. These reactions thus reduce the amount of UV radiation which penetrates the atmosphere and reaches the Earth's surface. UV radiation affects certain biological molecules, including amino acids, proteins and nucleic acids. Exposure to UV radiation can lead, for example, to alterations in the genetic code, damage to cells or disruption of metabolic processes. Exposure to high levels of UV radiation in humans causes sunburn and is likely to be associated with development of skin cancer. UV radiation also affects plant tissues and the process of photosynthesis, and hence the yield from crop plants. Because of the harmful effects of UV radiation on living organisms, the ozone layer has a protective role.

There is, however, increasing evidence that the ozone layer is becoming thinner, or in some places has almost disappeared. A region with reduced ozone is described as an **ozone hole**. The depletion of ozone is attributed to increases in certain pollutants which produce the free radicals that accelerate the destruction of ozone. There is particular concern that **chlorofluorocarbons (CFCs)** are responsible for breakdown of ozone and depletion of the ozone layer. The two CFCs which have attracted most attention are CFC11 ($CFCl_3$), used as a propellant in aerosol sprays, and CFC12 (CF_2Cl_2), used as a cooling agent in refrigerators and air conditioning. CFCs are also found in solvents, in foam materials used in furnishings and foam packaging that has become popular with the 'fast food' industry. Other pollutants which destroy ozone include oxides of nitrogen, derived in particular from motor vehicle emissions.

Evidence for the existence of ozone holes and for the thinning of the ozone layer in certain regions comes from research carried out since 1957 in Antarctica, by the British Antarctic Survey. The Survey first noticed a thinning of the ozone layer in October 1982. A general decline in ozone levels had been noted since about 1970, which coincided with increasing levels of CFCs in the atmosphere. The link between CFCs and their effect on the ozone layer has prompted international agreements which aim to reduce the use of CFCs. In 1987 a number of industrialised nations agreed to a programme which aimed to reduce CFC use to 50 per cent by 1999. However, it is predicted that some of these damaging substances are likely to persist in the stratosphere for 50 or more years, so a drastic reduction is needed if the former balance between the reactions of ozone to oxygen is to be restored.

At much lower levels in the troposphere, near to the Earth's surface, ozone becomes a pollutant with potentially harmful effects on both human health and plant growth. Ozone is one of the components of **photochemical smog**, produced by the action of bright sunlight on a mixture of reactive hydrocarbons and oxides of nitrogen. These originate particularly from motor vehicle exhausts, but also from a range of industrial processes, from paints, dry cleaning fluids, aerosol propellants and household cleaners. Photochemical smog was first recognised in California (USA), especially Los Angeles, but since the 1970s it has occurred more frequently in other large urban areas, including London, Sydney and Tokyo. Notable symptoms in humans are headaches and eye irritation (at about 0.15 ppm) and higher levels are linked with coughs, sore throats and chest discomfort, though deaths are unlikely except in those already suffering from chronic lung diseases. In Los Angeles, concern over health has been sufficient to lead to cancellation of competitive athletics meetings because of the smog. Injury to plants is seen as brown or white spotting on the leaves and a corresponding reduction in growth.

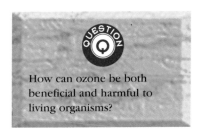

How can ozone be both beneficial and harmful to living organisms?

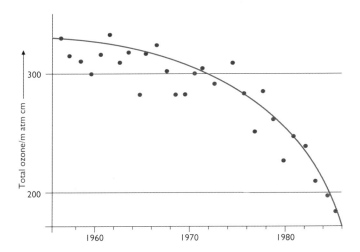

Figure 7.11 Changes in ozone concentration in the atmosphere over the Antarctic in October, between 1956 and 1986

The greenhouse effect

The greenhouse effect is a natural phenomenon in our global atmosphere and plays an important part in maintaining life on Earth as we know it. Without a greenhouse effect, global temperatures would be some 30 °C lower. In Europe, for example, summers would be more like winter and the winter temperatures would approach those in the Arctic and Antarctic regions. There would be much less liquid water on the surface of the Earth, tropical rainforest would not exist and the crops that could be grown around the world would be severely limited.

To understand the greenhouse effect, we need to trace the pathway of the energy received from the sun, through the atmosphere of gases around the Earth, to the surface of the Earth. Radiation from the sun is emitted in a range of wavelengths. Those that are important to biological systems range from short-wave ultraviolet (less than 400 nm), through visible (400 to 700 nm) to infrared (more than 700 nm). Gases in the atmosphere are relatively transparent to this incoming radiation, although about 30 per cent is reflected back into space. Part of the ultraviolet is absorbed by the reactions between oxygen and ozone so the UV radiation is effectively filtered out in the stratosphere. Some of the remaining energy which passes through the atmosphere warms the surface of the Earth. Energy is then radiated back away from the Earth's surface as longer wavelength infrared (4000 to 100 000 nm). Part of this escapes through the atmosphere into space, but some energy is absorbed by gases in the troposphere because the gases are less transparent to this longer wavelength infrared. As a result of absorbing this energy, the troposphere warms up. This warm layer then re-radiates the heat energy it has gained. Some is radiated back to the Earth, again providing warmth. The effect of these gases in the atmosphere is to keep the surface of the Earth warmer than it would be without the gases. This way of trapping the heat is known as the **greenhouse effect**.

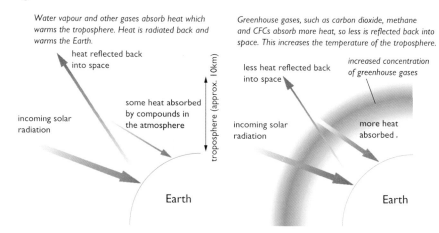

Figure 7.12 The greenhouse effect

We can see a parallel (but not identical) situation in a glasshouse (greenhouse). The glass is transparent to incoming radiation, so both the ground and the air inside warm up. The panes of glass help retain the warmth, partly because they help prevent the warm air escaping by convection. The ground and air are thus kept warmer than they would be without the glass covering. The gases in the troposphere are equivalent to the panes of glass, though they retain the heat in a different way.

Several different gases absorb the long infrared radiation and these are known as **greenhouse gases**. The main greenhouse gases are listed in Figure 7.13, in order of their estimated contribution. With the exception of CFCs, all these greenhouse gases occur naturally, but all of them are also produced as a result of human activities. Increases in the levels of the greenhouse gases have become apparent over the last 150 years, alongside development of industrialisation. The rate of increase has accelerated in recent decades, towards the late twentieth century. The concern is that these anthropogenic sources of greenhouse gases are leading to an exaggerated or enhanced greenhouse effect, and that the acceleration will continue unless positive steps are taken to limit the accumulation of greenhouse gases.

The **carbon dioxide** level in the atmosphere is maintained primarily by the balance between respiration and photosynthesis. Levels of carbon dioxide at different times have been estimated from carbon dioxide trapped as bubbles in ice in Antarctica and Greenland. These show a continuing and steady rise of carbon dioxide from pre-industrial level of about 280 ppm (parts per million) to

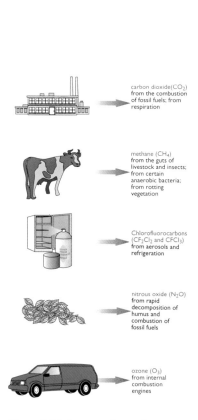

carbon dioxide(CO_2) from the combustion of fossil fuels; from respiration

methane (CH_4) from the guts of livestock and insects; from certain anaerobic bacteria; from rotting vegetation

Chlorofluorocarbons (CF_2Cl_2 and $CFCl_3$) from aerosols and refrigeration

nitrous oxide (N_2O) from rapid decomposition of humus and combustion of fossil fuels

ozone (O_3) from internal combustion engines

Figure 7.13 Sources of greenhouse gases.

about 315 ppm in 1958 and 353 ppm by 1990. These figures show how the increase has accelerated in the three decades from 1960 to 1990. The annual fluctuations reflect seasonal changes in the rate of photosynthesis, but the trend is clearly upwards.

Increases in carbon dioxide are attributed mainly to the burning of fossil fuels (coal, oil and natural gas). In the early 1990s this was estimated to be in excess of 6 billion tonnes per year. The carbon in fossil fuels was fixed by photosynthesis millions of years ago when the vegetation was growing, so burning the fuels now releases carbon dioxide which had effectively been removed from circulation. Deforestation on a large scale may also upset the contemporary balance between respiration and photosynthesis. When forest trees are cut down and removed, the land is often converted to agricultural land which is then used for grazing or growing other crops. The amount of carbon dioxide used in photosynthesis is likely to be very much less than when trees were growing, resulting in a net increase of carbon dioxide in the atmosphere. Clearing the land during deforestation involves burning of residues in the forest and disturbance of the soil may release further carbon dioxide.

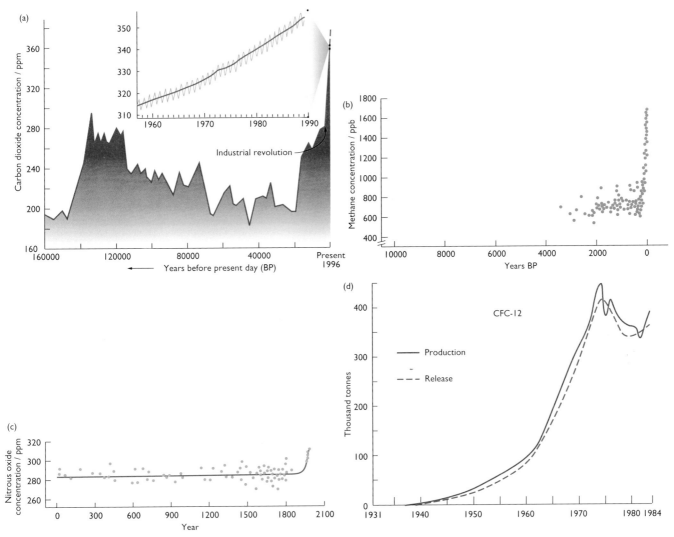

Figure 7.14 Changes in concentrations of greenhouse gases in the atmosphere (note the different time scales); (a) carbon dioxide over 160 000 years; (a) carbon dioxide detail from 1956 to 1990; (b) methane over 10 000 years; (c) nitrous oxide over 2000 years; (d) CFC12 for 50 years (from 1931 to 1984)

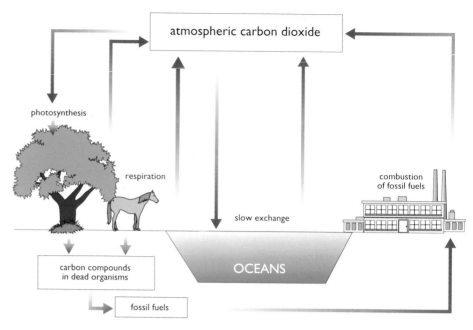

Figure 7.15 The carbon cycle

Chlorofluorocarbons are entirely anthropogenic. Their origin and their contribution to depletion of ozone in the ozone layer is described on page 136. The two which make the most significant contribution to the greenhouse effect are CFC11 ($CFCl_3$), and CFC12 (CF_2Cl_2). Even though present at very low concentrations, CFCs are far more efficient than carbon dioxide in absorbing infrared radiation. (They are considered to be more than 12 000 times more effective). CFCs persist in the atmosphere for 60 years or more, hence the concern over their potential long-term contribution to the greenhouse effect.

Methane is a hydrocarbon which is produced naturally by bacteria in wet, marshy, anaerobic conditions, hence its name 'marsh gas'. Considerable quantities of methane are produced from the activities of these bacteria in rice paddy fields. Another significant source of methane is from the guts of animals, particularly ruminants such as cattle, sheep and camels. A domestic cow can produce about 200 dm^3 of methane a day, but only 12 dm^3 of milk! Methane may also seep out from landfill sites where it is produced from rotting vegetation and some methane leaks from gas pipes and coal mines. With increasing demand for food, larger areas of rice paddy are cultivated and more cattle are reared. Numbers of cattle, for example, doubled between 1960 and 1980. These sources have certainly contributed to the doubling of methane levels over the last 150 years. A global rise in temperature is likely to release further methane currently trapped in frozen northern tundra areas.

Nitrous oxide is produced naturally in the soil from nitrates by the activity of denitrifying bacteria. Increases in nitrous oxide, from a pre-industrial level of about 288 ppb to a 1990 level of 310 ppb, are attributed partly to the burning of fossil fuels and other forms of biomass, but also to the enormous increase in use of nitrogen fertilisers. Some nitrous oxide is produced from disturbance of soils and also from animal and human wastes.

The amount of **water vapour** in the atmosphere is affected by human activities, through burning of fuels and other industrial processes. Any rise in temperature would cause more water to evaporate from the sea, which would add yet more water vapour to the existing greenhouse gases in the atmosphere. Formation and destruction of **ozone** is described on page 135 and an indication of its contribution as a greenhouse gas is given in Figure 7.13.

It is well established that levels of greenhouse gases have increased over the last 150 years. It is also recognised that human activities produce greenhouse gases and that these activities have increased during the same period. This suggests that, as a result of these increases in greenhouse gases, there may be an enhanced greenhouse effect, leading to a rise in temperature on the surface of the Earth. This is described as **global warming**. Climatic patterns since the mid-nineteenth century do show a rise in global temperature as a general trend. We cannot, however, say categorically that the temperature rises have been *caused* by the human activities, even though we accept that there are strong links between the two. Events which determine climatic patterns are highly complex. Reconstruction of past climates shows that there have been considerable fluctuations in global temperatures at least over the past 20 000 years, dating back to the last glacial period. The recent rises in temperatures may be part of a general fluctuation, such as has occurred before, though the extrapolation of present trends suggests that, during the early part of the 21st century, global temperatures are likely to rise to a level higher than at any time in recent history.

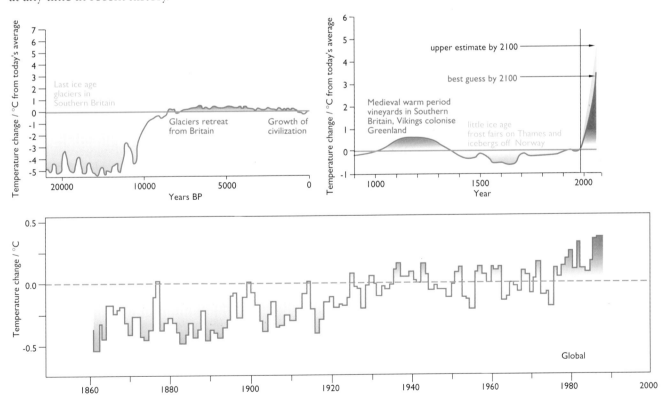

Figure 7.16 Generalised global temperatures. (a) Estimated global temperatures during and just before the history of human civilisation, shown as a comparison with today's average temperatures; (b) detail of the last thousand years, with prediction of global temperatures into the next century; (c) global temperatures from 1860 to 1988, compared with 1950 to 1979 (taken as the reference period)

PESTS AND POLLUTION

Why is it wrong to say that the greenhouse effect is *caused* by human activities? How far do you think that the increased human population has contributed to the enhanced greenhouse effect? Try to link specific human activities with the greenhouse gases they produce.

There is considerable concern about the implications of human activities and their possible effects on global climate but it is difficult to predict precisely what will happen in the immediate future. People may respond by reducing those anthropogenic activities which cause emission of greenhouse gases. This could be supported by policies agreed on an international scale, though this cannot have an immediate effect because certain persistent gases have already built up in the atmosphere.

If we assume that production of greenhouse gases continues at the present rate, a best estimate suggests that by the year 2030, carbon dioxide concentrations will reach double the pre-industrial levels. Computer models predict that this doubling would lead to a global warming of between 1.5 °C and 4.5 °C. This may not seem a very large increase in temperature, but the effects could be far reaching. One certain result would be a rise in sea levels, partly due to thermal expansion of the seawater in the oceans and also resulting from melting of glaciers and the ice sheets of Greenland and in Antarctica.

Climatic changes would also lead to changes in patterns of rainfall and of temperature. There would probably be shifts in both natural ecosystems and the agricultural crops compared with their distribution today. Yields of crops might benefit from the higher temperatures and from higher carbon dioxide levels. The zones of successful cultivation may extend beyond their present limits. As an example, maize could probably be grown 200 km further north in Europe. However, weeds and crop pests would also benefit, and vectors of disease might alter their range, also affecting crop losses. Perhaps with the warmer climate we would use less fuel to provide heating, but this could be offset by more air conditioning to help cope with the hot summers. We cannot predict the answers with certainty.

Acid rain

Rain, snow and other forms of precipitation are naturally mildly acidic, with a pH of about 5.6. (Remember that pH 7 represents neutral pH and that change by one unit on the pH scale represents a ten-fold change in acidity.) The acidity is due to carbon dioxide in the air, which dissolves to form carbonic acid. We use the term **acid rain** to describe precipitation of pH 5 or lower. This increased acidity is attributed mainly to the production of sulphur dioxide (SO_2) and oxides of nitrogen (NO_x) during the burning of fossil fuels.

Glaciers and ice sheets can be used to trace the history of the acidity of precipitation. For thousands of years up to the start of the Industrial Revolution, the pH was near to 6, or even higher, whereas by the middle of the 20th century, pH values of 4 to 4.5 have been commonly recorded in north America and Europe. Other evidence similarly shows a history of increasing acidity. As an example, diatoms have preferences for certain pH ranges, so the profile of diatoms deposited in sediments in certain lakes can give an indication of the history of the pH in the lake. Detailed monitoring of the acidity of rain in Europe over the 10-year period from 1956 to 1966 showed that the rain had become more acidic and that the area affected had expanded.

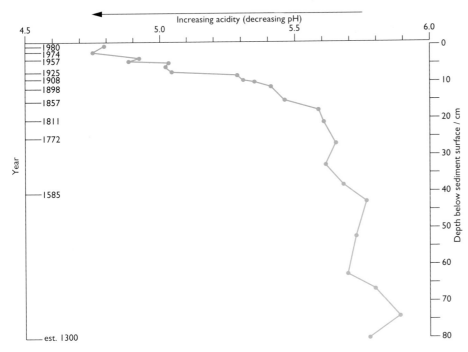

Figure 7.17 Fossil diatoms can give an indication of the history of the pH in a lake. These data show the analysis of diatoms from Round Loch, a lake in southwest Scotland. Despite its isolation, the lake shows sharp increases in acidity from about the time of the Industrial Revolution in the mid-nineteenth century

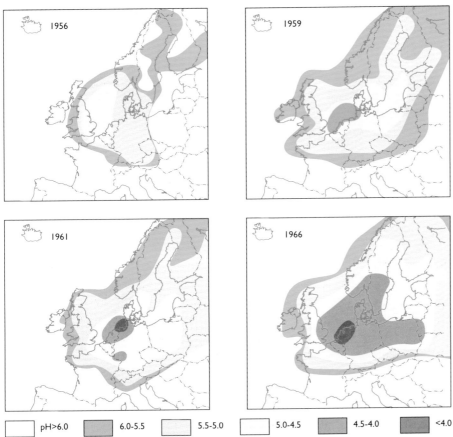

Figure 7.18 Increases in the acidity of precipitation in Europe, between 1956 and 1966

A complex series of reactions produces the acidity, mainly in the form of sulphuric and nitric acids with some hydrochloric acid. The proportions of the acids differ, depending on the origin of the pollutants. The deposition of the acid rain may be in areas that are considerable distances (even hundreds of kilometres) away from the source of pollutant, though it may be difficult to identify the exact source. Compared with lowland regions, mountainous areas are likely to receive a relatively high dose of acid rain, simply because of their higher rainfall. Scandinavia appears to have received a disproportionate deposition of acid rain. Some of this probably arises from industrial activities in Britain, but a substantial contribution is made by industries in Europe, including the Eastern European countries. During the 1950s, the policy in Britain was for the introduction of tall chimneys for power stations. This was an attempt to reduce ground level pollution by lifting the emissions to a higher level so that they were dispersed away from the source. While at least partially successful on a local scale, this practice has undoubtedly contributed to acid rain deposition at more distant localities.

The effects of acid rain on animal and plant life in both aquatic and terrestrial ecosystems, and also on buildings, are now well recognised. In freshwater ecosystems, particular attention has been paid to fish populations (or lack of them). A study made in the late 1970s, of 1679 lakes in Norway, showed a strong relationship between the pH of the lakes and the abundance of fish in them. The majority of lakes with a pH below 4.5 had no fish whereas virtually all those with a pH higher than 6.0 had good populations of fish (Figure 7.20). Similarly, there is evidence of loss of fish, because of the increased acidity, from rivers in Canada and the United States, and in lakes in Scotland and Wales.

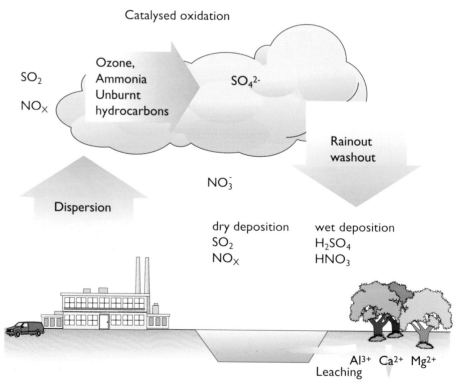

Figure 7.19 Formation and deposition of acid rain

Changes in pH affect fish in various ways. Below pH 4.5, trout do not produce the enzyme which breaks down the outer coating of eggs, so the larvae get trapped inside. This prevents successful reproduction. Acidity leads to reduced calcium concentration and some toxic metals are mobilised. One of these is aluminium, which appears to make the fish produce an excess of sticky mucus on their gills. This leads to reduced intake in salt through the gills which is part of the fish's osmoregulating process. Because of the mucus, the gills become clogged and gas exchange becomes difficult. Attempts to reverse the effects of acidity by adding lime to affected lakes may have some short-term benefit to the stock of fish in the lake.

Increased acidity in aquatic ecosystems is linked to a loss in phytoplankton. Water in acidified lakes is noticeably more transparent because there is less microscopic material in suspension. The effects spread through the food chains, some species becoming more abundant but others being lost. In some lakes, where the pH is below 6, the moss *Sphagnum* shows vigorous growth, covering the bottom of the lake and pushing out other plant life.

In terrestrial ecosystems subjected to acid rain there is clear evidence of damage to plant life. In large areas of coniferous forest, the trees show poorer growth, lower productivity, discolouration of the needles, shallow roots and die-back of the crown. Many have been killed. In Europe, deterioration has been observed in some important tree species, notably Norway spruce (*Picea abies*), white fir (*Abies* sp.), Scots pine (*Pinus sylvestris*) and beech (*Fagus sylvatica*). Generally deciduous trees suffer less than conifers. Acid deposition affects other organisms in terrestrial ecosystems, and there is evidence that populations of animals have declined in areas affected by acid rain.

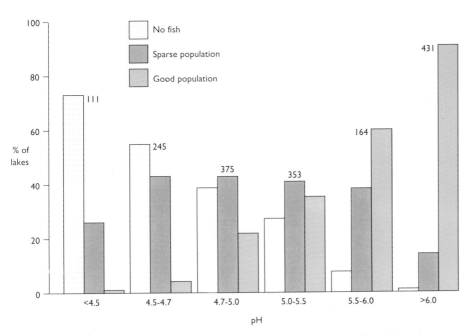

Figure 7.20 The relationship between pH in the water and abundance of fish. These data were obtained from 1697 lakes in Norway and show how low pH (high acidity) appears to have a detrimental effect on fish. The numbers above each group of bars indicate the number of lakes observed within the particular pH range

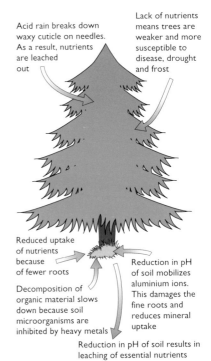

Figure 7.21 The effects of acid rain on trees

Acid rain may also affect human health. A low pH in the soil releases ions of certain heavy metals, such as cadmium, lead and mercury, and these may contaminate drinking water supplies. Pregnant women are at risk with water originating from affected areas, because the fetus is particularly sensitive to mercury poisoning. Acidity may cause leaching of copper and lead from water systems. In Sweden, high levels of copper and lead have been detected in drinking water. Copper from this source may have caused outbreaks of diarrhoea in young children and account for green colouring to baths and even hair! Release of aluminium may be harmful because of the possible link between aluminium and Alzheimer's disease (pre-senile dementia) in humans.

Water pollution

Water on the Earth continually circulates through the processes of evaporation, transpiration, condensation and precipitation. The main events of this **hydrological cycle** are summarised in Figure 4.3 (see page 62). Human activities can interfere with or pollute the water at any stage of the cycle. Pollutants may alter the physical conditions in the water, thus disturbing the balance of organisms living in aquatic habitats, or they may do direct harm to the living organisms which would normally be present. There is increasing concern over contamination of drinking water for humans, which is extracted from groundwater and also from rivers. Some pollutants, such as nitrates, have reached unacceptably high levels in these water sources in certain areas. Harmful effects of pesticide residues are discussed later in this chapter. Polluted water may also encourage spread of disease.

An outline of the ecology of aquatic habitats

A brief consideration is given of the physical factors of importance in an aquatic environment, with an indication of how these factors influence living organisms. This will help us understand the changes brought about by pollutants. We will focus on freshwater habitats, though similar principles apply to marine or brackish habitats. Even under so-called normal conditions, these factors fluctuate, on a daily or seasonal basis.

In lakes and ponds the water is relatively static, and mixing of water in different parts may be a relatively slow process. In rivers and streams, the water moves in one direction, though the flow rate may vary from very slow (as in canals or pool stretches of rivers) to extremely fast (in some mountain rivers). In a given stretch of a river, there is continual exchange of water, so that events at some distance upstream can affect the river much lower down (Figure 7.22). Movement of water sometimes creates turbulence so mixing occurs more readily than in static waters. Often this movement considerably increases the oxygenation of the water.

The normal **nutrient status** depends closely on the bedrock, soil and vegetation of the surrounding catchment area. The water may be relatively poor in nutrients (**oligotrophic**), through intermediate nutrient levels (**mesotrophic**) to rich in nutrients (**eutrophic**). This classification of nutrient status refers mainly to levels of inorganic nitrogen and phosphorus, although other mineral elements are also important to aquatic organisms. Higher nutrient levels are likely to support high productivity in terms of biomass within the water environment. **Organic compounds** found in water are derived mainly from decomposition of plant and animal material (detritus). They may include proteins, carbohydrates and fats as well as more complex particles of organic matter, which are sometimes suspended rather than dissolved in the water.

Table 7.3 *Comparison of nutrient levels (phosphorus and nitrogen) and productivity in oligotrophic, mesotrophic and eutrophic lakes*

Nutrient level and biological productivity	Type of freshwater lake		
	Oligotrophic	Mesotrophic	Eutrophic
total phosphorus / ppb	<1–5	5–10	10–30
inorganic nitrogen / ppb	<1–200	200–400	300–650
net primary productivity / g dry mass m^{-2} yr^{-1}	15–50	50–150	150–500
phytoplankton biomass / mg dry mass m^{-3}	20–200	200–600	600–1000

ppb = parts per billion

Oxygen dissolves in water and its concentration is a critical factor in determining the types of living organisms (plants, animals and microorganisms) which are present in the aquatic environment (see Chapter 4). The solubility of oxygen decreases with increasing temperature. Aeration is increased by turbulence, so shallow rivers flowing fast over a rocky substratum acquire more oxygen than stagnant pools. The oxygen level is also dependent upon the balance between photosynthesis and respiration. There is a diurnal fluctuation with an increase in oxygen during daylight linked to its production by the photosynthetic activity of plants. Seasonal changes in temperature and light intensity influence oxygen availability through their effect on photosynthetic activity. Organisms use oxygen in respiration. The demand for oxygen varies considerably, depending on the number of organisms present and on their activity.

Table 7.4 *Decrease of oxygen solubility with increase in water temperature*

Temperature / °C	Oxygen solubility / mg dm^{-3}
0	14.6
10	11.3
20	9.2
30	7.6
40	6.6

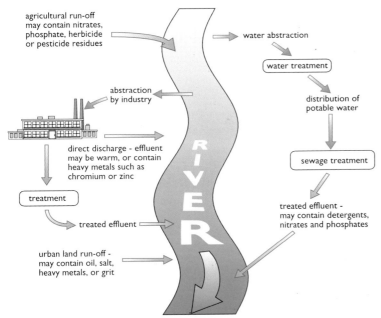

Figure 7.22 Events affecting quality of water in a river

The amount of oxygen being utilised by living organisms in the water is often expressed as the **biochemical oxygen demand** (**BOD**). The BOD of a sample of water is measured under standard conditions at 20 °C over a period of 5 days. Presence of polluting organic material in the water leads to increased activity from microorganisms involved in the decomposition of this material, resulting in a high demand for oxygen. Measurement of BOD can, therefore, provide an indication of the level of pollution.

Carbon dioxide, like oxygen, dissolves in water and is more soluble in cool than in warmer water. Carbon dioxide forms carbonic acid which is a weak acid, so even in non-polluted areas, the rain is likely to be mildly acidic. Carbonic acid dissociates into hydrogen ions (H^+) and hydrogencarbonate ions (HCO_3^-). The H^+ ions contribute to the acidity of the water, or make the pH lower. The hydrogencarbonate ions can dissociate further to produce hydroxide ions (OH^-) and carbonate ions (CO_3^{2-}). The hydroxide, hydrogencarbonate and carbonate ions contribute to the alkalinity of the water. Through these reactions, changes in levels of carbon dioxide lead to fluctuations in **pH**. Some organisms are sensitive to the pH of the water, particularly when acid (see *Acid rain*, page 142). Carbon dioxide is utilised by plants in photosynthesis. When photosynthetic activity is high, the pH also rises because the carbon dioxide is used up, so there is less carbonic acid and the H^+ ion concentration falls.

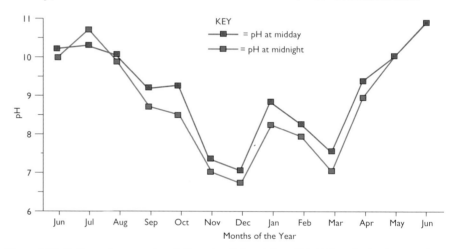

Figure 7.23 Daily and seasonal variations in pH in a lake. pH is higher during increased photosynthetic activity because carbon dioxide is used in photosynthesis. The pH is generally higher at midday than at night and higher during summer than winter

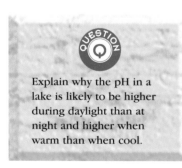

Explain why the pH in a lake is likely to be higher during daylight than at night and higher when warm than when cool.

Temperature affects the density of water, which is greatest at about 4 °C. Sometimes, particularly in large lakes or relatively still water, there may be a sharp separation between layers of water of different density. The boundary between the layers can act as a physical barrier to small organisms, such as plankton, whereas larger organisms may move from one temperature layer to another where conditions may be more favourable. Compared with air and land, water takes longer to heat up and cool down. Water is thus a relatively stable habitat, though living organisms have their own tolerance limits of extremes of temperatures as well as preferred optimum temperature ranges. The effect of temperature on dissolved gases, particularly oxygen, is described above.

Light intensity decreases with depth in the water, and is also reduced by material suspended in the water. Such material may be inorganic, derived from bedrock or particles of soil washed into the water, or there may be organic matter suspended in the water. The intensity of light at different depths within the water is a critical factor in determining the distribution of photosynthetic organisms.

Plant life in water ranges from microscopic organisms through to larger plants (**macrophytes**). Most of the microscopic organisms (known as **phytoplankton**) are found in the surface layers, floating or swimming freely. The phytoplankton includes bacteria and blue-green bacteria as well as diatoms and other algae. Through the photosynthetic activity of its organisms, the phytoplankton plays an extremely important part in the primary productivity of aquatic ecosystems. Population numbers of these organisms can fluctuate dramatically in response to seasonal changes or availability of nutrients. Macrophytes include large visible algae, such as blanket weed (*Cladophora* spp.), as well as Angiosperms such as pondweed (*Elodea canadensis*) and water lilies (*Nymphaea* spp.).

Animal life in water ranges from microscopic **zooplankton**, through invertebrates to fish, amphibians, birds and mammals. The level of available oxygen is an important factor in determining the distribution of animal species. Certain invertebrate species can be used as indicators of pollution levels because some tolerate very polluted water with poor oxygenation whereas others are restricted to clean well-oxygenated water (see Chapter 4).

Sewage – inadequate treatment leading to pollution

Sewage from domestic sources consists mainly of human faeces and kitchen waste. The latter includes detergents and residues from food preparation. Sewage is also likely to carry some industrial waste which may incorporate stronger liquids used in cleaning or in a variety of industrial processes. Other components may be present, including toxic substances such as chromium from leather tanneries or copper and zinc from metal-plating industries.

In Britain and many other countries, much of the sewage produced is treated in sewage-treatment plants, which involves both physical and biological processes. The effluent (liquid) and sludge (solid) produced after treatment is completed can usually be used or dumped without harm to the environment. However, sometimes raw sewage is released into rivers or discharged into the sea without treatment, giving rise to pollution. In addition, occasional accidental leakage of sewage occurs, or treatment may be inadequate. Pollution may also result from leakage of animal slurries from farm wastes or from silage liquor.

Suppose some raw or inadequately treated sewage or slurry is discharged into a river. There will be instant changes in the physical factors of the aquatic environment and impact on the organisms in the river. Table 7.4 shows the main components of sewage and summarises some of the effects of such pollution in a river. By looking at the graphs in Figure 7.24, we can follow this in more detail through various stages downstream.

Table 7.4 *The main components of sewage and their effects in a river*

Component of sewage	Impact/consequences on environment
suspended solids	size ranges from large and visible to colloidal and dispersed usually organic, so are degradable and can be decomposed by microorganisms • reduce penetration of light • high demand for oxygen during breakdown of organic material by microorganisms

nitrogenous compounds → nitrates	originate mainly from proteins and urea
	often present in the form of ammonium compounds (NH_4^+) oxidised in stages by nitrifying bacteria (*see nitrogen cycle*) to nitrates
	    → nitrite (*Nitrosomonas*)
	    → nitrate (*Nitrobacter*)
	
	• NH_4^+ ions toxic to fish
	• excess nitrate leads to **eutrophication**
	• eutrophication may encourage **algal blooms**
	• respiration of algae during the night leads to an **increased BOD,** resulting in depletion of oxygen
	• death of masses of algae results in an increased BOD (while the algae are broken down)
	• some toxins are produced during growth of algal bloom (from certain blue-green bacteria)
	• high nitrate concentration damaging to human health if the water is used as a source for drinking water

phosphates	present in faeces and in modern detergents • excess phosphate leads to eutrophication (described above for nitrates)
toxins	heavy metals, such as Cu, Pb, Zn, may accumulate persistent pesticides, from agricultural run-off rather than sewage (see *Control of insect populations,* pages 127–135) • toxic effects on organisms in the water, or for humans if the water is used as a source for drinking water
microorganisms	may include viruses, bacteria, protozoa and fungi (some may be pathogenic) • health risk for humans, particularly if water is used for drinking without adequate treatment
detergents	'hard' detergents (used in the 1950s) create foam and are unsightly on the surface 'softer' detergents (used since the 1960s) are biodegradable, but rich in phosphate • foams on the surface interfere with aeration of the water • high levels of phosphate may lead to eutrophication

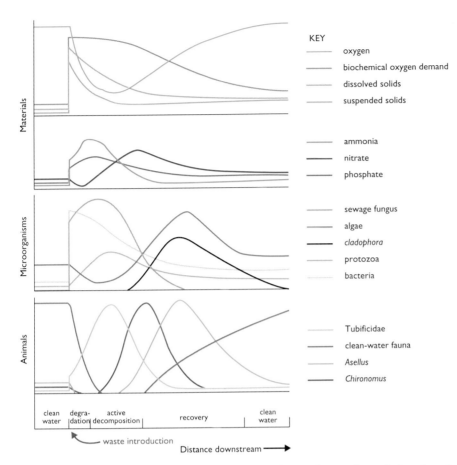

Figure 7.24 Typical changes of water quality and the plant and animal populations in a river after introduction of sewage. See text for discussion of events

In the initial stage, just below the point at which sewage enters the river, there is immediate decomposition of the organic material, mainly through the activities of aerobic bacteria. These bacteria may have been present in the river water or associated with the sewage material. The demand for oxygen is high, resulting in a rapid decrease in the concentration of oxygen in the water. This 'oxygen sag' occurs when the rate of utilisation of oxygen in respiration by the microorganisms exceeds the rate at which oxygen is replenished. In extreme cases, the oxygen level may fall so low that conditions become anaerobic. In very polluted water, visible slime-like growths known as 'sewage fungus' are likely to appear. The term sewage fungus is misleading as the slime consists of filamentous bacteria, protozoa and algae as well as fungi. Sewage fungus is tolerant of high levels of ammonia and anaerobic conditions.

As the organic substrate is utilised, the numbers of aerobic bacteria decrease, but protozoa which feed on other organisms show an increase. An increase in algae begins when the suspended solids begin to settle out or have been decomposed and the water becomes clearer. Improved light penetration allows algae and other plants to carry out photosynthesis and this helps in the recovery of the oxygen level in the water. Inorganic ions are released from the decomposing material. Ammonia, though initially high, is soon converted to nitrate by nitrification. The nitrate, phosphate and other ions released from the material are likely to be utilised by the increasing populations of algae.

PESTS AND POLLUTION

QUESTION

Sometimes, as a result of a discharge of raw sewage, a sudden mass mortality of fish occurs, known as fish kill. High levels ammonia are toxic to fish and there may be poisonous substances such as heavy metals in the water. What is likely to be the main reason for the sudden death of fish?

The lowest graph in Figure 7.24 shows typical fluctuations of invertebrate populations in the different zones of the river downstream from the sewage discharge. The *Tubifex* worms and *Chironomus* (midge larvae) are tolerant of low oxygen levels, whereas *Asellus* (water louse), snails, leeches and fish require progressively cleaner water. These population numbers reflect changes in the clean quality of the water and demonstrate how the water can effectively go through stages of self-purification as a result of the activities of microorganisms in the water. In highly polluted waters, there is a danger that no or only very slow recovery from the initial drop in oxygen concentration will occur. The distance over which such purification takes place depends on many factors, including the temperature of the water, the severity of the initial pollution and the existing microbial population in the water. To some extent, the effect of the pollutant is diminished by dilution as it passes downstream.

Fertilisers, eutrophication and algal blooms

Use of chemical fertilisers on agricultural land has increased markedly with changing agricultural practices aimed at increasing yields of crops. The main inorganic ions applied in fertilisers are nitrate (NO_3^-), phosphate (PO_4^{3-}) and potassium (K^+). These ions dissolve in soil water and are leached out by rain and water percolating through the soil. The water drains into lakes or rivers and this results in an artificially enriched nutrient status. In some cases the body of water may initially have been oligotrophic, with a relatively low initial level of nutrient, whereas in other situations the nutrient status may naturally be higher (mesotrophic or eutrophic). We use the term **eutrophication** to indicate the artificial nutrient enrichment of an aquatic system, regardless of its initial status.

Table 7.5 *Changes in fertiliser production in different countries between 1970 (taken as a score of 100) and 1988. The final column gives actual production in 1988*

Country	Fertiliser production					Tonnes / km²
	1970	1975	1980	1985	1988	1988
Canada	100	184	301	403	371	2.6
USA	100	128	147	129	132	5.1
Japan	100	102	101	112	109	13.7
France	100	114	142	156	166	13.3
Germany	100	109	139	136	138	20.6
Italy	100	145	198	206	182	7.6
Netherlands	100	116	122	121	101	46.7
Spain	100	134	161	171	201	5.5
Sweden	100	115	111	111	103	7.6
UK	100	121	143	179	169	20.9
North America	100	130	152	138	140	4.6
World	100	136	185	212	242	5.4

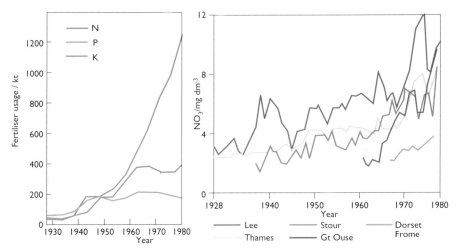

Figure 7.25 Increased use of fertilisers is linked with increases in the concentration of nitrate in rivers. (a) Increases in use of fertilisers containing nitrogen (N), phosphorus (P) and potassium (K) between 1928 and 1980. (b) Trend in nitrate concentration in five rivers over the same period. Both graphs refer to the UK

Excess or enrichment of nutrients in the water encourages rapid growth of algae and blue-green bacteria, resulting in a population explosion, known as **algal bloom**. Such growths may also occur in natural situations stimulated, say, by a seasonal increase in temperature. The phase of rapid growth may be followed by sudden and massive mortality, perhaps because the supply of one of the nutrients has become limiting. Dead algae then rise to the surface of the water as an unsightly green scum.

The most serious consequences of algal blooms arise from the depletion of oxygen. While growing, the algae have a high demand at night for oxygen used in respiration, though during the day, the net output of oxygen from photosynthesis is likely to be greater than that used in respiration. When the algae die, the mass of organic material is broken down by microorganisms (mainly aerobic bacteria), making further demands on the supply of oxygen. If the level of oxygen falls too low, there may be sudden death of masses of fish, a situation known as **fish kill**. Waters that are severely eutrophic may become anoxic and under anaerobic conditions unpleasant odours become apparent, due to production of hydrogen sulphide. Under these conditions methane (marsh gas) may also be produced. Certain blue-green bacteria produce toxins which may also cause the death of fish. The bacterium *Clostridium botulinum* flourishes in anoxic conditions. The toxins it produces cause paralysis known as botulism, which can affect birds and mammals (including humans swimming in the water). Overall, there is usually a drop in species diversity as well as reduction in numbers. Some species, however, prefer or can tolerate the anoxic conditions.

Excessive growth of algae as a result of eutrophication restricts penetration of light into the water. Other water plants suffer from a reduced rate of photosynthesis and this further diminishes the supply of oxygen. So we can see how the events following eutrophication can lead to a rapid deterioration in the quality of water and changes in the populations of plants, animals and microorganisms. Recovery is generally associated with improvement in oxygenation as well as dilution of the original source of nutrient enrichment.

Why do you think the effects of pollution are probably less in a fast-flowing mountain stream than in a slow-flowing lowland river?

Figure 7.26 (a) The oil tanker Braer, *wrecked off the Shetland Isles in 1993; (b) Devastating damage to seabirds as a result of oil spilled from the* Braer

Oil pollution

Damage from oil pollution is associated mainly with marine environments, though freshwater lakes and rivers can also be affected. Coastal salt marshes, mangrove swamps and inlets are particularly vulnerable when exchange of water or introduction of fresh water is slow or restricted. Large-scale disasters tend to hit the media headlines: most of us are familiar with the names of wrecked oil tankers and the reports are often accompanied by pictures of birds dragged from the sea or beaches, their feathers useless because of a coating of oil and tar (Figure 7.26 and Table 7.6).

Table 7.6 *Examples of wrecked oil tankers*

Name	Date	Area affected	Oil lost / tonnes
Torrey Canyon	1967	Scilly Isles, UK	119 000
Amoco Cadiz	1978	Brittany, France	227 000
Exxon Valdez	1989	Alaska	37 000
Braer	1993	Shetland Isles, UK	85 000
Sea Empress	1996	Pembrokeshire, UK	70 000

The main source of pollution, however, is from oil spills occurring from oil tankers in routine operations such as loading or discharging their cargo in ports, and seepage from oil installations, both on land and off-shore. Another source of oil spills comes from the emptying of ballast water, taken into the tanks to provide stability while returning to the loading terminal. Some oil remains in the tanks and is released with this water. Improvement in this area has been achieved by allowing the oil to rise to the surface in the tanks, then draining off the water but stopping before the oil layer is reached. Inadequate disposal of oil from motor vehicles also makes a contribution to seepage into the sea.

Oil consists of a variable mixture of substances, including long-chain hydrocarbons. After a spill, the oil initially spreads over the surface of the water, but as lighter fractions evaporate, more dense oils remain and sink. Chemical reactions produce tar which forms into balls. Tar may persist in the water or on the shore, where it is undesirable if the beaches are used by tourists. Over a period of time, weathering processes change the physical and chemical properties of the oil. A slick may be broken up into small droplets by wave action and gradually dispersed. Certain marine bacteria and microscopic animals (such as copepods and small crustaceans in plankton) digest organic compounds in the oil and initiate the processes of **biodegradation**. In the presence of light, further chemical reactions break down the oil, eventually into carbon dioxide and water. The time for these processes to remove the oil successfully varies from just a few weeks to many years.

Pollution from oil affects living organisms in various ways. A layer of oil on the surface reduces the exchange of oxygen between the water and air. This affects organisms dependent on breathing at the surface of the water. Oil is also drawn into holes made by worms burrowing in the mud so the layer of oil means organisms are denied access to oxygen. Plants also suffer, particularly in salt marsh areas; their roots may become coated with oil, preventing normal ventilation. In salt marshes close to a refinery near Southampton, the grass *Spartina anglica* has been killed over extensive areas.

Toxic substances from the oil may kill marine organisms, including plankton, shellfish (crustaceans and molluscs), fish and birds. There is evidence of reduced reproductive capacity in some bird species following oil spills. After the Torrey Canyon and Amoco Cadiz disasters, there was a marked decrease in breeding populations of puffins, guillemots and razorbills in the affected areas.

Oil trapped on feathers of marine birds causes considerable distress and the birds lose their ability to fly. Often sea birds are killed outright, but those that survive and attempt to preen their feathers suffer from ingestion of harmful substances in the oil. After the Torrey Canyon disaster an estimated 4500 seabirds died as a result of the oil pollution. Birds and mammals (such as sea otters) depend on the insulating properties of their feathers or fur to keep warm, but this is lost when the feathers or fur become clogged with oil. After the Exxon Valdez disaster, even though many sea otters were rescued and cleaned, they died from emphysaema after breathing toxic oil fumes or from liver or kidney failure. Another problem is that oil taints the flesh of fish and shellfish. This leads to economic losses for the fishing industry until the pollution has been cleared, or until public confidence has been restored.

After a major oil spill, different methods have been used to clean up the pollution, particularly when it affects a popular beach area, valuable fishing ground (including fish farms) or numerous seabirds. Mechanical ways of removing the oil include pumping and using booms to skim the oil or the 'mousse' off the water. (Sometimes when it has become thoroughly mixed and emulsified with the water, the oil forms a 'mousse' on the surface.) Contaminated sand can be shovelled from beaches to remove offending lumps of oil and tar. Chemical dispersants and detergents can be used to break up an oil slick, but in some cases these chemicals themselves have caused damage to living organisms, perhaps more than the oil itself. Improved detergents with lower toxicity are now used when clearing an oil spillage from a restricted area.

Many naturally occurring microorganisms can break down the complex hydrocarbons in oil to simpler non-polluting compounds. Microbial action (**biodegradation**) is an important stage in recovery of an area from the effects of oil pollution. In the technique known as **bioremediation**, a polluted area can be seeded deliberately with suitable microorganisms. In 1991, there was a huge oil pollution problem on the shores of Kuwait and Saudi Arabia, arising from the Gulf War. The loss of oil was in the order of 1 000 000 tonnes and the pollution spread along more than 700 km of the coastline. No immediate attempts were made to clean up the area physically, but some months later extensive blue-green mats had developed over some of the patches of oil on the beaches. These mats were made up mainly of blue-green bacteria, tolerant of the unusual conditions and capable of biodegrading the oil. It seems that the oil had killed off some of the other organisms which usually graze on the

Suggest why the time for complete recovery from pollution after an oil spill is likely to be much longer in sheltered areas than in those exposed to vigorous wave action.

"Pollution experts in bid to save coastline ... workers help nature in slow clean-up of blackened beach" (Headlines from the INDEPENDENT, Sat 17 Feb 1996, in report on oil tanker *Sea Empress*, which ran aground off the coast of Pembrokeshire).
What damage to wildlife might be expected as a result of the oil spill? What do you think the "workers" would try to do? What would "nature" do? In the long term, how might this accident affect the environment?

blue-green bacteria, allowing them to proliferate and, as it happens, carry out the useful task of degrading the oil. Given time, natural systems have considerable potential for 'self-cleaning' and recovery. In a similar way, population numbers of plankton, crustaceans, fish, seabirds and other organisms damaged by oil pollution, gradually rise as they re-invade the area.

Detergents

Detergents originate from both domestic and industrial effluents. Detergents used in the 1960s contained synthetic chemicals with multi-branched carbon chains which did not break down in water (hard chains). These reduced the effectiveness of processes in sewage-treatment plants and gave a problem with foam that developed on the surface of water in rivers and elsewhere. These foams reduced access to oxygen by organisms in the water: a concentration of detergent of 0.1 ppm reduces the availability of oxygen in the water to about 50 per cent, and at about 1 ppm is likely to kill freshwater fish, such as trout. The foams were extremely unsightly and there was the danger that pathogenic organisms in the original effluent would be blown about in the foam and spread disease. Since the 1960s, 'softer' detergents have been used, in which the components are biodegradable, but the main problem comes from their high phosphate content which can lead to eutrophication (see page 152). Detergents also contain toxic substances which can harm or kill living organisms in the water.

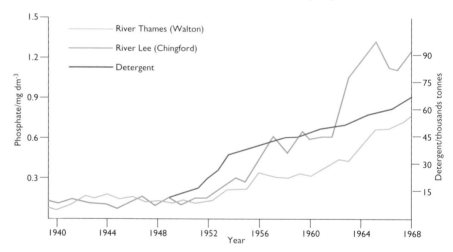

Figure 7.27 Increases in phosphate levels in rivers are linked to increases in use of detergents. Annual consumption of detergent between 1940 and 1968 and the phosphate level in two rivers over the same period

Thermal pollution

The term **thermal pollution** is used to describe the unintentional raising of temperature above normal, in a body of water as a result of human activities. The origin of the heat is usually effluents from industrial processes, particularly hot water discharged from the cooling towers of electric power stations. Increases in temperature resulting from thermal pollution are usually in the order of 5 to 10 °C, but the persistence of the raised temperature and the overall effect depends on the rate of flow or size of the body of water affected. Temperature of the water in River Severn, for example, is affected by the Ironbridge power station. When the flow of water is low, the temperature may rise as much as 8 °C whereas during floods, the temperature rise may be only 0.5 °C.

The effects of thermal pollution may be complex. In some cases the raised temperature may even be beneficial rather than harmful. Temperature changes affect the solubility of oxygen, which becomes less as the temperature increases (see page 147 and Chapter 4). Aquatic species have differing temperature tolerances: trout, for example, may be killed at temperatures above 25 °C. Temperature also affects development and growth rates of fish and influences behaviour patterns, such as migration and spawning. The time taken for fish eggs to hatch depends on temperature, expressed as day-degrees. Trout eggs, for example, require 300 day-degrees to hatch, so at 10 °C they take 30 days to hatch but if the temperature is raised to 15 °C, they require only 20 days. Some fish farms are deliberately located at sites close to effluents from power stations to exploit this enhanced temperature effect.

The communities of living organisms are likely to alter when exposed to thermal pollution. In response to a sudden temperature increase, some species are likely to be killed whereas others may grow more vigorously. The raised temperatures may encourage rapid growth of algae or, at higher temperatures, of blue-green bacteria, which may lead to the effects associated with algal blooms. If the raised temperature persists over a longer period, permanent changes in the species composition may result.

Even after quite severe incidents of water pollution, recovery and rehabilitation of the habitat often occurs. Usually the most important stage is for oxygenation to be restored and then the plant and animal species need time to re-establish and build up their numbers, though the species composition may have changed. We can illustrate this by the effects of pollution in the River Thames over the last 200 years. The River Thames was once known as a notable salmon river, but by about 1830 these fish had disappeared because the level of pollution preventing them from reaching their spawning grounds upstream. Many other species disappeared during the hundred years or more leading up to the 1950s, but have gradually reappeared after massive clean-up operations from the mid 1960s. Some examples, illustrated in Figure 7.29, will indicate the success of the measures taken. The key factors in bringing about this improvement have been introduction of measures to control discharge of effluents into the river and the recognition of the importance of maintaining an adequate level of dissolved oxygen in the water.

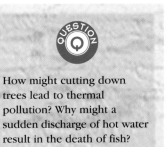

How might cutting down trees lead to thermal pollution? Why might a sudden discharge of hot water result in the death of fish?

Figure 7.28 Eutrophication is the artificial nutrient enrichment of an aquatic system. Here, excessive growth of water hyacinth fills channels leading into a lake in Kunming, south-west China.

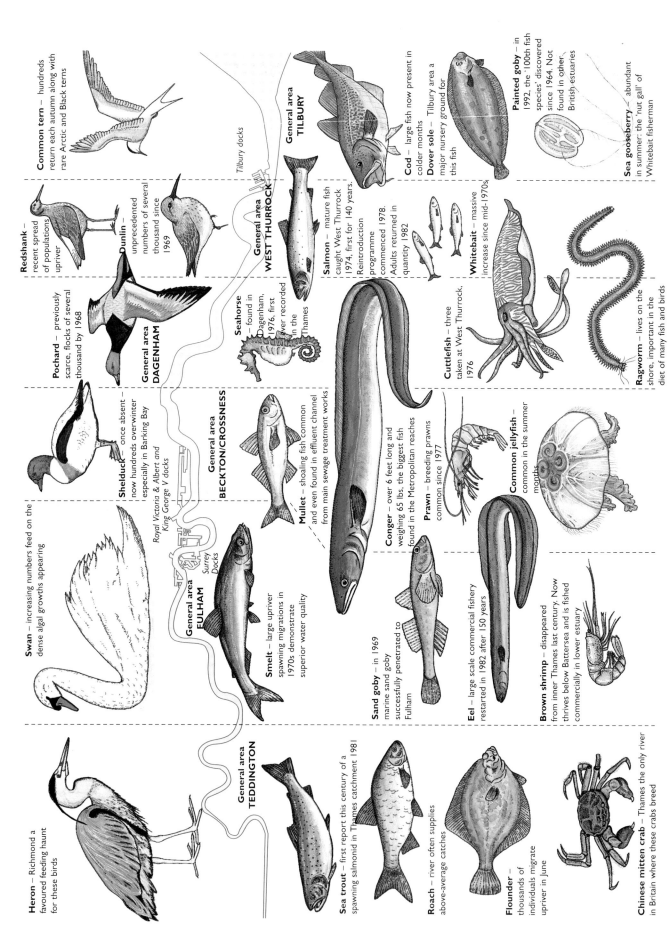

Common tern – hundreds return each autumn along with rare Arctic and Black terns

Redshank – recent spread of populations upriver

Dunlin – unprecedented numbers of several thousand since 1969

Pochard – previously scarce, flocks of several thousand by 1968

General area DAGENHAM

Shelduck – once absent – now hundreds overwinter especially in Barking Bay

Swan – increasing numbers feed on the dense algal growths appearing

Heron – Richmond a favoured feeding haunt for these birds

Tilbury docks

General area WEST THURROCK

General area TILBURY

Cod – large fish now present in colder months

Dover sole – Tilbury area a major nursery ground for this fish

Painted goby – in 1992, the '100th fish species' discovered since 1964. Not found in other British estuaries

Sea gooseberry – abundant in summer: the 'nut gall' of Whitebait fisherman

Salmon – mature fish caught West Thurrock 1974, first for 140 years. Reintroduction programme commenced 1978. Adults returned in quantity 1982

Whitebait – massive increase since mid-1970s

Cuttlefish – three taken at West Thurrock, 1976

Ragworm – lives on the shore, important in the diet of many fish and birds

Seahorse – found in Dagenham, 1976, first ever recorded in the Thames

General area BECKTON/CROSSNESS

General area FULHAM

Royal Victoria & Albert and King George V docks

Surrey Docks

Smelt – large upriver spawning migrations in 1970s demonstrate superior water quality

Mullet – shoaling fish common and even found in effluent channel from main sewage treatment works

Conger – over 6 feet long and weighing 65 lbs, the biggest fish found in the Metropolitan reaches

Prawn – breeding prawns common since 1977

Common jellyfish – common in the summer months

Sand goby – in 1969 marine sand goby successfully penetrated to Fulham

Eel – large scale commercial fishery restarted in 1982 after 150 years

Brown shrimp – disappeared from inner Thames last century. Now thrives below Battersea and is fished commercially in lower estuary

General area TEDDINGTON

Sea trout – first report this century of a spawning salmonid in Thames catchment 1981

Roach – river often supplies above-average catches

Flounder – thousands of individuals migrate upriver in June

Chinese mitten crab – Thames the only river in Britain where these crabs breed

Figure 7.29 Examples of wildlife which has returned to the Thames since the big 'clean-up'

158

Environmental change and evolution

Natural selection

The age of the Earth is estimated to be 4.6×10^9 years and the first forms of living organisms appeared about 4×10^9 years ago. These included bacteria-like and algae-like microorganisms. Now there are millions of different kinds of organisms on Earth, ranging from bacteria and single-celled fungi to complex flowering plants, insects and mammals.

Over this enormous period of time, organisms have changed dramatically. Whole groups, such as **trilobites** (early arthropods, Figure 8.1) and the **dinosaurs**, flourished and then became extinct. In general, as time passed, the structure of organisms became progressively more complex. For example, marine algae appeared about 590 million years ago, the first land plants about 400 million years ago and flowering plants first appeared in the fossil record 145 million years ago. Humans and their immediate ancestors have been on Earth for only about the last two million years.

Figure 8.1 A fossil trilobite, Olenoides

The process by which these changes have occurred over geological time is termed **evolution** and the branch of science which is concerned with the study of biological events in the past is known as **palaeontology**.

One of the major questions which has faced biologists has been to explain what has been responsible for evolutionary change. The idea that plants and animals have evolved is not new and is found in the writings of some of the philosophers in Ancient Greece. However, it was not until the 18th and 19th centuries that interest in evolution began to flourish and possible explanations for the causes of evolution were developed.

What features do trilobites show which enable palaeontologists to classify them as arthropods?

In 1859, **Charles Darwin** published a book which described the evidence for evolution which he had collected over twenty-five years. He called the book *The Origin of Species by Means of Natural Selection or the Preservation of Favoured Races in the Struggle for Life*. The ideas contained in this book are often referred to as **Darwin's theory of evolution by natural selection**, although another biologist, **Alfred Wallace**, working in Southeast Asia, had independently developed the same ideas.

The theory of evolution by natural selection can be summarised by three key points:
- The individual characteristics of an organism, such as its height, colour, or speed of movement, are vitally important for its ability to survive and to breed.
- Individuals within a given species of organism vary in many characteristics. Individuals with certain advantageous characteristics, such as ability to avoid a predator, are likely to live longer and to produce more offspring.
- Only a small proportion of the offspring will survive. If those characteristics which help the organism to survive are inherited by their offspring, then

individuals with those characteristics will gradually become more common, generation after generation, in other words they will be **selected for**. The numbers of individuals less well adapted to the environment will correspondingly decrease, that is, they will be **selected against**. This is the basis of evolutionary change.

In the following sections we will consider some aspects of natural selection and look at the ways in which changes in the environment are likely to influence evolution.

Differential mortality and natality

In Chapter 5, we described how populations, if unchecked, tend to increase in size. Some organisms are capable of producing enormous numbers of offspring, for example the weed, fat hen (*Chenopodium album*), may produce 400 000 seeds in one year. Of these, only very few will germinate and survive to reproduce. This is because growth of populations is normally kept in check by either increases in mortality (deaths) or by decreases in natality (births), as the population increases. The number of births or deaths is often expressed in terms of the number per adult member of the population. These figures are referred to as the birth rate, or **natality rate** and the death rate, or **mortality rate**. As an example, a natality rate of 2% would represent two births in a population of 100.

Availability of food is an important factor which can influence the natality rate. When food is plentiful, a population may increase dramatically in numbers as a result of an increase in the birth rate. This can lead to population explosions, as natality greatly exceeds mortality. Rapid increases in population size are frequently followed by an equally dramatic decrease as either the mortality rate increases or organisms move away from an area.

Increased numbers of prey organisms provide increased food for predators, and these can therefore be important agents which increase the mortality of prey. However, predators, such as birds, can be selective and this has a significant effect on the course of evolutionary changes in populations. For example, if some of the prey organisms are more conspicuous than others, they are more likely to be spotted by a predator and eaten. Less conspicuous organisms will therefore have a better chance of surviving to breed. This selective predation is an example of natural selection in action.

Natural selection in action

We will illustrate this by reference to the peppered moth, *Biston betularia*, which has been the subject of very detailed scientific investigation. Studies on the peppered moth were started in the 1950s by H.B.D. Kettlewell and have been ongoing since that time. There are several different coloured forms of the peppered moth. The typical form, called *typica*, is pale and speckled. Another form, *carbonaria*, is a sooty black colour. The moths are active at night, but settle on the bark of trees during the day. In rural parts of Britain where the air is cleaner, the bark of trees is lighter in colour than in the sooty areas of industrialised towns and is often covered by lichens and small plants such as mosses. These small plants are pale green or almost white in colour. More soot is deposited on trees in industrialised areas and fewer lichens and other small plants grow on these trees. Lichens and mosses are sensitive to sulphur dioxide in the air which prevents their growth.

Figure 8.2 The typica *form of* Biston betularia *seen on the bark of a tree from a rural area*

Figure 8.3 The carbonaria *form of* Biston betularia *seen on lichen from a rural area*

Figure 8.4 The typica *form of* Biston betularia *seen on the bark of a tree from an industrialised area*

Figure 8.5 The carbonaria *form of* Biston betularia *seen on the bark of a tree from an industrialised area*

The pale (*typica*) form of the peppered moth is difficult to see against the bark of a tree in a rural area, but the black (*carbonaria*) form is very conspicuous. However, when viewed against the bark of a soot-covered tree, the reverse is true. Here the *typica* form stands out but the *carbonaria* form is almost invisible. These are illustrated in Figures 8.2, 8.3, 8.4 and 8.5.

Moths of both colours are eaten by birds such as thrushes, robins and nuthatches. It might be supposed that the more conspicuous forms are more likely to be predated by these birds and will therefore be selected against. Kettlewell used several different methods to investigate whether the colour of the moths made any difference to their ability to survive in different localities. He used each method in two areas, one a soot covered woodland near Birmingham and the other an unpolluted woodland in Dorset. Two of these methods and the results are summarised below.

Method 1

Equal numbers of both the *typica* and *carbonaria* moths were released, which settled on trunks and branches in the woodlands. The moths were watched and the numbers of each form that were eaten by birds were recorded. If, during one day's observation, all the moths of a particular form were eaten, more of that form were released to ensure that both forms were always available for the birds.

The total numbers of the different forms that were eaten by birds during two days' observation in each woodland are shown in Table 8.1.

Table 8.1 *Total numbers of peppered moths eaten by birds in two woodlands*

Colour form	Woodland near Birmingham	Woodland in Dorset
typica	43	26
carbonaria	15	164

Method 2

Moths were captured in each woodland and the numbers of *typica* and *carbonaria* forms were counted. Two different trapping methods were used, a **light trap** in which moths are attracted to a fluorescent lamp, and an **assembler trap**. In an assembler trap, female moths are placed in a gauze container and pheromones released by the females attract males, which can then be caught in a net. Using two different trapping methods in each woodland ensured that the samples caught were representative, as it is possible that one colour form might be attracted to a light more than the other form. The proportions of each form captured by each method were very similar.

The total numbers of moths caught by both trapping methods in the two woodlands are shown in Table 8.2.

Table 8.2 *Total numbers of peppered moths caught in two woodlands*

Colour form	Woodland near Birmingham	Woodland in Dorset
typica	55	359
carbonaria	422	34

The results of method 1 show that the more conspicuous colour form is eaten by birds in larger numbers in both areas. The results of method 2 indicate that the inconspicuous form is much more numerous than the conspicuous form in each woodland. Clearly the inconspicuous form is less likely to be eaten by birds than the conspicuous form.

The main conclusion to be drawn from these experiments is that peppered moths whose colour contrasts with their background are less likely to survive than peppered moths whose colour is similar to their background.

The story of the peppered moth illustrates an important principle in evolutionary biology, that is, that **the phenotype of an organism affects its chances of survival**. As a general rule, the ability of an organism to survive will influence its ability to produce offspring, in other words, the longer an organism lives, the more likely it is to reproduce. The term **fitness** means the ability of an organism to survive and produce offspring which themselves can survive and produce offspring. In the case of the peppered moth, different phenotypes in one locality will differ in their fitness.

There have been many other studies of predators acting as selective agents, for example the influence of the song thrush on populations of snails. The brown-lipped snail, *Cepaea nemoralis*, exists in a number of different coloured forms (Figure 8.6). The shell can be almost white, yellow or pink and there can be a variable pattern of brown bands around the shell. These forms, like the different forms of the peppered both, are determined genetically. The main predator on *Cepaea* is the song thrush. Studies have shown, for example, that in beech woods, where the ground is covered by dark brown leaf litter, the pink-brown forms of *Cepaea* are better adapted to survive predation by thrushes than are the more conspicuous yellow forms.

Thrushes break the shells of snails by holding the snail in their beak and cracking it against a stone, referred to as an anvil. If you have access to such an anvil, in a garden, park or woodland for example, you could collect the shells and compare the numbers of different coloured forms which have been selected by thrushes.

Predation by thrushes is not the only selective agent acting on populations of *Cepaea*. In some exposed places for example, the light coloured forms predominate although they are conspicuous. These forms, however, reflect the sun's rays during the day and radiate less heat during cold nights, so climate can be a selective agent for shell colour.

Figure 8.6 Variation in the brown-lipped snail, Cepaea nemoralis

Natural selection can therefore act in different ways on the same species in different parts of the organism's distribution. This is one way in which a species can diverge and, ultimately, develop into separate species in different places.

Stabilising, directional and disruptive selection

We have already stated that individuals within a particular species show variation in a number of inherited characteristics. This variation is important because natural selection acts against some individuals, leaving others to survive and reproduce. Variation in a characteristic, such as height or mass, shows **continuous variation**. When plotted as a histogram, continuous variables produce a bell-shaped curve, often showing a **normal distribution**. As an example, Figure 8.7 shows the distribution of lengths of a sample of 86 leaves. Notice that the lengths are expressed as ranges (e.g. 120 to 130 mm) and the numbers of leaves (or frequency) in each range are plotted on the y-axis.

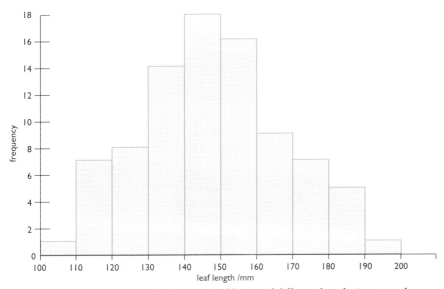

Figure 8.7 Histogram showing the frequency of leaves of different lengths in a sample

There are three ways in which natural selection can act on a population showing continuous variation, known as **stabilising**, **directional** and **disruptive** selection. These are illustrated in Figure 8.8.

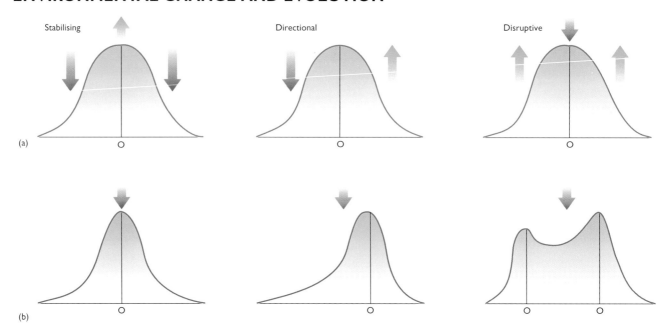

Figure 8.8 Three types of natural selection: (a) selection beginning; (b) distribution after selection. O = optimum value. Upward-pointing arrows indicate a selective advantage; downward-pointing arrows indicate a selective disadvantage

Stabilising selection

This type of selection favours the mean of the distribution. One example of this is seen in the study by M.N. Karn and L.S. Penrose on the relationship between human birth weight and mortality. Karn and Penrose collected the birth weights of 13 730 babies born in a London hospital over a period of 12 years together with data on the survival of the babies. Figure 8.9 shows the relationship between mortality and birth weight, mortality being determined by the percentage of babies failing to survive for 4 weeks. This study shows that the optimum birth weight (that is, at which mortality is lowest) is very close to the mean value. On either side of this value, the expectation of survival with either increasing or decreasing birth weight decreases rapidly, reaching a minimum at the two extremes of the distribution.

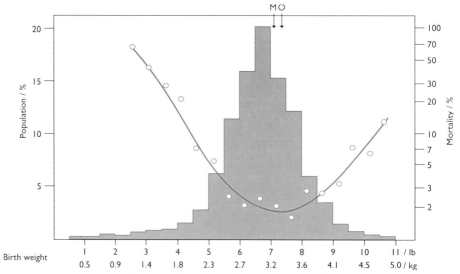

Figure 8.9   Relationship between infant mortality and birth weight. M = mean birth weight; O = optimum birth weight

Directional selection

This type of selection favours one extreme of the range of characteristics, as the other extreme is selected against. Examples of directional selection include selection against light-coloured peppered moths in an industrialised area, and selection against grasses which are intolerant of heavy metals in soil contaminated with lead, tin, or zinc.

Disruptive selection

Disruptive selection favours both extremes of a distribution, with selection occurring against the mean. This eventually results in a bimodal distribution, as shown in Figure 8.8. Disruptive selection has been demonstrated in both natural populations and laboratory experiments. For example, populations of the fruit fly, *Drosophila*, show variation in the amount of spontaneous activity when each fly is placed in standardised conditions. Figure 8.10 shows the result of an experiment in which males and females from the low activity end of the activity range were mated together, and males and females from the high activity end of the distribution were mated. After fourteen generations, two distinct populations emerged.

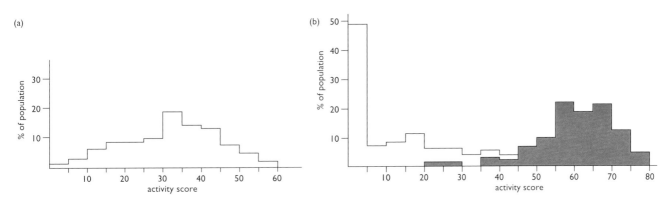

Figure 8.10 Spontaneous activity in Drosophila*. (a) frequency distribution in original population before selection; (b) frequency distribution after selection for fourteen generations*

In natural populations, disruptive selection may occur to select against intermediate forms. For example, some individuals of the European swallowtail butterfly (*Papilio machaon*), pupate on leaves or stems which are brown; others pupate on green leaves or stems. Two distinct colour forms of the pupae are found, namely brown and green, but very few intermediates.

Explain why intermediate colour forms of the pupae of *Papilio* would be at a selective disadvantage.

Isolation and speciation

Populations of organisms are genetically very variable, and one population may diverge to form two separate species in the following way. Suppose that there is a population of moths living in all parts of a tropical forest. The conditions in the forest have been very similar for millions of years and the moths are well adapted to these conditions. Now suppose that the climate changes and generally becomes much drier. Much of the forest will die out and be replaced with scrub or desert. Some of the forest, however, survives as two separate areas on mountain sides, which remain sufficiently wet to maintain the forest. What was previously a widespread, single population of moths is now separated into two populations, confined to the forest regions on the mountains. The moths which previously lived in the areas of forest which have disappeared have died out.

It is likely that the climatic conditions on the two mountains will be different. One might be colder and wetter than the other as illustrated by Figure 8.11. Over many generations, natural selection would favour genetically new kinds of moths which would be better adapted to these new conditions. Where previously there was just one kind of moth, there would now be two, which are genetically distinct.

Suppose that the climate now changes again and the low-lying desert area becomes wetter and tree covered. The moths in the two populations are now able move into the newly grown forest and meet up. If the two populations have not diverged too much, they might mate with each other, produce fertile offspring and merge into a single population again. It is possible, however, that the two populations will have diverged so much that they are unable to interbreed; in other words they have become **reproductively isolated** from each other. Once two populations become reproductively isolated they evolve along their own separate paths.

In the example we have described, the cause of speciation of the two populations of forest moths was **geographical isolation** arising from a change in the climate. Speciation as a result of geographical isolation is also referred to as **allopatric speciation** (different places). Divergence and speciation in populations may also arise as a result of **behavioural isolation**. For example, small differences in the feather patterns or songs of some birds may change their attractiveness as mates to members of their original population. Similarly, plants living in the same area may be reproductively isolated if they flower at different times of the year. Small differences can therefore isolate a group from the rest within one population, and natural selection will increase the divergence until they constitute a new species.

When two populations become reproductively isolated they are said to belong to different species. The process by which new species are formed is termed speciation.

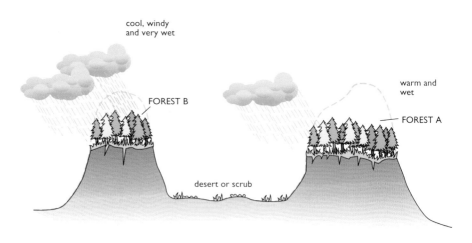

Figure 8.11 Different climates in two forest areas which were once part of a single area of forest

Sympatric and parapatric speciation

Sympatric speciation occurs within an existing population (in the same area) and arises if, for example, mutant individuals preferentially mate with each other, or if sterile hybrids are formed. In the latter case, the parental types will diverge. As an example, polyploidy in plants may immediately isolate a polyploid plant from others. Plants are capable of asexual reproduction, so the numbers of polyploid plants may increase until a large, interbreeding population is formed. Cross breeding between, for example diploid and tetraploid plants would result in the formation of sterile triploid hybrids, so isolation is produced between diploids and tetraploids.

This can be illustrated by reference to the evolution of a new species of cord grass. Cord grasses are a group of plants which colonise mud flats in estuaries. The original native British species is the small cord grass, *Spartina maritima*. In 1878, there was a report of a new, interspecific hybrid between *S. maritima* and an accidentally introduced American species, *S. alterniflora*. Spontaneous doubling of the chromosomes of the hybrid resulted in a new species, *S. anglica*, which is self-fertile, but sterile with both of its original parental species. *S. anglica* is well adapted to living in estuarine conditions, having a vigorous growth and a good ability to withstand fluctuating levels of salinity. The evolution of *S. anglica* is shown in Figure 8.12.

Parapatric speciation may occur if individuals of one population, although occupying a different area, remain in contact with individuals of the same species along the borders. Natural selection in both areas can result in genetic differences and separation of the two populations. This type of speciation occurs in organisms with low dispersal powers, such as land snails.

Prezygotic and postzygotic reproductive isolation

Genetic isolation is an essential requirement for speciation. Isolation of one population of individuals may result in the accumulation of different allele frequencies and they may eventually behave as a separate species. The mechanisms for maintaining genetic isolation between populations of one species are known as reproductive isolating mechanisms. Prezygotic (or premating) isolation means that zygotes are not formed because gametes do not meet. However, if zygotes are formed, they may not develop as a result of postzygotic (or postmating) isolation.

Table 8.3 lists some of the ways in which reproductive isolation may be brought about.

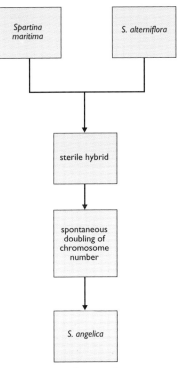

Figure 8.12 Evolution of a new species of cord grass by interspecific hybridisation

Table 8.3 *Reproductive isolating mechanisms*

Time when isolation is effective	Nature of isolation	Notes
prezygotic	geographical	populations inhabit different areas
	ecological	populations use different habitats within one environment
	temporal	populations inhabit same area but are active or reproduce at different times
	behavioural	courtship displays prevent mating between individuals of different species
	mechanical	reproductive parts may not fit each other
postzygotic	hybrid inviability	hybrids are produced but they fail to live to maturity
	hybrid sterility	hybrids fail to produce functional gametes
	hybrid breakdown	F_1 hybrids are fertile but F_2 generation fails to develop or is infertile

Human evolution

What makes and made us human?

If we watch a group of chimpanzees, we easily become captivated by aspects of their behaviour which seem remarkably like that of humans. Chimpanzees live and interact in social groups in which they move, groom themselves and feed together. Bonds develop between related individuals; they share food; the young play with each other and with their parents. They exhibit a variety of facial expressions which seem to depict happiness, even laughter, anger and other human-like emotions. To some extent, chimpanzees can use tools, selecting suitable objects from the environment then manipulating them in different ways, maybe using a twig to dig out termites or to extract honey from a hive, or hammering with a stone to crack open nuts. They even show elements of foresight and planning in selecting the right tool for the job. In captivity, chimpanzees have been trained to ask for tools and show the beginnings of communicating through language and other signs. But there is a limit as to how far chimpanzees can achieve the special attributes that still seem to separate humans from apes and other animals, in terms of mental capabilities, behaviour and cultural activities.

The story of how humans have evolved to become the dominating mammal on Earth is both fascinating and frustrating. The fascination lies in a natural curiosity about our origins and how life of any form began. The same curiosity compels us to unravel the biological events which have led to the emergence of humans as animals capable of using tools, able to cultivate land for production of their own food and with brains that can create music and paintings, electronic circuits and motor cars. These human animals also walk around on two legs; they talk, chatter and argue with each other, they sing songs and write down their language. Humans exert considerable control over their environment; they create a microclimate around their own bodies by wearing clothes and by building houses to live in. They look after their children for the first decade or so of their lives and organise themselves into social groups with elaborate hierarchies and interactions.

The frustration is that we shall never know the complete story. We use evidence from different sources to reconstruct the story of human evolution, but this evidence comes in fragments, or at best is indirect. We sift and evaluate the evidence systematically and scientifically, but we cannot do an experiment which might confirm our findings as with many scientific investigations. Research workers use modern techniques to analyse material, even at the molecular level, but the answers are limited by the range of material available. As more evidence is obtained and interpreted, it is added to the story. Sometimes new evidence does not fit the existing pattern, but rather than reject it we may have to re-assess earlier views.

Figure 9.2 Activities that are distinctively human: using the hand; upright posture; speech and making music; wearing clothes and building houses; living in social groups. No other mammal has achieved a similar level of evolutionary success.

Make a list features or behaviour patterns that you think are special to humans but poorly developed or non-existent in other animals (particularly mammals).

We are reasonably secure in our understanding of the evolution of the structures of our ancestors, but less sure about how they lived and interacted both with each other and with their environment. Attempts to build up a picture of their everyday lives is often pieced together by inference from present day situations, linked to scattered clues, such as artefacts found close to fossils. Even so, it is like taking a handful of present-day people from different countries, letting them represent the immense variation which exists in all the populations of living humans and using just a few of their possessions to tell us about their behavioural and cultural patterns.

Fig 9.1 A group of chimpanzees. Some aspects of their behaviour seem close to that of humans

Origins and evolution of living organisms

The age of the Earth is estimated to be 4.6×10^9 years and the first forms of living organisms appeared about 4×10^9 years ago. These included bacteria-like and algae-like microorganisms. Now there are millions of different kinds of organisms on Earth, ranging from bacteria and single-celled fungi to complex flowering plants, insects and mammals. Over this enormous period of time, organisms have changed dramatically. Whole groups, such as trilobites (early arthropods) and the dinosaurs, flourished and then became extinct. In general, as time has passed, the structure of organisms has become progressively more complex. Humans and their immediate ancestors have been on Earth for only about the last 2 million years.

In attempting to reconstruct the past we need to establish a timescale though, inevitably, some events are dated only approximately. For early **geological** events, we are dealing with millions of years (Myr) and it is often hard to visualise such enormous spans of time. But for the more recent **biological** events, particularly when human-like animals appear, we can begin to have a more realistic feel of the time dimension. We can relate the timescale to a typical human generation time: assuming a generation spans about 25 years, going back 100 years in a family tree would give about four sets of parents (or child to great grandparents). Extending further backwards would mean about 40 sets of parents in 1000 years. This suggests there were about 400 generations between us and the beginning of the Neolithic or New Stone Age (10 000 years ago) and about 80 000 generations to the earliest known evidence of humans as *Homo habilis* (approximately 2 million years ago).

Table 9.1 *Some major events in the evolution of plants and animals, from the origin of the Earth nearly 5000 million years ago (Myr) up to the beginnings of humans, probably about 2.5 Myr. The times (in Myr) indicate the approximate beginning of each period. Detail of primate evolution during the Tertiary and Quaternary, particularly of hominids, is given in Table 9.2.*

Era	Period	Myr	Events
Cenozoic	Quaternary	2	rapid hominid evolution (Palaeolithic and Neolithic cultural stages); appearance of elephants, modern horse
	Tertiary	65	modern genera of plants evolving; age of mammals (including primates) and birds of the air; first hominids towards end
Mesozoic	Cretaceous	145	flowering plants appear and become dominant; large dinosaurs radiated; many groups of insects; mammals and birds
	Jurassic	215	gymnosperms (especially cycads) prominent; age of reptiles, early fossil birds
	Triassic	250	conifers and ferns dominated forests; early dinosaurs and mammals
Palaeozoic	Permian	285	early conifers; reptiles diversified
	Carboniferous	360	rich vegetation of clubmosses and horsetails forming forests; amphibians radiated; early reptiles
	Devonian	410	early clubmosses; insects and vertebrates on land; various fish
	Silurian	440	first fossil plants; first jawed fish
	Ordovician	500	first fungi; molluscs diversified; annelids
	Cambrian	600	large red and green algae; some aquatic invertebrates (sponges, Trilobites, molluscs and echinoderms)
Precambrian	Proterozoic	1000	simple organisms resembling bacteria and cyanobacteria
	Archaeozoic	3000	
	Azoic	4600	

The first relatively simple marine algae appeared at about 590 Myr and the first land plants about 400 Myr. Flowering plants first appeared in the fossil record at about 145 Myr. Similarly, records of animals show how they evolved from simple marine organisms to those which colonised the land and became progressed larger and more complex. Often after the first appearance of a major group of plants or animals, the form would radiate and diversify, sometimes becoming dominant, but later many groups become extinct. The events given in the table indicate early appearance and sometimes the radiation or dominance, but not the disappearance of major groups. The names of the eras and periods need not be remembered but are given to help to relate the biological events in the evolution of living organisms to the geological events.

The process by which these changes in living organisms have occurred in geological time is termed **evolution**, and the branch of science which is concerned with the study of biological events in the distant past is known as **palaeontology**. This includes the study of fossils and their evolutionary relationships and, as far as possible, the ecology of the organisms of the past. The study of the more recent events in human prehistory is known as **archaeology**. One of the major questions which has faced biologists is to explain what has been responsible for evolutionary change. The idea that plants and animals have evolved is not new, and is found in the writings of some of the philosophers in Ancient Greece. However, it was not until the 18th and 19th centuries that interest in evolution began to flourish and possible explanations for the causes and mechanism of evolution were developed.

Table 9.2 *Timescale of the recent events associated with the evolution of hominids, including humans. One stage of hominid evolution merges into another and overlaps with the next, but it is convenient to associate the main events with the successive geological epochs. The times (in Myr) indicate the approximate beginning of each period or epoch. As stated for Table 9.1 the names of the periods and epochs need not be remembered.*

Period	Myr	Epoch	Events
Quaternary	0.01	Holocene	modern *Homo sapiens*; Neolithic culture
	2	Pleistocene	*Homo erectus*, Neanderthal man, *H. sapiens*; Palaeolithic cultures
Tertiary	5	Pliocene	Australopithecines, *Homo habilis*
	23	Miocene	Hominoids, origin of hominids
	35	Oligocene	Anthropoids, origin of hominoids
	56	Eocene	early Old World simians
	65	Palaeocene	Prosimians

A turning point in the development of the theory of evolution came about 150 years ago. In 1859, Charles Darwin published a book called the *Origin of Species by Means of Natural Selection or the Preservation of Favoured Races in the Struggle for Life* (or *The Origin of Species* as it is better known). This work set out Darwin's detailed arguments in favour of 'descent with modification' which laid the foundation for our modern views on evolutionary processes. For Darwin, the publication represented many years of indecision and uncertainty underlying his growing conviction about the processes of evolutionary change. Darwin was not alone in holding such views and he was finally able to collaborate with Alfred Wallace who independently had developed similar ideas on evolution. Darwin and Wallace set out their ideas in a joint paper which they presented to the Linnean Society in 1858. Predictably there were strong reactions opposing their theory of evolution and difficulty in gaining support for it in the scientific world. There was even more hostility from the 'public' who were reluctant to accept the notion of the 'descent of man . . from apes'.

Figure 9.3 Charles Darwin, author of The Origin of Species

The essential steps in the theory of evolution as put forward by Darwin are that organisms show **variation**, and that more are born than can survive so that there is a **struggle for existence**. Some of the variants are favoured so have a greater chance of surviving against the selection pressures of the environment and therefore of reproducing. This results in **natural selection**. These selection processes lead to an accumulation of favourable variants which, over a long period of time, may produce new forms of life resulting in the **origin of species**.

Darwin's theory has been refined as a result of our modern understanding of genetics. Darwin was unaware of Mendel's work in genetics but it is through the mechanism of heredity that we can see an explanation of the changes in the characteristics of a population from one generation to another. In the late 20th century, molecular biology has provided a detailed molecular mechanism for evolutionary change and reinforced rather than destroyed Darwin's theory of evolution.

Over a period of time, groups of individuals (or populations) of a species may become separated from each other. If this **isolation** prevents breeding between the populations, gradually the genetic make-up (gene pool) of these separated populations begins to differ. At some point, the divergence may be such that a new species has formed, branching away from the original species. Isolation may, for example, result from geographical barriers. Rising sea levels could separate two land masses that had been connected, so populations of animals that previously mixed freely might become separated on two or more islands. The genetic make-up of a population may also drift as a result of random effects and this genetic drift may have had an influence on evolutionary change, particularly in small populations.

Biologists accept that essentially the same processes have been responsible for evolutionary changes in all living organisms including those that have led to the emergence of human-like animals and ultimately of *Homo sapiens*, today's human species. Biologists see the origin of species as a dynamic process, rather than being static and finite with the present array of species being fixed for all time. Some people, usually because of conflicts with religious beliefs, find it difficult to accept the principle of evolutionary change, particularly when applied to humans. **Creationists** believe that each living species was created by a supernatural being rather than by organic evolution, though there is overwhelming biological evidence which supports the facts of human evolution. Acceptance of evolution by change from common ancestors is central to our understanding of human evolution.

A **classification** of living organisms can help us see the relationships between groups and so trace their evolutionary history. In a biological classification, organisms are grouped together if they show similarities and they are put in separate groups if they show differences. Following on from the Darwinian theory of evolution, in a classification scheme we should expect to find patterns of close (or distant) relationships which reflect the evolutionary history of the organisms. Structures which have a similar origin and therefore are closely related in evolutionary terms are described as **homologous**, even though they may now carry out different functions. By contrast, **analogous** structures may show resemblance in their functions, but have completely different origins and hence different evolutionary history. Homologous structures can be used to indicate common ancestry. The term **phylogeny** describes the way we attempt to reconstruct the evolutionary history of any group, by means of features and relationships used in a classification.

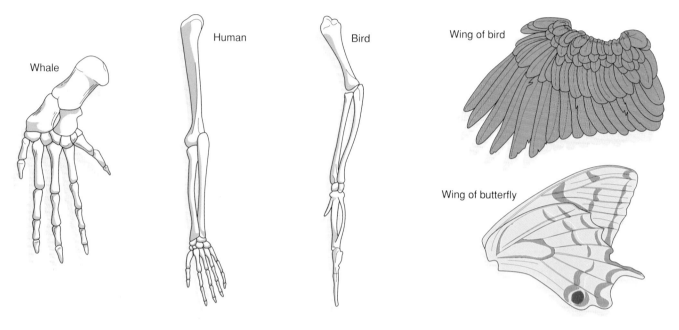

*Figure 9.4 The wing of a bird is closely related to or **homologous** with the human arm or forelimb of a whale. The same wing of the bird is **analogous** to the wing of an insect*

Primates: our nearest living relatives

Humans belong to the classification group or Order known as the Primates. We can use the outline classification of some living primates (Table 9.3) to help trace the phylogenetic relationships and see trends which suggest changes that have taken place in their evolutionary history. Initially we will consider some external features and behavioural patterns, then look at certain internal structures, including the skeleton, skull, teeth and brain and finally review evidence from the biochemistry and molecular structures of living primates to see how far we can derive support for the relationships suggested by this classification. The assumption is that similarities indicate a closeness of relationships and the degree of difference is a measure of how long ago the groups diverged.

Humans are in the **hominid** family (Hominidae). The closest relations to the hominids are the great ape family (Pongidae), which includes orang-utans, gorillas and chimpanzees. Gibbons are placed in a separate family (Hylobatidae). These three families are grouped into a 'superfamily', the Hominoidea, which includes apes and humans.

Apes and humans are distinguished from the rest of the primates by being tailless and by having the ability to swing their arms freely, around and up and over their heads. Gibbons inhabit the forests of Thailand and other parts of south-east Asia.

They show a characteristic graceful arm-swinging locomotion, a habit described as **brachiation**, in which they swing adeptly from branch to branch through the trees, one arm alternating with the other and hanging from the branches as they go. On the ground they walk or run a little uneasily on their two hind legs (bipedally), often waving their arms in the air. Their long slender fingers, with delicate nails, are remarkably human-like as they explore and pull apart potential foods.

Table 9.3 *Some living primates and how they are classified. [Note that the scientific (latinised) names are used to give a complete picture, but in most cases you will find that the common names can be used.]*

Suborder	Infraorder	Superfamily	Family	Common name
Prosimii (prosimians)	Lemuriformes (lemuriforms)	Lemuroidea (lemurs)	Lemuridae	lemurs
Anthropoidea (simians or anthropoids)	Platyrrhini (New World simians)	Ceboidea (New World monkeys)	Cebidae (true monkeys)	capuchins
				spider monkeys
			Callitrichidae	marmosets tamarins
	Catarrhini (Old World simians)	Cercopithecoidea (Old World monkeys)	Cercopithecidae	cheek-pouched monkeys
				baboons
				leaf monkeys
		Hominoidea (apes and humans)	Hylobatidae	gibbons
			Pongidae (great apes)	orang-utans gorillas chimpanzees
			Hominidae (hominids)	humans

Look first at the **families** (in the highlighted box) to see how these are grouped together. In the text we work from the familiar (humans), to the less familiar (lemurs). Use the table to see the underlying basis for the major divisions in the classification.

The suborder Prosimians ('before the simians or monkeys') are the oldest group and have retained certain primitive features. Relatively few modern families exist. The suborder Anthropoidea includes the monkeys (simians) and apes (anthropoids), considered to be the higher or more advanced Primates. The suborder Anthropoidea is further divided into two infraorders: the Platyrrhini (New World simians), and the Catarrhini. The name for the Platyrrhini is derived from their flat noses with well separated nostrils (think of the horn on the nose of a *rhino*ceros) and they have prehensile tails. The Catarrhini have downward facing noses and nostrils and their tails (in monkeys which possess them) are not prehensile.

The great apes (chimpanzee, orang-utan and gorilla) are progressively heavier (male gorillas can weigh up to 200 kg) and, compared with gibbons, move less freely through the trees. On the ground, they walk on all fours (quadrupedally), rolling their fingers into the palm of the hand and using their knuckles for support. When upright, they waddle rather than walk. Figure 9.1 shows how chimpanzees seem very human-like in some of their behaviour patterns and facial expressions, so it is not difficult to appreciate that apes and humans are closely related. Orang-utans are found in south-east Asia; chimpanzees and gorillas inhabit parts of central Africa.

Moving up the table we find the **Old World monkeys** (superfamily Cercopithecoidea), such as leaf monkeys, baboons and cheek-pouched monkeys.

Figure 9.5 Different primate families (from top to bottom): ring-tailed lemur; long-haired spider monkey; olive baboon; male silver back mountain gorilla

The Old World monkeys inhabit Africa and Asia; they have tails, and walk on all fours on the ground or in the trees, using the palm of the hand for support. Leaf monkeys feed almost exclusively on leaves (hence their name) and other Old World monkeys are mainly fruit eaters. The Old World monkeys have not developed the free arm-swinging movements of apes and humans (Hominoidea).

The **New World monkeys** (superfamily *Ceboidea*) are found in Central and South America. This is a fairly diverse group, whose features make them a little hard to classify satisfactorily. Marmosets and tamarins are small (the pygmy marmoset weighs only 100 g) and they have claws rather than nails. An unusual feature is their diet which often consists of gums scraped from the surface of trees. Spider monkeys are larger and noted for their long prehensile tail which is used for grasping branches and thus helps with locomotion. Spider monkeys feed mainly on fruit and leaves. Another group of New World monkeys includes capuchins, squirrel monkeys and owl monkeys. These are larger than marmosets and tamarins, and some use their tails in a prehensile way for grasping.

Present day **lemurs** (superfamily *Lemuroidea*) live only on the island of Madagascar. The several existing species are clearly relics of a formerly more widespread and primitive family. Fossil representatives have been found in North America and Europe. In their history, the group appears to have been overtaken by the more successful monkeys and apes, but the few remaining species have probably survived in Madagascar because here they were isolated from competitors. Lemurs run around on the branches of trees and are herbivorous. They have tails which are not prehensile. They show a certain amount of social grouping but less manual dexterity than monkeys or apes. In a number of features, they show themselves to be the most primitive of the primates.

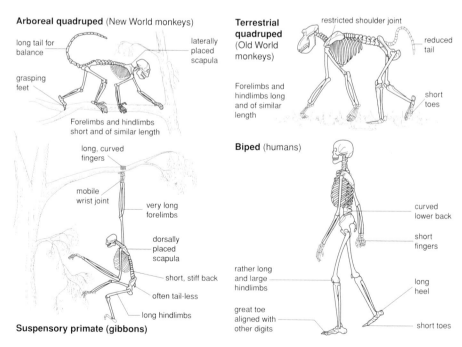

Figure 9.6 Skeletons and movement in living primates. These primate skeletons show progression from an arboreal quadruped through to humans as upright walking bipeds

Figure 9.7 World distribution of major groups of living primates

As a group, primates show adaptations to an arboreal way of life and are mainly vegetarians. Associated with this life is well-developed stereoscopic vision, which is essential for monkeys and apes for accurate judgement of distance when they leap rapidly through the trees. Looking at primates as a whole, this classification allows us to see trends within the order, from lemurs and monkeys which are representatives of earlier and more primitive forms, through to hominoids (apes and humans) which are considered to be more recent and more advanced types. Similarities and gradual changes observed between different families give evidence of shared ancestry, but not necessarily of direct descent.

Skeletons, skulls and teeth: a closer look at living hominoids

A comparison of the anatomy of different living primates gives an indication of how closely they are related. Notable trends within the primates are summarised in Figures 9.8 to 9.14. Particular attention is paid to the differences within the hominoids, illustrated by apes and humans. Later in the chapter we see how fossilised skulls, bones and teeth provide important evidence for evolutionary changes which occurred in human ancestors, so the study of living primates gives a firm basis for comparison of past with present.

In these diagrams, look at the **skeleton** in relation to locomotion and trend towards upright posture, at the length and form of the **limbs** and **feet** in relation to freedom of movement and at the **hands** with their development of manipulative skills. Then look at the shape of the **skull** and its ability to accommodate a larger brain, and at the changing shape of the **jaw** and the **teeth** inside the mouth.

Reorganise the classification chart of the primates (Table 9.3) into a family tree, from suborder through to family. Then include in it the features which can be used to separate the groups.

Then add the features to do with skeletons, skulls and teeth to the information in the family tree you devised for the primates. Use some of these features and devise a dichotomous key which could be used to identify members of the main groups within the primates. List the features which characterise and distinguish humans from other living primates.

HUMAN EVOLUTION

Skeletons, locomotion and posture – features and trends

***Trend** – quadrupedal walking → bipedal walking*

- **Limbs** – quadrupeds have limbs about the same length, short if the quadruped is arboreal, long if terrestrial; apes have proportionately longer arms → humans have relatively longer legs.
- **Vertebral column** – arched vertebral column (Old World monkey) → straight vertebral column (chimpanzee) → human vertebral column with reverse curvatures in lumbar (lower back) and cervical (neck) regions.
- **Centre of gravity** – S-shaped vertebral column of humans brings head and trunk above centre of gravity; body weight (mostly in vertebral column and bones of legs) acts vertically through centre of gravity at hip joints.
- **Pelvis and shoulder girdle** – (apes) for walking and climbing on all fours + swinging in branches → human: short wide pelvis allows bipedal walking + broad, flexible shoulder blade allows free arm rotation.
- **Big toe** – opposable in apes → non-opposable in humans (note that humans cannot grasp with the big toe and how this feature gives a characteristic human footprint) + human has arch.
- **Knee** – leg bones straighten when walking (femur in straight line with lower leg bones).
- **Ribcage** – barrel-shaped in humans (no longer uses arms in locomotion).

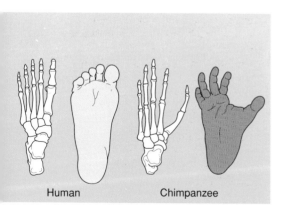

Figure 9.10 comparing feet – humans and chimpanzees

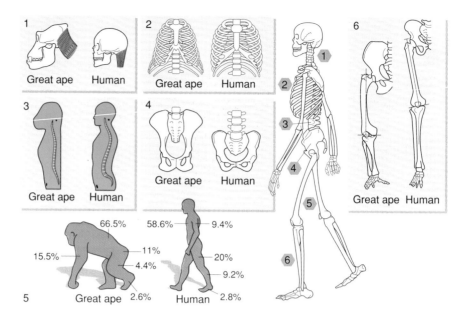

Figure 9.8 Getting upright and walking on two legs – comparing humans and great apes

(1) The human head is well balanced on top of the vertebral column; a great ape's head is held in position by powerful neck muscles. (2) The human ribcage has become more barrel-shaped than that of great apes, probably because human arms are not used for locomotion. (3) The human vertebral column has additional forward curves in the neck and lower back. These reverse curves help to bring the head and trunk above the centre of gravity in the upright position. (4) The human pelvis is broader and lower than that of the great apes. (5) Human legs are longer than their arms so a greater proportion of the body weight (hence their centre of gravity) is lower. (6) The human femur is angled outwards towards the knee, allowing the knee to be brought under the body and close to the line of action of weight. Humans extend the leg fully when walking and the bones of the leg form a straight line

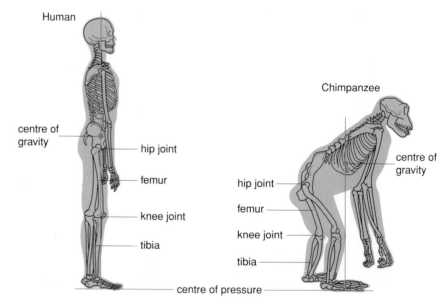

Figure 9.9 Getting upright and shifting the balance – comparing humans and chimpanzees

(1) In humans, the position of the centre of mass is in the pelvic region and when standing upright acts downwards along the line shown, down through the legs to the feet. (2) In chimpanzees, the centre of mass lies above a line between the four legs, so when the chimpanzee stands on two legs, the tendency is to tip forward to the more stable position on four legs

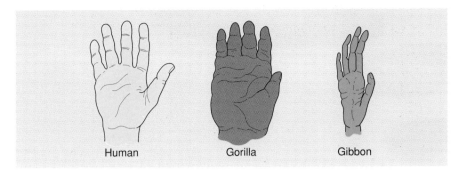

Figure 9.11 Thumbs and fingers – comparing humans, gorillas and gibbons. Note how in gibbons the relatively short thumb does not interfere with arm-swinging locomotion (brachiation) in trees. Gorillas have relatively short fingers (compared with gibbons and humans) and their fingers curve towards the palm, allowing them to use their knuckles for support when walking.

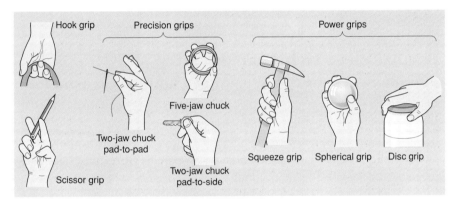

Figure 9.12 Human manipulative skills. The ability of the thumb to rotate and the mobility of the fingers allow humans to show enormous versatility in their hands and carry out many precision movements. Some of the most commonly used grips are illustrated in these diagrams

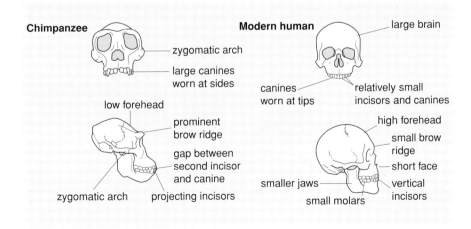

Figure 9.13 Skull and jaws – comparing chimpanzees and humans

Hands – features and trends

Trend – adapted for locomotion → human dexterity and ability to manipulate objects with precision

- **Prehensile** – able to grasp objects and pick up objects between thumb base and finger tips.
- **Opposable thumb** – can rotate so that fleshy tip can touch other fingers, giving highly mobile thumb joint (from Old World simians only).
- **Tactile sensitivity** – on palms, explore and hold objects for closer examination by eye.
- **Ridged patterns (fingerprints)** – convey sensory information to brain.
- **Rotation at wrist** – proximal end of radius can rotate, hence forearm twists – enables hand to be flat on ground or turned upwards.
- **Length of fingers** – relatively short fingers in gorilla allow knuckle walking (c.f. gibbons and humans).
- **Length of thumb** – relatively short thumb in gibbons does not interfere with arm-swinging locomotion (brachiation) in trees
- **Manipulative skills** – extreme versatility in manipulative skills and precision movements, e.g. hook grip, precision grip, power grip; flexion (bending) of fingers at joints, rotation of fingers (especially thumb); mobility of hand (rotation at wrist).

Skulls and brain size, jaws and teeth – features and trends

Trends – adjustment → upright walking, relative increase in size, jaw shape, teeth

- **Position of skull when upright** – relationship with joints of neck (centre of gravity), position of foramen magnum (hole through which spinal cord goes) further back in apes, eyes at front.
- **Size of skull** – (in relation to body size) enlarged cranium accommodates increased brain size, humans: largest brain volume in relation to body weight.
- **Shape of skull** – loss of prominence of brow ridges, protruding jaw → shortened face.
- **Jaw features** – U-shaped jaw → more V-shaped.
- **Teeth modifications** – loss of large conical canines, reduced molars.

HUMAN EVOLUTION

Figure 9.14 *Changes in teeth and jaw shape – comparing chimpanzees, australopithecines and humans. Compared with modern chimpanzees, the jaw shape in modern humans is less protruding and more rounded and the canines are noticeably reduced. (a) Side view of the jaw in the skull; (b) Lower jaw. Jaws of fossil* Australophithecus afarensis *and* A. africanus *have been included to show the probable evolutionary transition from a chimpanzee-like ancestor to form in modern humans. (c) Surface view of the upper jaw in chimpanzees and humans.*

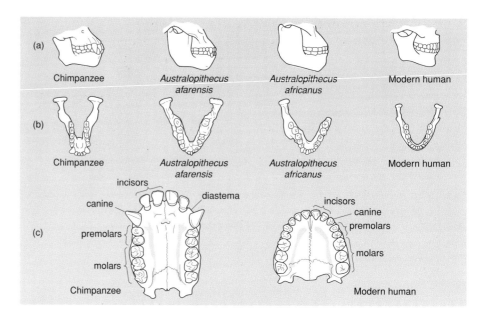

Table 9.4 *Relationships shown by immunological studies of blood sera – comparison of the antigen / antibody reaction of a number of mammal species*

Species tested	Percent reaction
human	100
chimpanzee	95
gorilla	95
orang-utan	85
gibbon	82
baboon	73
spider monkey	60
ruffed lemur	35
dog	25
kangaroo	8

Molecular clues to human ancestry
Immunological studies – the antigen/antibody reaction

The immune system protects the body against invasion by foreign material, such as protein, which might be harmful to it. The foreign substance is known as an **antigen**. The body responds to antigens by producing **antibodies** which react against the antigen to destroy it or at least neutralise its effect. Antibodies are proteins and specific to the invading antigen. The serum of blood contains proteins but no clotting factor. Thus proteins from one animal are foreign and behave as antigens to an animal of another species. If the antibodies produced are mixed with the original protein material, they react against it and produce a precipitate. This response has been used to investigate the closeness of relationships in a wide range of animals. A high percentage reaction indicates the species are closely related whereas a low percentage suggests greater divergence, both in time and in relatedness (Table 9.4).

1 Blood sample taken from human.

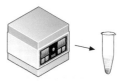

2 Blood cells are separated from plasma.

3 Serum from human is injected into rabbit. Rabbit develops antibodies to human serum.

4 Blood sample from rabbit is separated to produce sensitised rabbit serum.

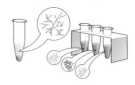

5 Sensitised rabbit serum is reacted with serum from other animals.

6 Level of precipitation is recorded.

Figure 9.15 *Stages in the antigen/antibody reaction*

Amino acid sequences in proteins

Automated biochemical techniques can be used to work out sequences of amino acids in proteins or specific parts of them. Comparisons have then been made of equivalent parts of cytochrome c and of the α and β chains of haemoglobin from different species. In the haemoglobin molecule, the α chain is composed of 141 amino acid residues and the β chain of 146 residues. Humans and chimpanzees are identical in these two chains; orang-utans differ from humans in two amino acids and gorillas differ in four. This suggests that humans are more closely related to chimpanzees than to orang-utans or gorillas.

How far do these data for the antigen/antibody reaction match the relationships suggested by the classification of primates, given in Table 9.3?

(a)

	β80	β87	β104	β125	α12	α23	α113
Human	N	T	R	P	A	E	L
Common chimpanzee	N	T	R	P	A	E	L
Pygmy chimpanzee	N	T	R	P	A	E	L
Gorilla	N	T	K	P	A	D	L
Orang-utan	N	K	R	Q	T	D	L
Gibbon	'D	K	R	Q	T	D	H
Old World monkeys	N	Q	K	Q	A	E	L
New World monkeys	N	Q	R	Q	A	D	H
Tarsier	N	K	R	Q	A	D	H
Lorises	N	K	R	Q	A	D	H
Lemurs	N	Q	T	A	T	E	H

Key to acid residues:
A=alanine; R=arginine; D=asparagine;
N=aspartic acid; C=cysteine; E=glutamic acid;
Q=glutamine; G=glycine; H=histidine;
I=isoleucine; L=leucine; K=lysine; M=methionine;
F=phenylalanine; P=proline; S=serine;
T=threonine; W=tryptophan; Y=tyrosine; V=valine

Using DNA

The information which determines the genetic make-up (genome) of an individual, and of a species, is contained in the DNA molecules. DNA is a double-stranded molecule made up of nucleotides; the bases of one strand are paired with specific bases in the complementary strand. The base cytosine pairs with guanine and adenine pairs with thymine. The link between the complementary strands is through hydrogen bonds. The sequence of bases in the DNA codes for sequences of amino acids, which are assembled to make proteins. Over a period of time, changes in the base sequence may occur, so that DNA sequences in an ancestral form may show differences in later generations. The longer the time of divergence, the greater the number of differences which are likely to have accumulated.

(b)

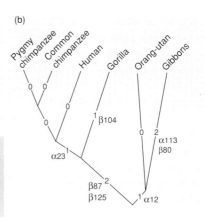

Figure 9.16 Comparing amino acid sequences in chains of α-haemoglobin and β-haemoglobin in some living primates. (a) In the primates listed there are only seven positions in the chains which show amino acid differences. (b) Possible evolutionary tree for apes and humans, based on these amino acid differences in the haemoglobins. The numbers within the branch lines indicate the number of amino acid residues that are different and the numbers beside the lines show the positions at which the differences occur

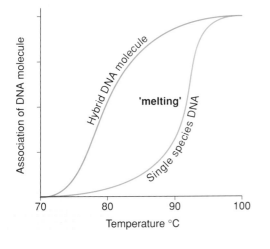

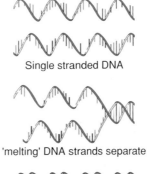

Single stranded DNA

'melting' DNA strands separate

Double stranded DNA

Figure 9.17 DNA hybridisation, a technique used to compare the genetic relatedness of DNA from different species

In the technique known as **DNA hybridisation**, an extracted length of a DNA molecule is denatured or 'melted' by a slight temperature rise. This causes the weak H-bonds to break, then the two complementary strands separate or unzip. The effect can be reversed by cooling the DNA to the same temperature, a process known as **re-annealing**. If a single strand of a DNA molecule is allowed to re-anneal with the complementary strand from another species, a hybrid is formed. The stability of this hybrid DNA molecule reflects how closely the two strands were able to match up their complementary bases. The greater the number of bases that paired, the higher the temperature required to separate (melt) the hybrid molecule. DNA–DNA hybridisation shows the closest relationship to be between humans and chimpanzees, then between humans and gorillas and orang-utans. Gibbons and Old World monkeys are more distant.

Figure 9.18 DNA profiling, a way of comparing DNA from different species to show how closely they are related. The polymerase chain reaction (PCR) can be used to amplify very small amounts of material obtained from fossils

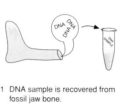

1 DNA sample is recovered from fossil jaw bone.

2 The recovered DNA is mixed with ingredients needed to amplify DNA. The process is called the polymerase chain reaction (PCR).

3 The thermal cycle heats and cools the DNA sample producing millions of copies.

4 PCR provides enough DNA to carry out analytical techniques like DNA profiling. The DNA is separated by a process called gel electrophoresis on an agarose gel.

5 The DNA is transferred to a nylon membrane where it is hybridised (made single stranded)

6 A probe is used to visualise the DNA. It binds to DNA sequences on the membrane. The DNA pattern obtained can help determine phylogenetic relationships.

The actual sequence of bases along certain lengths of the DNA molecule can be revealed by the technique known as **DNA profiling**. The DNA is extracted then cut into lengths using enzymes. The enzymes which cut DNA are known as **restriction enzymes**; these enzymes recognise particular sequences so can be chosen to be very specific in where they cut the DNA molecule. The fragments of DNA are loaded onto an agarose gel, and then separated by electrophoresis. DNA is negatively charged so moves towards the positive electrode. Large fragments move more slowly than small fragments when the electrical field is applied. The resulting patterns of banding of the DNA fragments can be revealed by treatment with radioactive labels or dyes. This allows the sequence to be 'read' and different genomes can be compared. Studies of DNA sequences confirm that the genomes of humans are very similar to those of chimpanzees and gorillas, with greater differences from orang-utans and even more from gibbons.

Fragments of DNA can sometimes be recovered from fossil material, such as part of a jaw bone. If present in very small amounts, it may be necessary to amplify this material (make many copies) by means of the polymerase chain reaction (PCR). This then provides enough material to carry out the fingerprinting procedure or to match the DNA sequence with other known samples and thus provide evidence at the molecular level for phylogenetic relationships of extinct species.

How far do these relationships worked out from molecular evidence agree with the classification scheme given in Table 9.3?

DNA in the mitochondria **(mitochondrial DNA or mtDNA)** has been used to trace phylogenetic relationships. Mitochondrial DNA contains relatively few genes, which give rise to recognisable patterns when sequenced. Because mtDNA is inherited through females, it is effectively isolated and retains its own genetic material. The only changes which occur are due to mutations, so mtDNA can be a useful means of tracing the inheritance of the particular sequences contained in this DNA. Attempts to use mtDNA to trace the origin of all humans back to a single mother (known as 'Eve') are a bit optimistic, but mtDNA can be used alongside other evidence to indicate the closeness of relatedness (see Figure 9.19).

Attempts have been made to link fossil evidence (considered in the next section) with molecular evidence and to establish a timescale for the divergence of the different hominoids (apes and humans). Several recent studies suggest the following:

- between 13 and 22 million years ago, the gibbon line diverged;
- between 12 and 19 million years ago, the orang-utan line diverged from the other two;
- between 4 and 10 million years ago, humans, chimpanzees and gorillas diverged from each other.

(a)

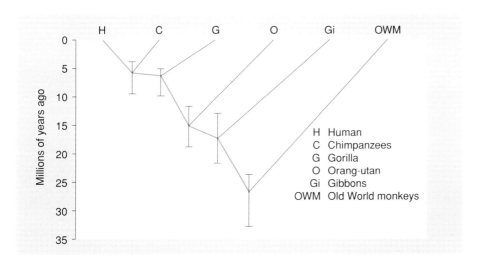

Figure 9.19 Possible evolutionary tree for primates, based on globin and mitochondrial DNA sequences. For each of the branching points, the range of estimated times is shown by the vertical bar

Fossils: evidence of ancestral forms

The preserved remains of living organisms are usually described as **fossils** when they are more than about 10 000 years old. In plants, the fossilised material is likely to be from the wood, veins of leaves, pollen grains or seed coats, and in animals mainly from bones and teeth. Soft parts are usually not preserved. The fossils may be the actual remains of the living material, or may be in the form of imprints, moulds or casts of the former living material. The hard parts of the material may have undergone changes. For example, protein in bones may be replaced by minerals containing calcium or silica in a process known as **mineralisation**. This makes the fossil harder and stronger than the original material. In some cases, the original material is completely replaced during these mineralisation processes, leaving a cast or mould of the original shape.

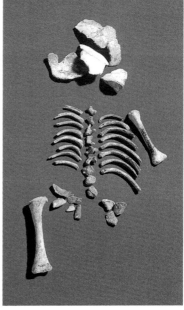

(b)

Figure 9.20 (a) Laetoli footprints, a trail of hominid footprints, fossilised in volcanic ash, found at Laetoli, Tanzania. The trail dates from over 3.0 million years ago and probably belongs to Australopithecus afarensis. *The footprints show a well-developed arch to the foot and no divergence of the big toe. They indicate upright walking and probably represent two adults and a child (b) Fossil bones of the skeleton of a Neanderthal infant*

HUMAN EVOLUTION

Suggest conditions in which decay of dead organisms would be very slow.

Formation of fossils is a chancy business. When living organisms die, the processes of decay by microorganisms usually set in rapidly. In addition, the effects of weathering by heat, cold, wind and water (particularly when acid), make corpses of animals (both large and small) disappear quite fast. Parts of animal corpses would have been consumed or dragged elsewhere by scavengers. Just occasionally, conditions around the time of death may allow the living material to be buried so that some parts of the original structure become preserved without undergoing decay.

Where fossils have formed, it is usually because sediments settled over the remains of the organism and buried them. Active flowing rivers carry various particles which may be deposited in the river bed, on the river banks or flood plain beside the river. Large particles of sand and gravels tend to alternate with finer silt and clay. Over a period of time, the composition and depth of the layers which form may be irregular, because of variations in the activity of the river. Sometimes a considerable build-up of sediment may occur in quite a short time. Sediments build up in layers in a similar way in lakes, as a result of rivers flowing into the lake. Deposits of sand, silt or clay also accumulate in caves, though sometimes collapse of the cave brings a rapid end to the opportunity for further burial of remains. Caves may have been places where carnivores dragged their prey and were certainly used by early humans for shelter or burial, resulting in accumulation of material suitable for fossilisation. Sediments are usually soft when formed, but over a period of time they harden to form rocks. The processes are due partly to pressure as more sediments are laid down on top and water is squeezed out from between the original particles, and partly as a result of chemical processes. Thus mud becomes shale, sand is converted to sandstone and calcium carbonate to limestone.

Active volcanos eject materials which can be favourable for the formation of fossils. Volcanic ash may fall over wide areas and become incorporated into the soil where other deposits are building up. Closer to the volcano, a thick layer of ash may cover the landscape, and hot lava and mud flowing out from the volcano may trap remains of organisms underneath. Fossil footprints probably originated as footprints in soft mud, which were then rapidly filled in with another contrasting material, such as volcanic ash.

Any fossil discovered is likely to represent only small fragments of the original organism. Its location is recorded, then the fossil is compared with other relevant material and its significance interpreted. An essential step is to date the material, both in relative terms and, as far as possible, on an absolute time scale. The age of material found at different sites must be correlated. The term **geochronology** refers to the methods used to find the age and sequence of geological events. The techniques used depend on both the nature and the age of the material. Often the dating is applied to material in the rocks immediately below and above the fossil rather than to the fossil itself.

Ideally, fossils are found between the layers of sedimentary rocks, with the oldest at the lower depths and the youngest at the top. Careful removal of the material, working down through the layers, could unearth a series of fossils which may reveal a sequence of changes in particular organisms over successive time periods. As a result of geological movements, the positions of rocks have often been disturbed so that the original pattern of the layers of the sedimentary rocks and any fossils within them is altered. It is perhaps fortunate that earth movements have occurred as this allows buried fossils to be brought to the surface, then exposed through weathering by heat, cold, wind or water. Without this, people would be very unlikely to find and uncover fossils from the rocks. The Rift Valley in Africa has become a particularly important fossil-hunting region, its richness being due to the combination of past events related to its lakes, rivers, volcanos and spectacular earth movements.

Formation of fossils and their exposure

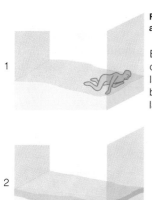

1

Bones of a living organism lying on a lake bed become buried by successive layers of sediment

2

3

4

Geological events may raise the rock above the water level, then weathering may erode the layers of sediment, thus exposing the fossil

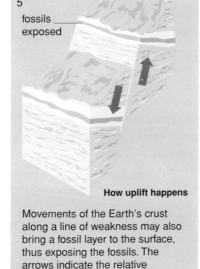

5

fossils exposed

How uplift happens

Movements of the Earth's crust along a line of weakness may also bring a fossil layer to the surface, thus exposing the fossils. The arrows indicate the relative movement of two blocks of land

6

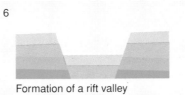

Formation of a rift valley

Figure 9.21 Fossils buried under layers of sediments can be brought to the surface when the layers of rock move relative to each other. Such geological movements certainly occurred in the Great Rift Valley in Africa, which has become an extremely important source of fossils relating to human evolution

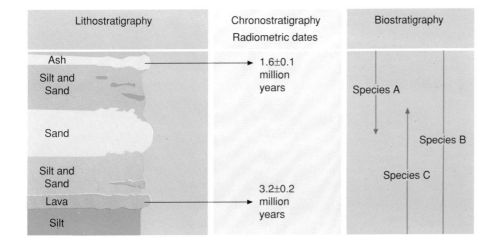

Figure 9.22 Ways of dating fossils in their geological context.
Lithostratigraphy *looks at the layers of sediments that have built up around the fossil.*
Chronostratigraphy *uses material to give an absolute measure of time (by techniques such as radiocarbon dating).*
Biostratigraphy *looks at fossils of other species associated or overlapping with the fossils, for which history may have been established elsewhere*

HUMAN EVOLUTION

Relative dating places phases in sequence relative to one another and where possible to other events. **Lithostratigraphy** studies the sequence of different types of sedimentary layers and assumes that a series of events would leave similar patterns in the build-up of layers of sediments at different sites. This may, for example, show as a thick layer of one type of material, followed by a thin layer of another. Volcanic material can act as a useful marker indicating the timing of a particular event. Similar patterns in sedimentary rocks may help connect simultaneous events at distant locations. The appearance of fossils in successive sedimentary layers gives an indication of the relative age of the fossils.

Sometimes changes in other species found associated with a group of fossils or in the same layers of rocks can give an indication of the age of the fossils, particularly if the other species has already been well documented from finds elsewhere. This is illustrated by the horse, elephant and pig, three species with fossil records which reveal a fairly complete record of stages in their evolution. Remains of these species, when found alongside human fossils, can help give a relative date to the associated human material.

Absolute dating methods attempt to express the occurrence of events in calendar years. **Radiometric dating** methods are based on measurements of radioactive decay. Radioactive forms (isotopes) of elements are assumed to decay at predictable rates, regardless of surrounding conditions. The proportion of related isotopes in any given material gives an absolute measure of the age at which the material originated. The two methods most useful in the study of human fossils and origins are **potassium-argon** dating and **radiocarbon** dating.

In the potassium-argon dating method, the element potassium (K) has an atomic mass of 39, though it includes a small proportion of a radioactive isotope potassium-40 (^{40}K). The radioactive potassium decays to give ^{40}Ar, a stable isotope of the element argon. The isotope ^{40}K is released in lava from active volcanos, though at this temperature, any argon released at the same time would be lost as a gas. Eruption of a volcano effectively sets the clock to zero. The ^{40}K (radioactive potassium) is incorporated into sedimentary deposits and there it changes slowly to ^{40}Ar. The ratio of ^{40}K to ^{40}Ar in the material examined gives an indication of its age. Calculation of its actual age is based on the **half life** of the isotope. For ^{40}K the half life is taken as 1.3×10^9 years (1300 million). Potassium-argon dating has been used mainly for materials in volcanic rocks older than 1 million years. The accuracy of this method is about 50 000 years.

The exceedingly abundant element carbon exists mainly as the stable isotope ^{12}C, but includes a small proportion of a radioactive isotope, ^{14}C. The ^{14}C is formed at a relatively constant rate by cosmic radiation. In radiocarbon dating methods, it is assumed that the living organisms (plants carrying out photosynthesis) took in carbon with a ratio of ^{14}C to ^{12}C equivalent to that in the surrounding atmosphere at the time. The plant material, when eaten, would then have been incorporated into animals. As with the potassium-argon method, changes in the ratio can be used to calculate the age of the material. For radioactive carbon (^{14}C) the half life is only 5730 years. Radioactive carbon dating has been used for estimating the age of much younger organic materials, such as bones or shells, up to about 50 000 years old.

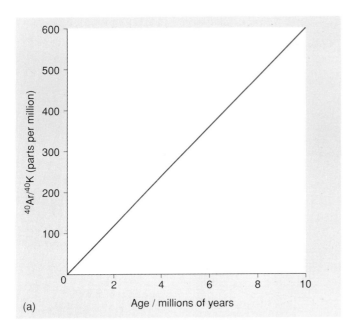

(a)

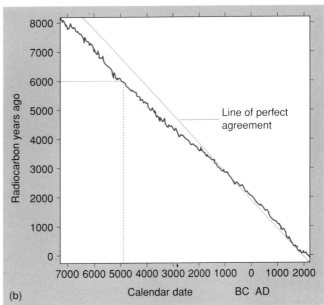

(b)

Figure 9.23 Radiometric dating is based on predictable rates of decay of radioactive isotopes.
(a) Potassium-argon dating is useful for fossils associated with volcanic rocks, about 0.4 million years or older. As ^{40}K decays, ^{40}Ar accumulates, so the ratio $^{40}Ar/^{40}K$ gives a measure of time since a particular volcanic event;
(b) Radiocarbon dating estimates age by measuring the proportion of ^{14}C to ^{12}C in a sample of material. Radiocarbon dating methods are sensitive but applicable only to the last 40 000 years. The graph shows radiocarbon dates in a sample of wood compared with age deduced from counting tree rings. A radiocarbon age of 6000 years corresponds to a calendar date of 4900 BC

In tracing human evolution, fossil skeletons have given a definite record of the size of the individual, its posture and the type of locomotion; fossil skulls show how the shape of the face and jaw have changed and how the brain has increased in size. Inevitably there are uncertainties surrounding the precise details of the evolution of human or of any other species and there is controversy and debate even amongst the experts who find and interpret the fossils. Dates are only approximate, partly because of limitations in the accuracy of methods used to estimate the date of the material and partly because finding a fossil in one location does not give information about how long the species might have existed, nor does it preclude the form being present elsewhere but where no fossil has yet been found.

Associated with fossils there may be **artefacts** or objects used by humans, such as shaped stones which would have been used as tools. When associated with fossil skeletons, artefacts have become an important source of evidence for building up a picture of how early humans lived and of their cultural development. Artefacts include flaked stones and bones; later evidence indicates use of fire, followed by more sophisticated expression in terms of jewellery or cave art. Artefacts continue to reflect the history and achievements of people, right up to the present day, through buildings, jewellery, paintings or even rockets launched into space.

The extinct australopithecines

Evidence from fossils has been used to build up a picture of the australopithecines, extinct primates which were undoubtedly related to early human ancestors. The australopithecines lived in Africa, between 4 million and 1 million years ago (4 to 1 Myr). The name *Australopithecus* is derived from the Greek for 'southern' (australo) and 'ape' (pithecus). They have been described as 'man-apes' and also as 'ape-men' and clearly show features of both groups. It is likely that they were a group of early hominids, sharing ancestry and overlapping with early human species, but not necessarily their direct ancestors. In Table 9.3 on page 175, the australopithecines would come just above humans, in the family Hominidae.

Figure 9.24 The australopithecines are ape-like hominids, and probably branched away from the African apes between 4 and 5 million years ago

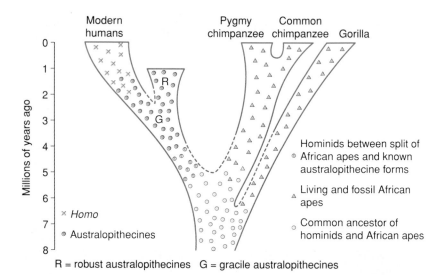

R = robust australopithecines   G = gracile australopithecines

Fossil skeletons assigned to the australopithecines have been found at different sites in Africa. They fall into two groups: a lightly built or **gracile** form and a more ruggedly built or **robust** form. The gracile form includes two species, *Australopithecus afarensis* and *A. africanus. A. afarensis* was found in the Afar region of Ethiopia, and the first partial skeleton found here was nicknamed 'Lucy'. Other *afarensis* fossils have been found at Laetoli in Tanzania, the site of the oldest known human footprints. The robust form includes *A. robustus* and *A. boisei*, though some workers feel these are sufficiently different from the gracile forms to justify a separate genus, *Paranthropus*. These fossil skeletons are far from complete, but enough can be pieced together to build up a picture of these primates when they were living.

Australopithecines show a number of features characteristic of humans in the genus *Homo*. Gracile australopithecines like Lucy probably walked upright, though perhaps a little awkwardly, and may have spent some time in the trees out of reach of predators. Lucy was about the same height and weight as a modern six-year-old girl. The narrow hips would give a narrow birth canal in females, consistent with a relatively small skull size, which in turn indicates a relatively small brain. *A. afarensis* represents the oldest australopithecine that has been found and probably resembled a small upright dark and hairy chimpanzee. The robust australopithecines, *A. boisei* and *A. robustus* were larger and heavier, with larger brain size. In height and weight *A. robustus* approached that of a modern human, though the brain size was much smaller. The arms were relatively long, but the teeth show reduced incisors and canines, a feature associated with hominids.

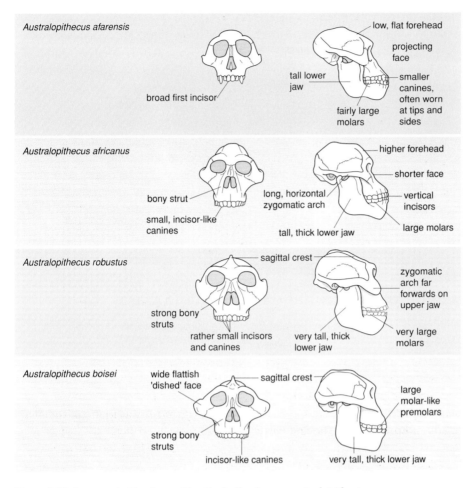

Figure 9.25 Front and side views of fossils skulls of some australopithecines

Australopithecines

Australopithecus afarensis and
A. africanus are the **gracile**
australopithecines and lived about
4 to 2.5 million years BP (before
present).

A. boisei and *A. robustus* are
robust australopithecines (or genus
Paranthropus) and lived about
2.6 to 1 million years BP.

Distinguishing features:
• upright, height 1 to 1.5 m, weight
  from 30 to 80 kg;
• relatively long arms;
• build – graciles: light build, some
  sexual dimorphism
  – robust forms: heavy build, some
  sexual dimorphism;
• brain size 400 to 530 cm³;
• skull form
  – graciles: flat forehead (higher in
  A. afarensis), projecting face
  – robust forms: prominent crest
  on top of skull, flatter face;
• brow ridges – prominent in
  A. afarensis, less in *A. africanus*,
  also present in robust forms;
• jaw and teeth
  – graciles: incisors and canines
  relatively large in *A. afarensis*,
  smaller in *A. africanus*
  – robust forms: very thick jaws,
  small incisors and canines, very
  large molars;
• location – eastern and southern
  Africa.

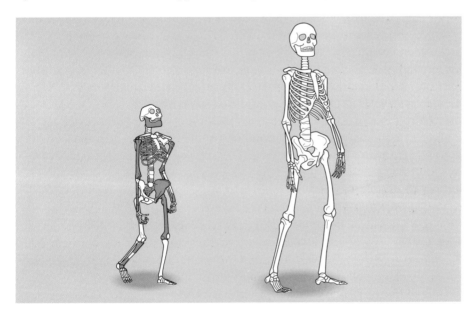

Figure 9.26 Skeleton of the australopithecine nicknamed Lucy, seen alongside modern human skeleton. The shaded portions show actual fossil bones, the remaining parts have been reconstructed from other available evidence. Lucy was only 105 cm high, though other australopithecines are taller.

Climatic changes occurring about 4 million years ago probably influenced the evolution of the australopithecines. The Ice Age at that time meant that a lot of water had been locked up as ice, and rainfall had dwindled. This led to a shrinking of forest areas and a spread of tropical grasslands, otherwise known as savannah. Such grasslands had relatively few trees or shrubs. But Primates were mainly tree dwellers and maybe the australopithecines were able to take advantage of the loss of forest and changes to grassland. Their fossil footprints suggest bipedal walking which would have left their hands free for making and using tools. If they were able to use even simple tools, such as stones or bones, they would have been less dependent on canine teeth for performing some tasks. Group co-operation might have helped survival on open grassland and establishment of a home base would have allowed food to be shared and helped encourage social ties. With a relatively stable home base would come the opportunity for the development of longer infant and childhood phases, with prolonged parental care.

The genus *Homo:*
where and when did early humans start?

Three or four stages or 'species' of early humans are recognised from fossils: *Homo habilis* ('handy man'), *Homo erectus* ('upright man') and *Homo sapiens* ('wise man'). *H. sapiens* sometimes includes the Neanderthals, treated as a subspecies of *H. sapiens* (*H. sapiens neanderthalensis*), though some people consider the Neanderthals to be a separate species (*H. neanderthalensis*). The name is derived from the Neanderthal river in Germany where the first fossils were found. In this chapter, we will use the term 'Neanderthal man' and avoid designation of species or subspecies.

Relatively few fossils of early hominids have been found, and those that have been discovered are rarely complete. Fossil evidence for *H. habilis*, found only in Africa, is particularly scanty. There is general agreement that *H. habilis* represents the earliest known form that can be called human and that it originated in Africa. The later *H. erectus* also developed in Africa and from there migrated to Asia. The oldest dated *H. erectus* fossils come from Africa but later ones have been found in Asia, at sites in Java and near Peking (Beijing) in China. *H. sapiens* diversified into a range of forms, known as archaic, early modern through to modern. These spread from Africa and Asia into Europe, and eventually in modern times, into most parts of the globe. Neanderthal man is known mainly from Europe, with an eastern distribution as far as central Asia.

We can see distinct evolutionary trends from the early *H. habilis* through *H. erectus* to Neanderthal man and the later *H. sapiens*, but the progression does not necessarily mean that one form evolved directly from the preceding type. Certainly *H. habilis* lived alongside the australopithecines, between 2 and 1.5 million years ago, before the australopithecines became extinct. *H.erectus* (from about 1.5 million years ago) also overlapped with the later australopithecines. *H.erectus* was probably a common ancestor to both modern *H. sapiens* and to Neanderthal man. Some people would support very much earlier dates for the origin of hominids, including the australopithecines and members of the genus *Homo*.

These critical developments in the evolution of *Homo* species coincided with considerable variations in climate, with cold alternating with warm due to successive Ice Ages. Fluctuating temperatures led to changing sea levels: during a glacial period sea levels were lower because water was locked up as ice and land connected some of the major continents as we know them today. These land bridges allowed migration of evolving forms from one area to colonise another. During an interglacial period when the ice melted, sea levels rose and some parts of the land masses became separated. Some branches of the *Homo* family tree may thus have become isolated on the different land masses and evolved along slightly differing lines.

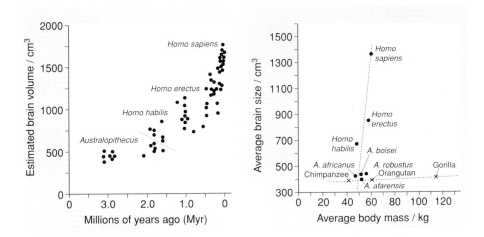

Figure 9.27 Brain size in the australopithecines and the genus Homo *compared with some modern primates. (a) Increase in size of brain over time; (b) Increase in size of brain in relation to body mass*

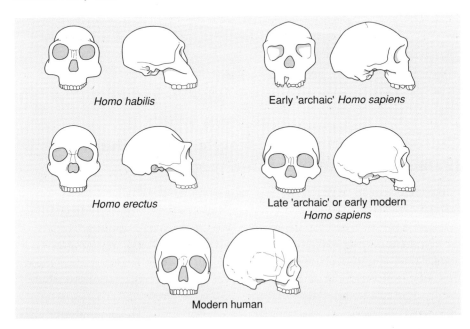

Figure 9.28 Homo *skull features - comparing* H. habilis, H. erectus *and* H. sapiens. *There is variation between different reconstructed fossil skulls assigned to each species, with respect to the size of the braincase, the brow ridges and shape of the jaw and face. Nevertheless, clear trends can be seen in the development of modern* H. sapiens *from the earlier* H. habilis

HUMAN EVOLUTION

Homo habilis
(about 2.4 to 1.6 million years BP) small and large forms recognised;
- upright, height 1 to 1.5 m;
- weight about 50 kg;
- relatively long arms;
- pelvis allowed birth of babies with big heads;
- brain size 500 to 650 cm^3 (small), 600 to 800 cm^3 (large);
- cranium shape (see Figure 1.28)
- brow ridge, forehead slopes backwards;
- jaw light to robust but protrudes; small canines, narrow molars (small to large);
- location – Africa.

Homo erectus
(1.8 to 0.3 million years BP)
- height 1.3 to 1.5 m;
- weight about 40 to 75 kg;
- brain size 750 to 1250 cm^3;
- skull long, flat and thick;
- brow ridge, receding forehead;
- protruding jaw, teeth smaller than *H. habilis*;
- location – Africa, E and SE Asia, spread to Europe.

Neanderthal man
(150 000 to 30 000 years BP)
- height 1.5 to 1.7 m;
- weight about 70 kg;
- brain size 1200 to 1750 cm^3;
- skull thinner, large nose, protruding mid-face region;
- brow ridge reduced, low receding forehead;
- some development of chin, teeth similar to archaic *H. sapiens* but smaller;
- location – Europe and western Asia.

Homo sapiens
(archaic 400 000 to 100 000, early modern 130 000 to 60 000 years BP)
- height 1.6 to 1.85 m;
- weight about 70 kg;
- brain size 1100 to 1400 cm^3 (archaic), 1200 to 1700 cm^3 (early modern);
- skull higher, higher and more upright forehead, face less protruding;
- almost no brow ridge in early modern;
- smaller jaws, chin developed, teeth smaller (small molars, vertical incisors, small canines);
- location – Africa, Asia, Europe.

Figure 9.29 Skull of Neanderthal (left) compared with that of modern human (right), together with a 'model' or reconstruction of a Neanderthal face. This emphasises the Neanderthal flattened, broad shape, swept-back cheekbones, large front teeth, large nose and brow ridge above the eyes

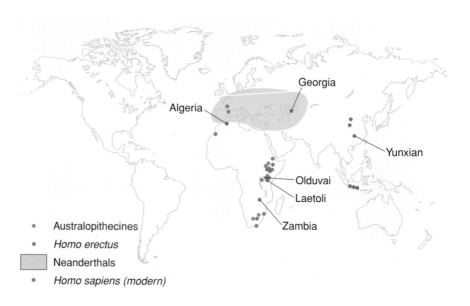

- • Australopithecines
- • *Homo erectus*
- ☐ Neanderthals
- • *Homo sapiens (modern)*

Figure 9.30 Main sites of Homo and Australopithecine fossil finds

Epoch	Estimated years ago	Climatic conditions	Probable stage of hominid evolution	Too
Middle Pleistocene	250 000–180 000	mainly temperate or cool, some glacial intervals • *large oceans, small ice caps*	• Europe – transitional 'archaic' *Homo sapiens*, early Neanderthals • Africa – late 'archaics' • China – late *H. erectus* and 'archaic' *H. sapiens*	Acheuli
	180 000–130 000	full glacial, some milder intervals • *small oceans, large ice caps*	• Europe – early Neanderthals, fixation of Neanderthal features • Africa – transitional to early Moderns • China – 'archaic' *H. sapiens*	
	130 000–115 000	last interglacial, warm conditions • *large oceans, small ice caps*	• Europe and Middle East – Neanderthals • Africa – first Moderns (and possibly Middle East) • China – 'archaic' *H. sapiens*	
	115 000–75 000	temperate / cool	• Middle East – Neanderthals and Moderns • China – 'archaic' *H. sapiens*	Mouste
Late Pleistocene	75 000–30 000	cool / glacial • *ice caps increasing*	• Europe – Moderns appear; Neanderthals extinct • Far East – Moderns appear	Microli

Table 9.5 Climate and sea level changes while hominids were evolving, with possible stages of evolution in different parts of the world. (See later in chapter for details of the different hominid forms.) The 'tools' give an indication of the stage of cultural development, but evidence of actual tools used depends on the area and sites excavated

Cultural evolution and lifestyle of early humans (Palaeolithic and Neolithic stages)

In tracing the main stages of the evolution of humans, we believe they adopted first a terrestrial rather than arboreal way of life, became truly bipedal and then developed the ability to use their hands. With this came enormous development of the capability of the brain. Cultural development progressed alongside the physical changes described in this chapter. Evidence for cultural evolution is mostly indirect. Sticks and other wooden objects were probably used as tools, but it is unlikely that these would be preserved. The earliest objects recognisable as tools were stones, shaped and worked in different ways. Such stones have been found associated with fossil skeletons and skulls and other evidence of human existence.

In the Palaeolithic (old stone age), from over 2 million years ago, there was considerable use of stone tools but no indication of settled agriculture. The Neolithic (new stone age), from about 10 000 years ago, was characterised by the domestication of plants and animals and the beginnings of true agriculture. These terms represent stages of cultural development which probably occurred at different times in different parts of the world. We shall never really know how early human communities interacted with each other, but we can speculate and build up at least a partial reconstruction of cultural life from about 2.5 million years ago. The next section describes the basic tools of the Palaeolithic stage, followed by a reconstruction of the lifestyle of these early humans, as they showed the early development of human lives including utilisation of tools and technology, and expression through art and religion.

Old stone age – Palaeolithic

The earliest stone tools were probably simple pebbles used for hammering, or knocked about to give a sharp edge which was useful for more complex tasks. Shapes similar to those found at early hominid sites have been produced recently by workers simulating the activities of early hominids. If one stone is hammered with another (the 'core'), pieces of stone known as 'flakes' break away. The flakes have sharp edges and can be used for various purposes, such as cutting up meat or shaping a piece of wood for use as a tool. The term **knapping** is used for a person working stone in this way.

The earliest examples of stone tools are described as **Oldowan**, after the Olduvai Gorge site in Tanzania. Similar stones have also been found at numerous other sites, in Africa, Asia and Europe. Several types are recognised in the Oldowan 'toolkit', including a chopper, scraper, hammerstone and flakes.

Draw up a chart to compare the australopithecines with the *Homo* species. List features which characterise and distinguish modern humans from their probable ancestors. In particular, note when the human brain size and shape of the modern face became established. Then look back to your earlier comparison of modern humans with other living primates. Compare them with descriptions given for the extinct australopithecines and for modern apes.

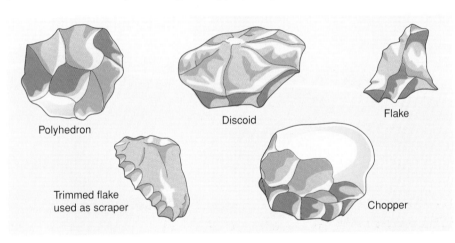

Polyhedron

Discoid

Flake

Trimmed flake
used as scraper

Chopper

Figure 9.31 Early stone tools of the Oldowan toolkit – the heavier stones can be used as choppers (for cutting meat or cracking open bones); the polyhedron has three or more cutting edges; the discoid has a sharp rim and the scraper can be used on the hides of animals. The flakes probably broke away while working the stones and would have been useful for various purposes, certainly for cutting meat off large animals. It is likely that H. habilis used the tools described for this Oldowan toolkit. The chopper is about the length of an adult thumb

More sophisticated tools in the form of hand axes have been found and described as the **Acheulian** culture, named after St Acheul, the place in France where they were first discovered. These axes have two worked faces giving two cutting edges and were elongated, tapering to a point. Making these Acheulian axes required a greater sense of purpose and represented a considerable advance in the skills of the humans making and using them. The earliest finds date from about 1.5 million years ago.

Neanderthals used an elaborate range of tools, with finer stone flakes worked to provide a means of skinning, scraping, sawing or sharpening other tools. These are known as the **Mousterian** toolkit, after Le Moustier in France where they were first found. Gradually the tools became smaller, sharper and generally more specialised. The flakes were worked into blades to produce a diverse range of implements. These refined small stone tools, described as **microliths**, appeared towards the end of the Palaeolithic culture, about 25 000 years ago, though some earlier finds have been reported. They could have been used in spears, harpoons, sickles and many other implements, showing potential for use in a wide range of activities.

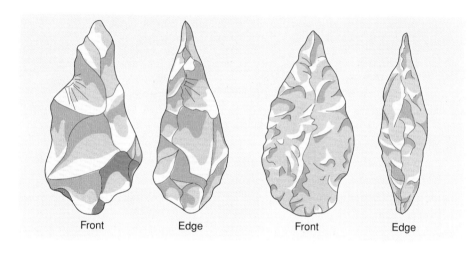

Figure 9.32 The Acheulian toolkit includes a wider range of shaped stone tools which were certainly used by H. erectus. The typical axes were elongated with two faces and were more difficult to make than the Oldowan chopper. The rounded end of the Acheulian axe could be held in the hand allowing the edge to be used for chopping and slicing (shown about one-third actual size)

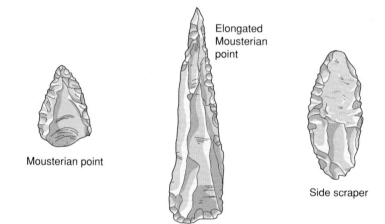

Figure 9.33 Some tools of the Mousterian toolkit, as used by the Neanderthals. Shown about half actual size

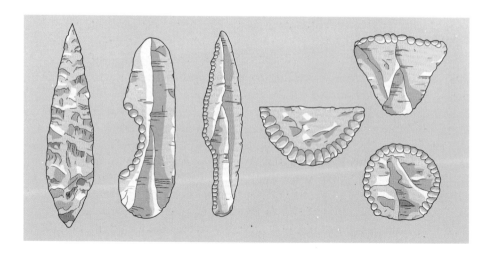

Figure 9.34 A range of skilfully worked blades, associated with the Upper Palaeolithic cultures

Homo habilis

The brain of *H. habilis* was larger than that of the australopithecines and markings inside the skull indicate development of a region known as 'Broca's area' suggesting the beginning of language ability. *H. habilis* is associated with Oldowan stone tools and the form of their hands and thumbs confirms rudimentary ability to use tools. The jaw and tooth pattern is consistent with their being fruit eaters and they probably also scavenged meat. The form of the foot shows they walked upright, but their limbs would also have enabled them to climb trees. They are thought to have lived by river banks in more wooded habitats (whereas the nearby australopithecines lived in more open areas of savanna grassland). *H. habilis* could have used the open areas for food gathering, but retreating to the wooded areas for protection. There is evidence that *H. habilis* made simple shelters but could also take refuge in trees from large predators. This indicates the beginnings of a settled home base, of living in groups with bonding between males and females and care of their children.

Homo erectus

The larger brain size of *H. erectus* gave potential for more complex activities. There must have been language ability and the position of the larynx indicates use of more elaborate sound patterns. Compared with *H. habilis*, *H. erectus* had a larger body size and was more muscular which would have allowed mobility over wide areas, consistent with an active hunting lifestyle. They used tools from the Acheulian toolkit, but neurological control of the arm and hand was still restricted so that manipulations with the hand were limited. This is deduced from the size of the canal in the skull for nerves entering the brain. They certainly would have used other tools, made from wood. In some areas (such as Java) no stone tools have been found but locally fashioned bamboo probably served the same purpose.

H. erectus were nomadic hunter-gatherers, hunting large game such as boar, bison, deer, horse and rhinoceros. This was made possible by their more elaborate tools which could have included spears or lassoos. The hunting required cooperation, probably between males, and the catch would have been brought back to the home-base to be shared with females and the rest of the families in the group. Their teeth suggest a mixed diet; increasing consumption of meat was supplemented with locally gathered fruits and other plant material. Evidence of use of fire comes from charred remains and ash. They would have used fire for roasting meat and probably for keeping warm in some of the colder parts of their geographical range. They built fairly elaborate shelters or often used caves for their living places.

Neanderthal man

Neanderthal man was probably a side branch in the evolutionary history of modern humans, which died out rather suddenly. Neanderthals were short, stocky and muscular, well built to withstand the colder climate of the Ice Age in Europe at that time. Their brain was actually larger than that of modern humans, though the Neanderthal skull was a little thicker, the jaw protruded and lacked a chin. The nasal cavity was enlarged and this would have given advantage in a cold climate by conserving heat and moisture from the freezing air outside. Their speech ability was probably still limited compared with *H. sapiens*.

HUMAN EVOLUTION

How is the short, heavily built body of Neanderthals better able to survive in cold climates when compared with a tall lean body of say present-day Masai people living in central Africa? You may need to refer to sections in Chapter 10 on temperature adaptations. See pages 204–214

Neanderthals used an elaborate range of tools from the Mousterian toolkit. The neck vertebrae show an enlarged hole, indicating that the spinal cord had become large enough to allow fine neurological control of manipulations with the hands. Their tools may also have included wooden implements and thongs, but such materials rarely survive with other remains. Their teeth were strong, and probably supplemented their toolkit in helping with some of the tasks they carried out.

Neanderthals lived in caves and also constructed large tents for protection. These would have been covered with skins from animals, captured in their hunting exploits. The skins were stretched over a framework of branches and weighted down with large bones, such as those from mammoths. Skins were also used for simple clothing, but with no evidence of needles for sewing. Clothing enabled them to create a microclimate around the body, important for survival in the cold climate. Fires would have helped to give heat to their caves or tent dwellings, and provided a means of modifying their food by cooking.

Neanderthals were big game hunters, capturing bison, horses, woolly cave bears, wild cattle, reindeer, mammoths and woolly rhinoceros. This would have required co-operation within the group suggesting rudimentary language skills. They may, for example, have used fire to drive herds over cliffs or into narrow gorges for mass slaughter. The bones of Neanderthal skeletons show frequent signs of wounds and damage, suggesting a fairly violent lifestyle, as might be associated with their big game hunting activities. Nearer the home-base, women and children would have gathered berries, honey, roots and smaller animals including insects. Excess meat or fruit was perhaps conserved for another season by drying, smoking or freezing.

There are signs of community care and beginnings of religious practices. Evidence for this is seen in burial sites: in one case a skeleton of an old arthritic person has been found, in another bones indicate deformities which would have rendered the person useless for hunting activities. Regrowth and healing of bone tissue suggests protection of the injured within the social group. Burial sites are recognised by the deliberate orientation of the skeletons, accompanied by remains of pollen grains, suggesting flowers had been buried with the corpse. The simplest form of art as something which has no practical use is seen in scratched pebbles, use of red colouring or a polished tooth. Neanderthals may have practised cannibalism, perhaps with the belief that by eating the brains or bone marrow of their dead relatives, strength would be given and their spirits would carry on in the living. Another view is that the marks seen on fossilised human bones, interpreted as signifying cannibalistic practices, have arisen from activities of big animals such as hyenas, or even rock falls in the cave. This uncertainty serves to reinforce the speculative nature of some of the interpretations given when reconstructing the lives of our ancestors. It is, however, clear that Neanderthals died out quite suddenly, and were replaced or overtaken by the modern form, *H. sapiens*.

Homo sapiens

With *H. sapiens* comes a rapid diversification and increase in complexity of the tasks which could be accomplished, consistent with greater mental capacity and improved motor control of activities, particularly with the hand. Cro-Magnon man, named after the site at Cro-Magnon in France, is taken to be early modern *H. sapiens*, very close in ancestry to modern living humans.

H. sapiens, typified by Cro-Magnon man, is a (late) stone-age human, but is clearly identifiable as an immediate ancestor to modern human populations. The species probably originated in Africa rather than being a direct descendant of Neanderthals, but Neanderthal man and *H. sapiens* overlapped for some time and there may have been some gene flow between the populations. Compared with Neanderthals, Cro-Magnon man was of slighter build in body frame though about the same height. Cro-Magnon man shows considerable advances over Neanderthals in capabilities (with respect to tools employed), sensitivity as shown by artistic expression, exploitation of the environment and independence of it in terms of shelter and protection from extremes of climate.

Cro-Magnon man created and used tools showing far greater diversity than previous types. They made specialised blade tools, worked from flints, then used them to shape bone, antler and ivory. They used bone needles with eyes for sewing, barbed hooks for fishing. They made harpoons and spears which could be thrown. They could thus attack animals at a distance, perhaps making greater use of the brain to find ways of trapping and ensnaring them. Catching fish added yet more variety to their already protein-rich diet, which was supplemented with fruits and nuts gathered and stored. Remains of baskets suggest they were used to help collect these items from the surrounding land. At certain times they lived in caves, well chosen for their aspect (to trap sun) and view over the open valleys. They also built large and elaborate shelters, sometimes with stone for the floor, the roof being propped up with mammoth bones. These homes may have housed several families. They made simple fur suits from animal skins, a way of retaining body heat and providing protection from climatic extremes.

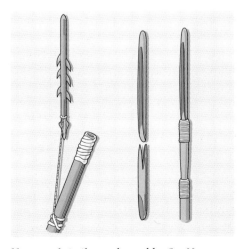

Figure 9.35 Examples of bone and similar tools used by Cro-Magnon man

Figure 9.37 Late Palaeolithic art on a reindeer antler

An outstanding advance associated with these late Palaeolithic humans was in their artistic culture. They carved patterns and pictures on bones and antlers, made bracelets from ivory and necklaces from coloured beads, pierced teeth from larger animals or sea shells. Remains hoarded in burial areas suggest accumulation of such ornaments was an important part of their lives. Cave paintings often depict large mammals, including bison, aurochs, deer, horses, mammoths and ibex. Examples of such cave art are found particularly in France and Spain.

The hunting-gathering lifestyle of the late Palaeolithic people seems to have been reasonably successful in terms of survival. Food was mostly plentiful, but if an area became depleted, these nomadic people moved on to harvest a fresh patch of countryside. There were large and small game as well as fish, and berries, nuts and other fruits from the bushes. At some stage, large seeds from wild grasses were deliberately harvested and added to their mixed diet.

There was a relatively short transition period, known as the **Mesolithic** (middle stone age), which spanned the end of the Palaeolithic into the beginning of the Neolithic. This coincided with the retreat of the last Ice Age, and with the melting of the ice, there was a general increase of temperature and rising of sea levels. Animals migrated northwards and plants became re-established in the temperate zones across Europe through to Asia. As tools became more refined, they were applied to a wide range of tasks. There is evidence of people using baskets and simple pottery. Quite elaborate huts and shelters were constructed and the lifestyle was becoming more sedentary. The Mesolithic stage merges easily into the Neolithic, characterised by the beginnings of agriculture and the settled life that goes with it.

New Stone Age
– Neolithic people and the beginnings of farming

The Neolithic culture dates from around 10 000 years BP, though similar patterns of development of agriculture occurred independently, at different times, in at least three areas of the globe. The oldest was probably in the 'fertile crescent', an area in western Asia, bordered by the Taurus and Zagros mountains to the north, lying in the basin of the Tigris and Euphrates rivers. Wild grasses found in the area today are related to modern wheat and barley, and in the same area the wild animals include sheep and goats which were among the earliest animals to become domesticated. This area also saw the development of the great early civilisations in Persia, Babylon and through to Egypt. Evidence of crops cultivated in this fertile crescent include wheat, barley, lentils and peas. Remains of cereal seeds show wheat similar to 'einkorn' and 'emmer' (*Triticum monococcum* and *T. turgidum* respectively), two species of *Triticum* believed to have given rise to modern varieties of wheat (*Triticum aestivum*). Artefacts from excavations show that Neolithic people had sickles with fine flint blades set into handles made from antlers. They used pestles and mortars for grinding their harvested cereal crops and had storage bins at their hut sites.

Wild animals that became domesticated include sheep (*Ovis orientalis* and *O. vignei*), goats (*Capra aegagus*) and aurochs (*Bos primigenius*), the now extinct ancestor of modern cattle. These animals had a natural herding instinct, which predisposed them to successful domestication. Their ability to digest cellulose meant that they did not compete directly with humans for crops. Domestication of sheep is known from about 11 000 years ago, of goats from about 10 000. Aurochs were first domesticated about 8 500 years ago and wild boar (*Sus scrofa*) at about the same time. Supplies of meat and later milk were thus secured from one year to the next. Other animals that were domesticated and used for transport include the camel, donkey and horse, while cattle were valuable as a dual purpose animal (food and ploughing).

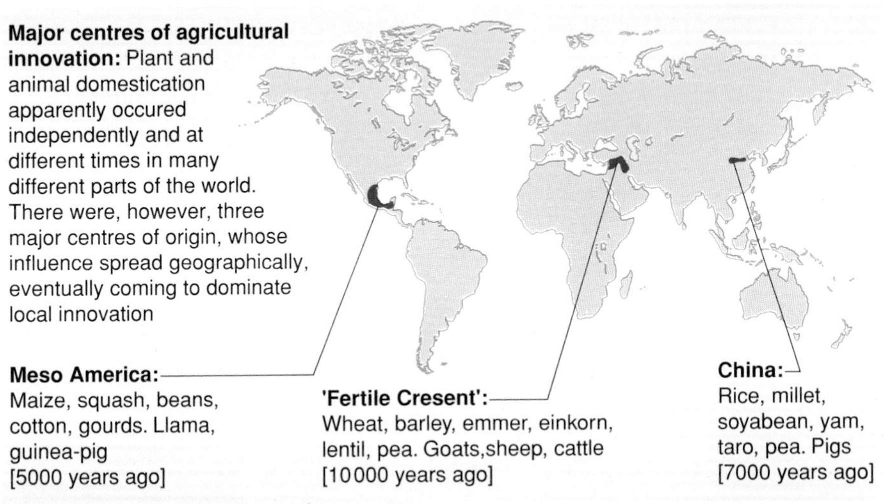

Major centres of agricultural innovation: Plant and animal domestication apparently occured independently and at different times in many different parts of the world. There were, however, three major centres of origin, whose influence spread geographically, eventually coming to dominate local innovation

Meso America:
Maize, squash, beans, cotton, gourds. Llama, guinea-pig
[5000 years ago]

'Fertile Cresent':
Wheat, barley, emmer, einkorn, lentil, pea. Goats,sheep, cattle
[10000 years ago]

China:
Rice, millet, soyabean, yam, taro, pea. Pigs
[7000 years ago]

Figure 9.38 There were at least three major centres for the origin of agriculture during the Neolithic cultural stage. Developments in the 'fertile crescent' about 10 000 years ago are described in the text. Two other important centres were China and Central America. There is evidence that settled agriculture existed in China about 7000 years ago, where the crops included rice, millet, soya beans, tea and yams. A later centre appeared about 5000 years ago, in central America, where the crops were maize, beans, squash and peppers

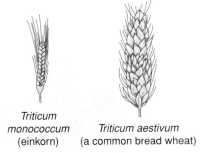

Triticum monococcum
(einkorn)

Triticum aestivum
(a common bread wheat)

*Figure 9.39 Wild wheat (*Triticum monococcum = *einkorn) compared with modern wheat (*T. aestivum)

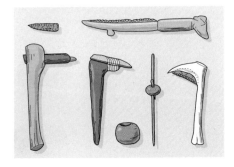

Figure 9.40 Some Neolithic agricultural tools

The practice of farming represents a major change in the relationship between early people and their environment. Nomadic hunter-gatherers take from their surroundings and their mobility allows them to move on and exploit fresh areas. Neolithic people, as they settled down, were beginning to interact in a different way with their environment. They manipulated plants and animals for their own benefit and exerted planned control over small areas in their production of food. They cared for their plants and animals, beginnings of the practice described as **husbandry**; they then harvested their crops and found ways of storing their produce for another season when food may have been less plentiful. They deliberately saved seed to plant for the next season's crop and ensured they kept stock animals for breeding. They are likely to have selected those with the most desirable characteristics; already there was probably a desire for higher yields or different varieties. They were developing the skills of selective breeding, changing the characteristics of the plants and

Figure 9.41 Domesticated animals and some of the ways humans use them: (top) Kutchis on the move in Afghanistan with domesticated camels, donkeys and dogs; (bottom) cattle used for ploughing in a rice paddy, Yunnan, south-west China

animals they used. The practice of settled farming meant that food had to be stored; tools were developed to become implements used in the working of the soil or harvesting and processing of the crop. Moving from place to place would have become complicated because of the accumulation of possessions, so that settled community life and agriculture went together in having a profound effect on the future development of people and society.

Finally, we will attempt to see in perspective the changes that have taken place in the evolution of modern humans from our ape-like ancestors. Molecular studies indicate the difference in genome between humans and chimpanzees to be a mere 2% yet the difference in both capability and achievement appears huge. Remember the timescale of human evolution and think first of the enormous rate of change and development of agriculture. The technology now utilised in modern intensive farming has developed in just 10 000 years since the start of the Neolithic culture (about 400 generations ago). Contrast this with the comparatively slow evolutionary progress from humans using the first Oldowan toolkits through Acheulian to Mousterian and finally the microliths in the late Palaeolithic era which spanned over one million years (40 000 generations). The biological story of evolution continues and diversifies into the history of human societies, of artistic expression and religions, of scientific discovery and increasing control by humans of the environment. It is appropriate to end this review of human evolution with a quotation from Chris Stringer, a research worker in the field of human evolution:

> "How on earth could an animal that struggled for survival like any other creature and whose time was absorbed in a constant searching for meat, nuts and tubers, and who had to maintain constant vigilance against predators, develop the mental hard-wiring needed by a nuclear physicist or astronomer?"

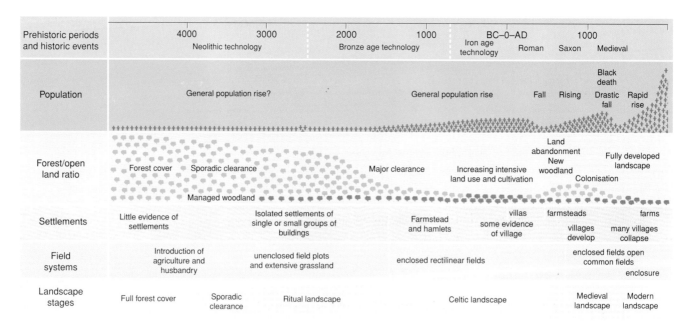

Figure 9.42 From Neolithic agriculture to the present day. A pictorial representation of further development of culture and human activities, and their effects on the moulding of the landscape in England (not to scale)

Detective work on a *Homo* hoax

The Piltdown Man was 'discovered' in Piltdown (Sussex) between 1908 and 1915. It was heralded with much enthusiasm as a find which represented a 'missing link' between ancestral apes and humans. The forgery was finally exposed in 1953, though the originators of the hoax have never finally been established. Certainly, the Piltdown skull misled workers in the field of human evolution for a number of years.

It seems that the skull had been put together quite cleverly, as follows:
- the skullcap was human, though older than the jawbone;
- the jawbone was from an orang-utan (ape);
- the molars were human and modern, and filed down.

It was also shown that the bones, fossils and stone tools found associated with the skull had been stained to match the Piltdown deposit and planted in the gravel pit.

1   What differences in the size and shape of the skull would you expect between modern apes and humans?

2   How have the changes in the skull from ancestral ape-like ancestors allowed further evolution of humans?

3   What differences would you expect in the jawbone and teeth between an ape and a human?

4   What features of the skull would tell you whether this ancestor walked upright? What other evidence would you look for to confirm this?

5   Give an indication of the relative length of the limbs in this imaginary ancestor. Refer to arms and legs in relation to each other and to the likely body size.

Suppose you are trying to establish the actual date and phylogenetic position of this skull.

6   What techniques(s) would you use to establish a date for the material in the skull? [You may need to use different techniques to cover the possible time periods up to the present day.]

7   About what age would you expect if the Piltdown skull had been a genuine ancestor of the genus *Homo*? [Think of the australopithecines.]

8   If the skull had genuinely been earlier than *H. habilis*, what sort of stone tools would you expect to find?

9   How could molecular evidence be used to reveal that the bones were from different species?

Human ecology

The human species - distribution and tolerances

Human beings (*Homo sapiens*) are remarkably versatile in their living or habitat tolerances. Other animal species are often quite precise in their niche or habitat requirements and this is reflected in their distribution on a local and broader geographical scale. To illustrate this, we can look at desert lizards. In a small area there will be certain lizard species associated exclusively with sandy areas, other species will only be found clambering in the spiny bushes and yet others will be seen in and out of the crevices in rocky outcrops. The daily behaviour pattern of these lizards shows that they differ in their tolerance to the heat of the sun with some being totally nocturnal. Their physiology enables them to conserve water allowing survival in situations of extremely low water availability. European lizard species probably would not last long in the deserts of Central Asia. We say that the desert lizards are successfully adapted to their environment. We could look at a wide spectrum of animal species and find similar limits to their tolerance of a range of physical factors, such as light, temperature, altitude, salinity, oxygen concentration and pH.

By contrast, humans exist in a wide range of terrestrial environments, having extended their territories by use of buildings and clothing which protect them from unsuitable or extreme environmental conditions. Historically, regions inhabited by human populations have been determined largely by accessibility to land for food production and by the need for water – for people, for their crops and their animals. Nevertheless, human populations show tolerance to wide temperature ranges, for example, from a hot 50 °C in July in the Sahara in Africa to a cold –65 °C in January in Yakutsk (in the former USSR). Most, however, prefer to live well away from these extremes. There are variations too in daylength in different parts of the globe. Arctic and Antarctic winters are totally dark for several weeks in the year, but have continuous light for a corresponding period in the summer. The upper limit for permanent settlements at high altitude is about 4500 m, though inside aircraft people are now able to cruise regularly at 10 000 m. Space travel has taken people to the moon, introducing space travellers to the experience of weightlessness. Travel in both aircraft and spacecraft is possible only by creating an artificial environment inside. Back on Earth, we spend a considerable part of our lives in artificially modified environments – of houses, shops and offices – and sometimes go a short depth underground in mines, in road and rail tunnels or under cities.

Figure 10.1 Humans live in and tolerate a wide range of physical conditions, from the heat and aridity of the Tunisian desert (top) to the sub-zero temperatures of the North West Territories of Canada (bottom)

Occasional exceptional feats have taken a few individuals outside the normally inhabited areas, but if the adverse conditions are too extreme or the human is subjected to them for too long, there comes a point when the body can cope no longer. We can look at the endurance shown by Ranulph Fiennes and Mike Stroud in the Antarctic during the winter of 1992 to 1993. For 97 days they journeyed on foot, dragging all their own supplies and equipment in their to walk across the Antarctic continent. We can contrast their success with the

Figure 10.2 Ranulph Fiennes and Mike Stroud completing their 2050 km journey on foot across the Antarctic

failure of the Antarctic expedition in 1912, led by Robert Scott. One by one the men failed physically or mentally and though they reached the South Pole, none survived. All were beaten by starvation, cold and frostbite before they could reach safety and the ship home.

The human species has thus colonised a wide range of habitats and shown its ability to tolerate considerable variation in the physical factors of the external environment. Nevertheless, the **internal environment** of the human body is maintained within quite narrow limits. The ability of an organism to control its internal environment is referred to as **homeostasis**. This is the result of internal physiological mechanisms which regulate factors such as body temperature, body water content, ionic composition, blood glucose concentration and oxygen concentration in the blood. The areas populated by the human species and the behavioural patterns which have evolved are such that the body regulatory mechanisms can operate satisfactorily.

This chapter on human ecology looks at humans in their environment and explores some of the ways in which the human body responds to variations in temperature and to the special effects of high altitude. It also looks at patterns of body functions in relation to the daily alternation of light and dark.

Temperature variations

Body temperature and thermoregulation

Human beings are described as **endotherms** and, like other mammals, are able to maintain a constant high body temperature independently of the external environmental temperature. In humans, normal body temperature is about 37 °C. With a clinical thermometer, we usually take the oral temperature (in the mouth) but we can also take a person's rectal temperature. The two values would be slightly different. The oral temperature is less reliable as it may vary according to recently consumed hot or cold foods and drinks, or because of breathing activities. If we were to use a thermocouple to measure temperature with greater precision, we would find that the body temperature is not the same throughout the tissues. The rectal temperature is close to the temperature of the deeper structures in the body, or the body **core**. Thus, when we speak of a constant body temperature, strictly speaking the term refers to the core temperature of the body, that is, the temperature inside the head, thorax and abdomen. Figure 3.14 on page 51 shows how the temperature distribution in a person varies in different parts of the body, with different environmental conditions.

In humans, the temperature of the core fluctuates within a narrow range, normally between 36 and 37.5 °C. There is a clear diurnal or daily rhythm, with temperatures reaching a peak from midday to early afternoon and a trough during the night after midnight. This rhythm gradually becomes reversed if daily activity patterns change, for example in workers on night shifts. In women the temperature is lowest during menstruation and rises noticeably at about the time of ovulation. Exercise may lead to a rise in core temperature with a temperature as high as 41 °C being recorded in an athlete after a marathon race. Core temperature may also rise in response to emotional situations.

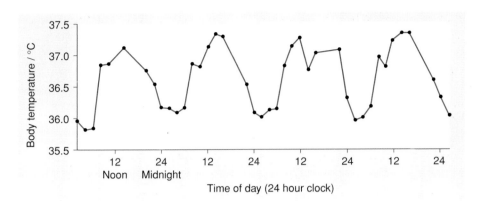

If a steady temperature is to be maintained in the human body, heat loss must equal heat gain. Most of the heat gained is generated internally from metabolic activities (thermogenesis), though there may be some gain from the surroundings. Heat is produced at a fairly constant rate from metabolic activities in various organs in the body, such as the liver and heart. For a person at rest, the heat produced is approximately 4 kJ per kg body mass per hour. This is equivalent to about 170 kJ per square metre per hour or about the same as a 40 watt light bulb. Activity in skeletal muscles increases heat production and short bursts of vigorous exercise may increase the heat produced by more than 10 times the level at rest. A brisk walk (or jog or game of squash) does a lot to warm the body on a cold day. Shivering is a specialised form of uncoordinated muscle activity which produces heat, to about five times the level at rest. Shivering thermogenesis may be initiated when the body becomes cooled and lasts for a few minutes at a time. Heat may be gained from the environment in situations where air temperature is higher than the temperature of the skin. The body also gains heat by direct radiation from the sun, from a fire or artificial heater, and indirectly from reflected radiation. Some heat is gained from the consumption of hot food and drink.

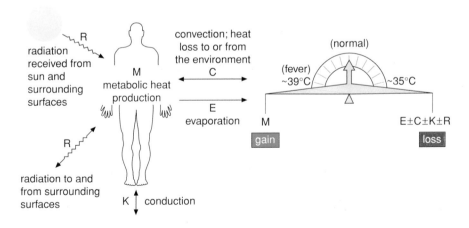

Figure 10.4 The balance between heat gain and heat loss in the maintenance of a stable body (core) temperature. Heat gain may be the result of internal generation (thermogenesis) from metabolism, including exercise and shivering, or external gain from the environment, hot food and drink.
Heat loss involves the skin, through conduction, convection, radiation and evaporation through sweating.

When the body is hotter than the surrounding environment, heat may be lost by direct radiation from the body, by conduction through areas of the body touching cooler objects or by convection to the surrounding air. The skin covers the surface of the body and plays a very important part in regulating the loss of heat from the body and thus in maintaining the required balance between heat gain and heat loss.

To understand the role of the skin in thermoregulation, it is important to look at its general structure, shown in Figure 10.5. Heat loss from the skin occurs by conduction, radiation and evaporation of sweat. Heat loss by conduction and radiation can be varied by altering the blood flow to superficial capillaries. The diameter of arterioles in the skin is controlled by sympathetic nerves originating in the hypothalamus. External cold results in a decrease in the diameter of these vessels (vasoconstriction) and the blood flow to capillaries is reduced. This reduces the loss of heat. Conversely, in warm conditions, the diameter of skin arterioles increases (vasodilatation), resulting in increased blood flow in the peripheral circulation leading to a considerable increase in heat loss.

Figure 10.5 The structure of human skin showing structures involved in thermoregulation.

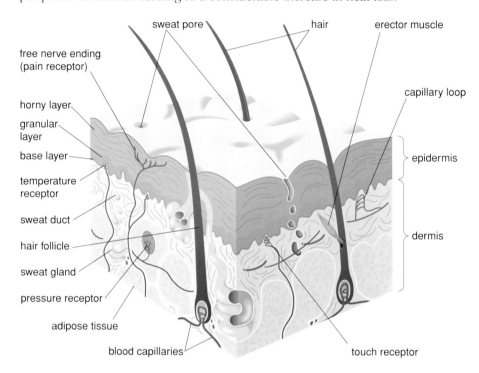

In hot situations, an important means of heat loss is through evaporation of moisture which may arise from the sweat glands or from the inside surface of the lungs and mouth. When 1 g of water vapourises it requires 2.42 kJ of heat and this latent heat is taken from the immediate surroundings, resulting in cooling. Thermal sweating occurs when the external temperature or body temperature rises. Sweat glands secrete a dilute solution containing sodium chloride, urea and lactic acid. Heavy sweating involves rapid loss of water and salt from the body. Dehydration and salt deprivation will occur unless enough water is drunk and adequate amounts of salt are taken to replace the losses.

In cold conditions, smooth muscles attached to hair follicles in the skin contract, raising the hairs. This is important in many mammals as it helps to trap a layer of air in their fur. Air is a poor conductor of heat so this helps to insulate the body against excessive heat loss. The insulating effect of hair is minimal in humans, except on the head, though lack of hair in bald adults and babies can lead to significant heat losses. We should all take heed of the advice to wear a hat in cold weather! The layer of adipose tissue also makes a contribution as a means of insulation as fat is a poor conductor of heat.

In most situations people wear clothes. This clothing has a protective effect which alters the exchange of heat between the body and the immediate environment. Clothes have an insulating effect when dry, so to some extent clothes modify the relationship of the skin with the immediate environment, by creating a 'microclimate' close to the body. We spend a considerable proportion of our time indoors and the design of buildings also serves to protect us from the natural environmental temperature. Clothing and buildings are behavioural responses which help maintain the required balance between heat loss and heat gain.

The **thermoneutral zone** (TNZ) refers to the range of external (environmental) temperatures over which heat production by metabolism is at a minimum. The thermoneutral zone has been defined as 27 °C to 31 °C for a naked 70 kg man. The lowest temperature (27 °C) is known as the **critical temperature**. The thermoneutral zone is a useful concept in the understanding of the operation of body temperature control mechanisms, but in reality people usually wear clothing so the critical environmental temperature would be much lower than the defined temperature of 27 °C. Figure 10.6 indicates the relative importance of the different thermoregulatory processes. Within the TNZ, any changes in body temperature are compensated by adjustment in blood flow in the peripheral vessels, by vasoconstriction or vasodilatation. When the surrounding environmental temperature falls below the critical temperature, non-evaporative heat loss increases, so if the core temperature is to be maintained, the balance is restored by increased production of metabolic heat. At environmental temperatures higher than the TNZ, you can see that evaporative heat loss by sweating is the most important means of thermoregulation.

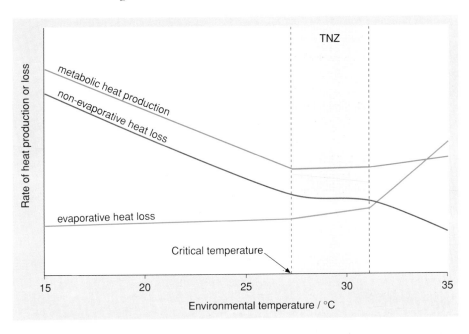

Figure 10.6 The thermoneutral zone (TNZ) refers to the range of environmental temperatures at which heat production by metabolism in the body is at a minimum. The graph gives values for a naked 70 kg man

The next sections look at how humans respond to extremes in environmental temperatures in both hot and cold environments. Physiological and behavioural responses are seen, but when conditions are such that thermoregulatory processes are inadequate or fail, the body suffers from stress.

Extremes of temperature

Response to high temperature – acclimatisation and heat stress

If you live in a cool or temperate climate and pay a visit to a much hotter place, in the tropics or hot desert, at first you feel very uncomfortable. For a few days you are likely to lack energy even for walking around, perhaps you find you cannot think very clearly and probably have little appetite. Then you begin to adjust, or **acclimatise**, and gradually find you can carry on your daily activities much as you did in the cooler place. You would, however, be wise to start your activities early in the morning when it is cooler, and be less active around midday and early afternoon when temperatures are at their highest. It would be sensible also to wear loose and generally light-coloured clothing, but not necessarily to abandon all and certainly not to sunbathe in the direct heat of the sun.

In the hot climate, reduction of physical activity and reduced food intake mean that less heat is generated internally as a result of muscular and metabolic activity. A preference may develop for foods with high water content, such as fruit and salads, and this may be related to the thirst mechanisms attempting to prevent potential dehydration in the heat. In response to the high temperatures, blood vessels in the skin dilate, bringing more blood near the surface. But because the environmental temperature may be close to or higher than normal body temperature, there is likely to be little or no heat loss to the surroundings as a result of conduction, convection or radiation. Evaporation of sweat thus becomes the most important means of losing heat and so cooling the body.

We will consider the effects of sweating in two different extremes of hot climate: hot and humid (as in tropical regions or in factories, mines and other enclosed environments) and hot and dry (as in desert conditions). When humidity is high, less sweat evaporates and sometimes sweating may cease altogether. Inability to lose heat through evaporation of sweat in these situations is serious and quickly leads to rise of the body core temperature. At the other extreme, if humidity is low, sweating is an effective way of losing heat. In temperatures as high as 40 °C a person may feel perfectly cool and comfortable, particularly if the clothing is loose, thus allowing sweat to evaporate from the body surface. The danger comes after prolonged exposure to high temperatures if the components of sweat are not replaced. Excessive loss of water through sweating leads to dehydration of body tissues. Loss of ions in the sweat, particularly of sodium (Na^+) and chloride (Cl^-), depletes these from body fluids and is likely to lead to painful muscular cramps.

Alteration of sweat production appears to be the main response to hot conditions as the body acclimatises. This effect is shown in Figure 10.7. Data for this study were obtained with people doing physical work in an artificial environment, subjected to hot temperatures for different lengths of time. Equivalent results are obtained when the body temperature is raised without the person indulging in physical work. It seems that similar increases in the rate of sweating occur in people of different ethnic groups, though females usually show lower rates than males. This may be because females tend to undertake less vigorous physical activity than men, though it is of interest that some female athletes do show rates of sweating which are comparable with those of men. People who normally live in hot climates appear to have an initially high rate of sweating compared with those from cooler climates, but show a similar response when exposed to higher than normal temperatures. Another effect of acclimatisation is that the sweat produced has a lower salt concentration, thus reducing the harmful loss of salt from body fluids. People living in hot climates may find it advisable (and desirable) to take extra salt with their food or drinks.

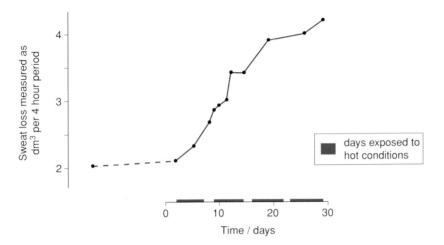

Figure 10.7 Acclimatisation and sweating – a group of people exposed to hot conditions for 5 or 6 days at a time over a period of 4 weeks showed a noticeable increase in sweat loss. The dashed line on the graph shows the average sweat loss of the people in the group for approximately 4 weeks before the experiment

Traditional clothing and housing which have evolved with people living in hot climates are usually highly effective in helping the people to tolerate heat. Clothing which covers the body minimises direct exposure of the skin to the radiation of the sun; loose clothing allows for evaporation of sweat and also keeps a still layer of air next to the skin which insulates against excessive absorption of heat; light colours reflect the heat. Clusters of compact, thick-walled mud houses with only narrow slits for windows, characteristic of villages in deserts, remain remarkably cool inside. Modern buildings use air conditioning, with controlled temperature and ventilation, in an attempt to provide comfortable working and living conditions, particularly for visitors to a hot climate.

Figure 10.8 Traditional clothing of desert people. Bedouin in Sahara Desert

Figure 10.9 Typical desert village in Afghanistan with thick-walled mud houses

When the body can no longer cope with excessively high temperatures, problems of **heat stress** arise. The skin may suffer from sunburn when exposed directly to excessive radiation from the sun. Redness, blistering and peeling of the epithelium occurs. Increased production of the pigment melanin, associated with a tan, gradually gives some protection from the sun's radiation. An irritation known as prickly heat may develop due to blocking or narrowing of the ducts from the sweat glands with the result that sweat cannot reach the skin surface.

In some situations a person may feel dizzy or suffer from **heat collapse**, probably because blood is diverted away from the internal organs to the skin and skeletal muscles. Recovery usually occurs quickly if the person is taken to a cool environment or lies down. **Heat exhaustion** is a more serious condition which may be due to dehydration because fluid lost in sweating has not been replaced. With a loss of 5 to 8% of the body fluid, fatigue sets in, and by a 10% reduction in body fluid, mental and physical deterioration occur. With further decrease in body fluid, intracellular fluid is withdrawn and this starts to cause damage to cells. Heat exhaustion may also result when losses of sodium chloride are not replaced.

The most severe condition is known as **heat stroke**, characterised by a breakdown of the thermoregulatory control mechanisms. This appears to be associated mainly with a failure or inadequacy of the sweating mechanism. At core temperatures of 42 °C or higher, irreversible damage to cells and proteins occurs and the person is likely to go into a coma. Unless immediate steps are taken to lower the body temperature, death may follow. It should be appreciated that the upper limit of tolerance to temperature is only a few degrees above normal core temperature.

Response to cold – acclimatisation and cold stress

The body's response to extreme cold is not simply a reverse of its response to heat. To begin with, 'coldness' may be due to very low environmental temperatures but people can also suffer from cold in more moderate temperatures, particularly in windy conditions. Moving air blows away the warm layer of air immediately surrounding the body and increases the loss of heat by convection. This **wind chill** factor is made worse if the body surface and any clothing being worn are wet, because further cooling occurs due to evaporation of water. Another difference when considering response to cold rather than heat is that the temperature gradient between the body surface and surrounding temperature is likely to be far greater in cold conditions. In many parts of the world, the surrounding temperature is well below the core temperature of 37 °C; indeed temperatures in the region of –40 °C may be experienced in some places normally inhabited by humans. The body can tolerate and recover from a lowering of core temperature by as much as 10 °C, whereas a rise in temperature of only 4 °C may prove lethal. If the body temperature is to be maintained close to 37 °C, the focus is on generation of internal heat and on protection and insulation of the body to **conserve** heat.

Estimated windspeed / mph	Actual thermometer reading / °C											
	10	5	−1	−7	−12	−18	−24	−29	−35	−40	−46	−51
↓	*Equivalent temperature / °C*											
Calm	10	5	−1	−7	−12	−18	−24	−29	−35	−40	−46	−51
5	9	3	−3	−9	−15	−21	−26	−32	−38	−44	−50	−56
10	5	−2	−9	−16	−23	−31	−36	−44	−50	−57	−64	−71
15	2	−6	−13	−21	−28	−36	−43	−50	−58	−66	−73	−81
20	0	−8	−16	−24	−32	−40	−48	−55	−64	−72	−80	−87
25	−1	−9	−18	−26	−34	−43	−51	−59	−67	−76	−84	−92
30	−2	−11	−19	−28	−36	−45	−53	−62	−71	−79	−88	−96
35	−3	−12	−20	−29	−38	−47	−55	−64	−73	−81	−90	−99
40	−4	−13	−21	−30	−39	−48	−57	−66	−74	−83	−92	−101

Windspeeds greater than 40 mph have little added effect

↑ Little danger for properly clothed person, maximum danger of false sense of security

↑ Increasing danger from freezing of exposed flesh

↑ Great danger

Figure 10.10 The wind chill effect – the table shows how increasing wind speed lowers the effective temperature on exposed skin. Look for the wind speed in the left hand column and the actual temperature across the top. From this you can find the equivalent temperature on exposed skin. Wind speeds are given in miles per hour (mph) in line with familiar usage

Increased metabolic rate in response to cold conditions leads to increased heat production. This can be stimulated by the hormones adrenaline and thyroxine. A person is likely to have a better appetite for food in cold conditions, thus increasing food intake. Voluntary muscular movements during physical activity and involuntary shivering also contribute to the generation of heat, though shivering is likely to be sustained only for short periods.

Diversion of blood away from the body surface by **vasoconstriction** is an important means of conserving heat. Reduced blood flow to the skin means that less heat is lost by conduction and convection from the body surface. Vasoconstriction for a prolonged period would restrict blood flow to the extremities of the limbs, i.e. the fingers and hands, toes and feet. This could cause damage to the tissues and is counteracted by occasional dilation of the blood vessels, in what is described as the 'hunting reaction'. Without this reaction, fingers tend to become numb or paralysed and loss of manual dexterity would lead to serious difficulties. The tendency to curl up the fingers or indeed the whole body with arms folded when it is cold reduces the surface area exposed, and illustrates another way of reducing heat loss to the surroundings.

Fat present as a subcutaneous layer provides **insulation**. This cannot be turned on and off at short notice, but development of a fat layer in certain people may be an advantage, for example, to long distance swimmers immersed in water for long periods. Young babies, up to about 6 months old, may benefit from 'brown fat', which generates additional heat by its metabolism rather than insulation effects. It is unlikely that brown fat plays a major part in the heat balance of adults.

Probably the most important way of protecting people from the cold is by conserving heat through **insulation** of the body, hence the importance of suitable clothing and buildings. In extreme temperatures and where wind chill might be significant, it is necessary for clothing to be windproof. Even though

Figure 10.11 Clothing in an environment of extreme cold – an Inuit in the Arctic region of Canada

windproofing is important it can lead to discomfort due to condensation of moisture produced from the body. In extreme cold, this moisture may freeze inside the garment. Air is a good insulator, so the most effective clothing includes a number of layers which trap air, provided the air is dry. The design of clothing needs to allow for adequate physical activity and in this respect very heavy clothing may become cumbersome. In temperatures below freezing, the face and limb extremities become vulnerable if inadequately protected. Inside buildings additional heat is supplied artificially from fires, electrically or by other systems of central heating. An important economic consideration is the adequate insulation of the building as a means of saving energy and this is given high priority in the design of modern buildings. Traditional housing in cold climates usually has thick walls and small windows, thus minimising loss of heat which is likely to be supplied from an open fire. The hard-packed snow used to construct Inuit (Eskimo) igloos has excellent insulating properties.

People living permanently in cold climates show certain physiological adaptations, though their main means of protection is through clothing and housing. One study showed that Australian Aborigines were able to sleep well, even though naked, in temperatures around freezing, whereas unacclimatised white people shivered violently and were unable to sleep properly in similar conditions. Inuit people have a higher blood flow through the hands and feet than do visitors in the same conditions, and the traditionally high protein diet of the Inuit may contribute to their high metabolic rate. Acclimatisation to cold by people who travel from warmer to colder climates does occur, but compared with acclimatisation to heat, it is less easy to define the mechanisms. It is likely that increased feeding, leading to increased metabolic rate, is an important response though the most important consideration is to ensure adequate protection by means of suitable clothing and housing.

When the body can no longer cope with the effects of cold, problems of **cold stress** may result. **Cold injury** is characterised by actual damage to tissues, most often the hands, feet and face, because of their direct exposure to low temperatures. Severely reduced blood circulation deprives the tissues of nutrients and their normal metabolic reactions. In mild form, chilblains may develop as the parts become tender and itchy. In more severe conditions, tissues actually freeze and this is known as **frostbite**. If only superficial, the tissues are likely to recover and injured skin is replaced by new growth. If prolonged or deeply frozen, underlying tissues including muscle and bone may suffer permanent damage. This is caused by the mechanical action of ice crystals on the cell structure and also by the dehydrating effects of removing liquid water from body fluids. Gangrene may set in, resulting in loss of toes or fingers. The condition known as **trench foot** is usually a result of prolonged cooling in cold water. Trench foot is characterised by blackening of the skin of the toes and foot. The main damage is to the muscles and nerves and can lead to gangrene if the affected limbs are not warmed up quickly.

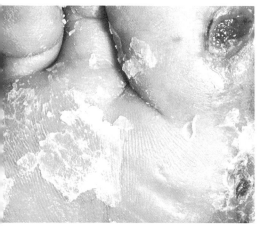

Figure 10.12 Frostbitten toes, post treatment. Frostbite refers to tissue damage due to freezing, from the destructive effects of the formation of extracellular ice crystals. Frostbitten parts need to be gently warmed in tepid water. Precautions are required against bacterial infection, to which frostbitten skin is susceptible

Hypothermia develops when the core temperature falls to 35 °C or below. Down to 35 °C shivering increases, but already the muscles are likely to be at a lower temperature. Below 35 °C there are signs of muscle weakness and the person shows difficulty in walking and coordinating movements. By 34 °C the person becomes mentally confused and vision is disturbed. Loss of consciousness occurs between 32 °C and 30 °C and death usually follows at between 28 °C and 25 °C if the person is not warmed up. Probably the most important effect of the lowering of core temperature is the reduction of the heart rate. Cooler blood flowing through the heart affects the **pacemaker** (sinoatrial node), which initiates the heart beat. Reduced blood output from the heart means that the coronary circulation and flow to muscles and the brain may be inadequate, hence the symptoms associated with hypothermia. People suffering from hypothermia, may also show a slower respiration rate and increased production of urine. The latter is linked to suppressed release of antidiuretic hormone (ADH). (The effects of the pacemaker and of hormonal control are described in more detail in *Systems and their Maintenance*.)

Recovery from hypothermia can be complete provided the body is warmed quickly. A number of deaths from hypothermia or **exposure** do occur each year among walkers and mountain climbers, even on the hills of Britain, usually through a combination of physical fatigue and inadequate protective clothing. Immersion in cold water can quickly lead to hypothermia and accidental deaths in water are often due to hypothermia rather than drowning. Compared with air, water has a higher thermal conductivity so the body cools faster when surrounded by water. Survival time of a naked unprotected person is about 90 minutes in water at 15 °C, but only 30 minutes at 5 °C. Attempts to swim to safety after an accident in cold water are probably misjudged, since the movement disturbs any remaining layer of warm air adjacent to the skin and the activity uses up valuable metabolic reserves.

Young babies and old people are particularly vulnerable to hypothermia. Up to about 1 year old, babies are unable to shiver and also their behavioural responses are limited. Old people generally show less physical activity and have a lower metabolic rate thus reducing internal heat generation. Their temperature control mechanisms, originating in the hypothalamus, are usually less effective with respect to the responses such as shivering, vasoconstriction and increase in oxygen consumption. In some cases the hypothalamus sets the temperature at too low a level. For those living on low incomes the situation may be made worse by inadequate heating and clothing.

In some surgical operations, involving the heart or brain for example, hypothermia may be deliberately induced by cooling the blood or body surface. This is used as a short-term means of reducing blood flow and use of oxygen in the tissues and allows the operation to take place. In these situations, body temperature can be lowered temporarily to about 25 °C.

A question of clothing

A manufacturer of specialist outdoor clothing advertises its garments by promoting the benefit of several layers of clothing. In their garments, they emphasise the following features:
- the 'wicking' properties of the inner layer (next to the skin);
- the choice of several different layers depending on external conditions;

- the importance of a windproof and/or waterproof outer layer;
- the overall light weight of their garments.

Imagine you are walking for several hours in the mountains, in wet, windy conditions with temperatures near to freezing, and you are wearing this type of clothing.

1 What do you think the manufacturer means by 'wicking' and why is this beneficial?

2 How do several layers keep you warm?

3 Why is it important that the outer layer is windproof?

4 Explain what features should be incorporated into the design of the waterproof material to make sure it is effective in these conditions.

5 Explain why it is important to pay attention also to the design of hats, gloves and boots worn in these conditions.

6 Why are lightweight garments preferable to heavy clothing?

7 How far do you think the manufacturer's claims about their garments are biologically sound?

8 Why don't you just stay at home? What benefits do you get from your walk? Is it really worth the effort?

High altitude

Mount Everest, or *Chomolungma* ('Goddess Mother of the Earth') in Tibetan, was first climbed in 1953. The climbers were part of a large-scale expedition, with many people involved in the team. The whole expedition required considerable planning and meticulous attention to detail with regard to routes to be taken, specialised equipment required, food stores and oxygen supplies. When Hillary and Tenzing stood at 8848 m, on the summit of the highest peak in the world, their success showed that humans can overcome the extremes of physical conditions associated with the highest mountains, in terms of exposure to cold, low levels of oxygen and dangerously hazardous terrain.

Since then, very few people have climbed Mt Everest solo and without the help of extra oxygen: two who have achieved this are Reinhold Messner in 1980 and Alison Hargreaves in 1995. An increasing number of other climbers have now reached the summits of Everest and of the remaining peaks over 8000 m. Many, however, have endured the hardships without the success. Some have lost their lives in the attempt (including Alison Hargreaves a few weeks after her success on Everest); others have become victims of frostbite and sacrificed toes or fingers. Nobody can stay for long at those heights: realistically a few hours at the most. The urgent need is to summon both the physical energy and mental concentration to make a safe descent to lower altitudes.

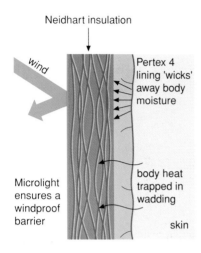

Figure 10.13 Clothing (from Rohan) designed to provide "one stop insulation" – an outer layer of 'microlight', a middle layer of 'Neidhart insulation' (thin, stable, featherweight) and an inner layer of 'pertex'. Successful insulation is achieved by trapping still air next to the body

Figure 10.14 Mountaineers face difficult terrain and are exposed to harsh physical conditions

People in the mountains

Native high-altitude people are seen as rugged, robust and sturdy, able to live successfully and in equilibrium with the particular conditions characteristic of high-altitude. These qualities may be due partly to the outdoor lifestyle associated with the mountain environment, but also indicate that the natives are well-adapted to the rigours of life at high altitude. For convenience, we will define 'high altitude' as being above 3000 m. Using this as an arbitrary limit, we find the main native high altitude populations in the Himalayan region of Asia (notably Tibetans and Nepalese) and in the Peruvian Andes of South America (Quechua Indians). The upper limit for permanent settlements is about 4500 m in both regions though to some extent this is a reflection of historical events and of accessibility and not determined solely by the altitude itself.

Figure 10.15 Natives in high mountain regions: (left) Tibetan woman, Western China (3200 m); (centre) shepherd in Himalayan region of north-west India (4000 m); (right) Quechua Indian woman and child, Peru (4000 m)

Visitors to these high-altitudes from the lowlands may suffer varying degrees of discomfort and illness. With increasing numbers of tourists visiting high mountain regions for short periods, these effects are becoming more familiar. If, on a quick trip up Mount Kenya, you spend a night below the peaks in a hut at about 4500 m, you are likely to have a persistent headache, probably slight nausea and an extreme disinclination to continue with your journey the next day. Trekkers in the Himalayan mountains, from altitudes of about 3000 m and higher, find that their rate of moving becomes slower. They are likely to stop frequently and also experience difficulty in maintaining steady breathing, or may become very conscious of it. Similar difficulties have been encountered by military personnel unaccustomed to life at high altitude, for example, when controlling border disputes in the Himalayan region. The problems are worst during the first few days at high altitude and can be much more severe if the ascent is rapid, perhaps arriving at high altitude in a motor vehicle or even by plane. Bus passengers in the Andes are sometimes supplied with oxygen when their journey takes them over high passes. Certainly, the best way to get into the mountains is a slow plod on foot. The effects usually disappear rapidly on descent to lower altitudes.

These symptoms described for visitors to high altitudes can be attributed primarily to the lower levels of oxygen, though some effects are a result of unfamiliar physical demands of mountain terrain and, at extreme heights, of the cold. The symptoms, collectively known as **acute mountain sickness**, include headaches (mild to severe); lack of concentration and giddiness; coughing and

Figure 10.16(a) Visitors on Mount Kenya at about 4000 m, already struggling and losing interest with still another 1000 m of height to climb to reach Point Lenana, one of the three summits

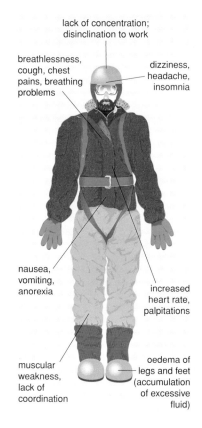

(b) Mountain sickness and its symptoms

Figure 10.17 Changes in air pressure with altitude

difficulty with breathing; palpitation and very rapid beating of the heart; loss of appetite, feelings of nausea and vomiting; muscular weakness, exhaustion and poor coordination; reduced production of urine, swelling particularly of the legs but also development of excessive fluid in the lungs; a general disinclination for exertion or coherent thinking and frequent difficulty in sleeping. At extreme heights or in severe cases, loss of consciousness or death may occur, unless the sufferer is brought rapidly to a lower level. Development of symptoms of mountain sickness depends on the rate at which the ascent is made but the severity is unpredictable, and fit young males are often more prone to suffering than middle aged people or females.

The environmental conditions at high altitudes

The physical conditions in high mountains which lead to the symptoms of mountain sickness are dominated by low levels of oxygen, known as **hypoxia**. In addition, the body has to withstand low temperature, high winds, low humidity and increased solar radiation. These physical conditions impose considerable stress on the human body.

Oxygen and hypoxia

The percentage of oxygen in the atmosphere remains constant, at about 21%, from sea-level to the peaks of the highest mountains and up to at least 110 000 m. However, air is a compressible gas, which means that at sea level, there are more molecules per unit volume than there are at high altitude. Thus, at high altitude, there are fewer molecules per unit volume so the atmospheric pressure is less. The part of the pressure due to the oxygen molecules is called the partial pressure, or pO_2. It is the partial pressure of oxygen that effectively determines the amount of oxygen available at the lung surface for uptake and transport into the body tissues.

Figure 2.18 shows that at sea-level, atmospheric pressure is approximately 100 kPa so the partial pressure of oxygen is 21% of this, or 21 kPa. At 3500 m, the atmospheric pressure is 65.5 kPa, giving a partial pressure of oxygen of 13.8 kPa. At 8500 m, approaching the summit of Everest, atmospheric pressure is reduced to 31.9 kPa and the partial pressure becomes 6.7 kPa. This is only 31% of its value at sea level.

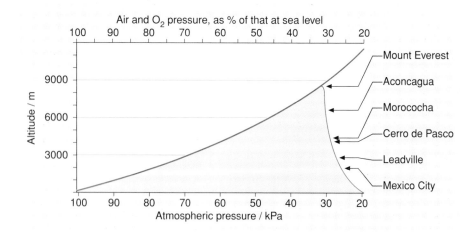

Cold temperatures, low humidity, high winds and solar radiation

Air **temperature** falls at approximately 1 °C for each 150 m of ascent, though this varies in different mountain ranges around the world. However, a rough estimate indicates that, compared with sea level in a nearby area or equivalent latitude, the temperature at 3500 m is likely to be 23 °C lower, and at 8500 m 56 °C lower. This emphasises the necessity, at very high altitudes, to conserve body heat by providing good insulation in clothing worn and in houses, or portable tents in the case of mountaineers.

Humidity of the air is often low at high altitudes. This leads to an increased loss of heat from the body as a consequence of evaporation of sweat. Mountaineers frequently suffer from dryness and cracking of the lips or other exposed parts of the skin. There is also a tendency to develop a persistent cough, because breathing is usually through the mouth and the air is cold and dry. Loss of water from the body by evaporation brings a danger of dehydration. Plants found growing at high altitudes tend to show xeromorphic (water-conserving) features, giving further evidence of the low humidity in these regions. In addition, strong winds are frequently associated with mountain regions, which adds to the potential heat loss from the body due to the wind-chill factor.

Solar radiation increases at high-altitudes because there are fewer molecules (of oxygen, nitrogen and ozone) per unit volume of the atmosphere. These molecules effectively absorb and scatter radiation at various wavelengths. Ultraviolet radiation at 3000 m is about double that at sea level because of the reduced ozone at high altitude. Excessive ultraviolet radiation may damage the cornea of the eye, a condition known as **snow blindness**, and mountaineers frequently use dark goggles as a means of protecting their eyes. Exposure to ultraviolet radiation may lead to a higher incidence of skin cancer; reflection from snow intensifies the effects of radiation on exposed skin. During the daytime the high solar radiation can provide additional heat for the body, though temperatures fall rapidly at night. A climber on Mt Everest at 8530 m was able to remove his down-filled clothing without suffering from cold when in the full sun and at low wind velocity.

Physiological responses of the body to hypoxia at high-altitude
Obtaining oxygen for respiration

Oxygen is used in cells in the process of respiration, which releases energy for the many metabolic and physical activities of the body. Any reduction in available oxygen leads to impairment of a range of body functions, and, in extreme cases, to death. To reach the mitochondria in the cells where the respiratory reactions take place, the oxygen molecules must pass through a series of barriers. There is effectively an oxygen **cascade**, down a gradient of pressure from the atmosphere to the inside of the cells. The stages for transfer of oxygen are summarised in Figure 10.18.

The partial pressure of oxygen at each stage is a measure of its availability. For a healthy person at sea level, the difference between the partial pressure (pO_2) of inspired air and the final pO_2 in the blood capillaries surrounding the body cells is sufficiently great for there to be no difficulty for the oxygen to reach the mitochondria. However, at higher altitudes, because of the lower pO_2 of oxygen in the atmosphere, this gradient is reduced considerably. The body responds in several ways to overcome or compensate for the effects. Native high-altitude people have already adjusted to the lower pO_2 of oxygen, whereas visitors beome **acclimatised** and usually adjust after a number of days.

Figure 10.18 The cascade of oxygen from alveoli to body cells – summary of respiratory processes

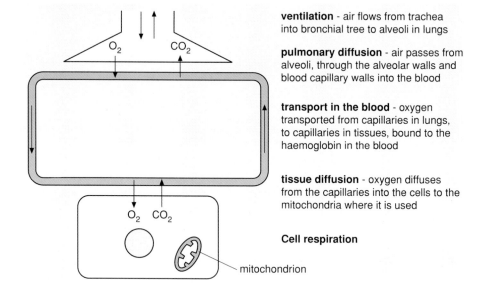

ventilation - air flows from trachea into bronchial tree to alveoli in lungs

pulmonary diffusion - air passes from alveoli, through the alveolar walls and blood capillary walls into the blood

transport in the blood - oxygen transported from capillaries in lungs, to capillaries in tissues, bound to the haemoglobin in the blood

tissue diffusion - oxygen diffuses from the capillaries into the cells to the mitochondria where it is used

Cell respiration

mitochondrion

The gradient in pO_2 at these different stages is shown in Figure 10.19. It compares the fall in pO_2 for a person living at sea-level, with that of a native living at 4540 m and a 'visiting' climber at 6000 m. When air enters the bronchial tree inside the lungs, it becomes saturated with water vapour which itself exerts a partial pressure. Expired carbon dioxide in the lungs also exerts a partial pressure and these two gases account for the immediate steep fall in pO_2 inside the alveoli.

Figure 10.19 Oxygen cascade from inspired air to venous blood. Changes in the partial pressure of oxygen (pO_2) for three groups of people: natives living at sea-level, natives living at 4540 m and visitors (climbers) at 6700 m. The small gradient at high-altitude makes it more difficult for oxygen to be taken up by the cells

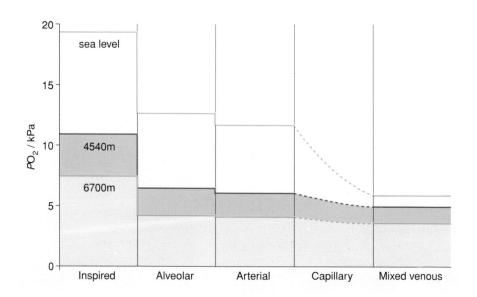

Haemoglobin in the red blood cells combines with oxygen to form oxyhaemoglobin. The reaction between haemoglobin and oxygen is summarised in the equation

$$Hb \quad + 4O_2 \; \rightleftharpoons \quad HbO_8$$
haemoglobin $\qquad\qquad$ oxyhaemoglobin

The term **percentage saturation** is used to give a measure of the amount of oxyhaemoglobin in relation to (deoxy)haemoglobin in the blood. In practice, saturation of haemoglobin with oxygen is rarely higher than about 95%. The percentage saturation varies with the partial pressure of oxygen but is not directly proportional to it. As a graph, we might expect a straight line, but the typical sigmoid shape of such curves (known as dissociation curves) is because of the way the affinity of haemoglobin for oxygen alters after the binding of the first oxygen molecule.

From the dissociation curve shown in Figure 10.20, we can see that at high partial pressures of oxygen the haemoglobin takes up oxygen, but at low partial pressures, oxygen will be unloaded or released from the haemoglobin. Thus oxygen is collected by haemoglobin in the blood flowing through the capillaries in the lungs at the alveolar surface where partial pressure are high. Oxygen is then off-loaded at low partial pressure in the tissues at the surface of the cell; there it passes to the mitochondria where it is used in respiratory reactions. The steepness of the curve at low partial pressure's indicates that a very small drop in partial pressure's allows large amounts of oxygen to be released. The critical partial pressure of oxygen at the mitochondria for oxidative reactions to take place is less than 0.4 kPa. If we look at the flat part of the curve, we can also see that the partial pressure of oxygen in the alveoli can drop from 13.3 kPa to 8 kPa without making any appreciable difference to the degree of saturation of the haemoglobin. Thus at moderate altitude, people do not suffer from a reduced uptake of oxygen by haemoglobin.

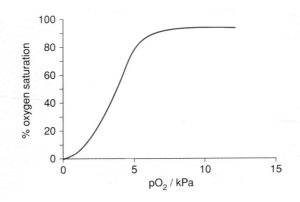

Figure 10.20 Oxygen dissociation curve of haemoglobin showing the percentage saturation with increasing partial pressure of oxygen, at 37 °C and pH 7.4

Overcoming shortage of oxygen
Hyperventilation

Visitors to high altitudes notice an immediate increase in their rate of breathing, even at rest. This is known as **hyperventilation**. The response may be observed from about 3000 m upwards and is also found in native high-altitude people. Quechua Indians in the Peruvian Andes hyperventilate at a rate about 30% higher than people at sea level, though visitors maintain a rate higher than that of the native highlanders. As well as more rapid breathing, the breaths may be deeper. This hyperventilation increases the ventilation inside the alveoli and increases the pO_2 of oxygen at the boundary between the alveoli and the blood.

Another consequence of hyperventilation is increased exhalation of carbon dioxide, which results in a degree of **alkalaemia**, or abnormally high pH in the blood. (Carbon dioxide dissolves in the blood to form carbonic acid, which increases the acidity, or lowers the pH.) The acid–alkali balance in the blood must be restored to avoid further physiological effects. A complication arises because the level of carbon dioxide in the arterial blood is one factor which controls the rate of breathing (hence ventilation). As the carbon dioxide level in the blood falls, ventilation rate decreases until a high carbon dioxide level builds up and ventilation increases again. This may lead to irregularity in breathing. The low carbon dioxide in the blood (or high blood pH) triggers the symptoms of nausea and dizziness.

Increased pulmonary diffusing capacity

The pulmonary diffusing capacity is a measure of the rate of exchange of gases between the alveoli and the (pulmonary) capillaries surrounding them. The capacity can be improved by an increase in surface area of the lung and by an increase in blood flow in the surrounding capillaries. There is some evidence that native highlanders have a lung volume larger than that of comparable lowlanders. This enlargement gives an increased alveolar surface area, thus allowing a higher rate of diffusion of oxygen from the alveoli into the surrounding blood capillaries. Native highlanders also show a relatively higher total volume of blood, which results in a greater flow through the pulmonary capillaries. Visitors to high altitudes do not seem to develop a similar increase in lung capacity: if anything at first there appears to be a decrease in vital capacity (maximum volume that can be exhaled) though this change is reversed after a few weeks at high altitude.

Increased transport of oxygen in the blood

The total quantity of oxygen transported in the blood to the cells depends on three factors: the **cardiac output**, the **haemoglobin concentration** and the **saturation of the haemoglobin** with oxygen.

Cardiac output may be altered both by changes in the **heart rate** (number of heart beats in a given time) and by changes in the **stroke volume** (volume of blood pumped out at each beat). Increases in one or both of these attributes will result in more blood being pumped through the pulmonary capillaries, allowing an increase in the collection of oxygen from the alveoli, and increased delivery to the cells. Visitors to high altitudes show an immediate increase in heart rate, though stroke volume remains steady then appears to fall. Overall cardiac output thus increases for the first few days then returns to about the same as that at sea level. In native highlanders the resting cardiac output is about the same as that for comparable lowlanders.

Native highlanders, compared with equivalent lowlanders, show higher levels of both the haemoglobin concentration in the blood and of the total number of red blood cells. A similar increase is noted when lowlanders visit high altitudes. The response is due to increased production of red blood cells, which takes place in the bone marrow stimulated by the hormone erythropoietin. Increased secretion of erythropoietin occurs in conditions of low oxygen. There is also an overall increase in blood volume. At 4500 m, these values are about 25 to 30% above the corresponding values at sea level. In visitors, the higher red blood cell count is noticed about 3 to 4 days after arrival at high altitude, though the increase in haemoglobin and blood volume continues for several weeks. With more haemoglobin in the blood, the capacity for carrying oxygen is increased, provided the pO_2 of oxygen is high enough for the haemoglobin to collect the oxygen (become saturated). The increased blood volume noticed at higher altitudes is due to the increased red blood cell volume. Plasma volume actually decreases. A potential danger of the raised red blood cell count is that the blood becomes more viscous (thicker), which in extreme situations may reduce the flow of blood through the capillaries.

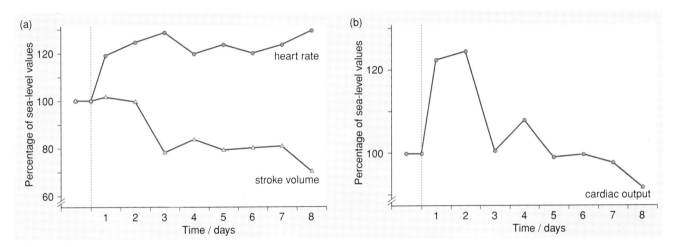

A closer look at the oxygen-haemoglobin dissociation curves reveals how readily the haemoglobin collects oxygen and how easily it gives it up again to the cells. If the curve shifts to the right, it means that oxygen is released more easily, but there is some loss in percentage saturation of the Hb with oxygen. This shift would provide some advantage at moderate high altitude (3000 m to 5500 m). Quechua Indians living in the Peruvian Andes do show such a shift to the right, as do visitors to high altitudes. At very high altitudes, a shift to the left would be favoured because it increases the saturation of haemoglobin with oxygen, though there is a corresponding disadvantage in terms of the point at which oxygen is released to the cells. A shift to the left has been found in the Sherpa people at high altitudes in the Himalayas. It is also of interest that some mammals characteristic of high altitudes, such as the llama, show a shift to the left of the oxygen-haemoglobin dissociation curve.

Figure 10.21 Changes in cardiac output of a group of people during the first 8 days after exposure to an altitude of 3800 m. (cardiac output = heart rate × stroke volume).
(a) heart rate and stroke volume;
(b) cardiac output

Table 10.1 *Red blood cell (rbc) counts and haemoglobin concentration at different altitudes*

	Rbc count / dm^{-3}	Hb concentration / g dm^{-3}
Sea level (normal values)	5.0×10^{12}	148
4540 m (natives in the Andes)	6.4×10^{12}	210
5790 m (mountaineers on Mt Everest)	5.6×10^{12}	196

All values are for male subjects.

Figure 10.22 Oxygen dissociation curve for high-altitude inhabitants (Quechuas and Sherpas). A shift to the left suggests adaptation; a shift to the right indicates acclimatisation. Visitors to high altitude are likely to show a shift to the right

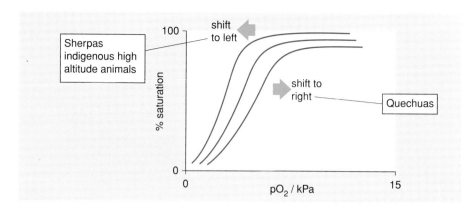

Redistribution of body fluids

Distribution of fluids in the body is influenced by a number of interacting factors, including blood pressure, blood volume, salt balance and hormones. The changes are complex and some are affected by hypoxia as experienced at high altitude.

One of these hormones is antidiuretic hormone (ADH), secreted by the posterior pituitary gland. ADH increases the reabsorption of water by the kidneys. This results in more water being retained in the body and a smaller volume of urine being produced. In **mild hypoxia**, there is probably a decrease in secretion of ADH, which will result in more urine being produced, a situation known as **diuresis**. However, in conditions of **severe hypoxia**, increased secretion of ADH has the opposite effect, leading to retention of water in the blood and reduced urine production. At high-altitude there is a redistribution of the circulation of the blood, with a reduced flow to the extremities. The excess fluid retained in the body tends to accumulate outside the blood vessels, particularly in the lungs and brain. The condition is known as **oedema**. In severe **pulmonary oedema** fluid collects in the lungs and the sufferer may froth at the mouth and become very breathless. In **cerebral oedema**, the brain swells with fluid and then presses against the cranium (the part of the skull which contains the brain). This induces severe headaches, leading to loss of consciousness and sometimes to death. To recover, in both situations, it is essential for the person to be taken rapidly to a lower altitude. Milder forms of oedema show as a puffiness in the face, particularly around the eyes, and as swelling in the legs and feet.

Mental reactions

Visitors to moderately high altitudes frequently show slower mental reactions and weaker decision-making ability than expected from the same people at lower altitudes. This can be tested by a variety of memory tests. There are many stories of mountaineers making foolish mistakes at high altitudes, which may often have been the cause of serious accidents in this difficult terrain. These mental reactions may have been influenced by a combination of the direct effect of hypoxia on brain function as well as stress imposed on the body by the onset of symptoms of mountain sickness. At altitudes over about 5500 m, the hypoxia has yet more severe effects on the activity of the brain, leading to loss of concentration, inability to carry out calculations or make reliable judgements. At still higher altitudes, hallucinations may become apparent, often manifesting themselves as the presence of another companion on the mountain. Habeler and Messner, when they climbed Mt Everest without oxygen in 1979, experienced a feeling of euphoria, somewhat dangerous at a height of 8848 m, surrounded by snow, ice and precipitous slopes. To quote from Habeler, 'I felt somehow light and relaxed, and believed nothing could happen to me. Undoubtedly, many of the men who have disappeared for ever in the summit region of Everest have also fallen victim to this treacherous euphoria'.

Acclimatisation or adaptation?

Native highlanders, such as the Quechua Indians in the Andes or Tibetans and Sherpas in the Himalayas, are able to lead active and physically demanding lives at altitudes in excess of 3000 m. Newly arrived lowland visitors to high altitudes notice an immediate decline in their normal physical activity, and though their performance gradually improves, it may never reach that of the native highlander or at least not for a considerable time. Athletes competing in the 1968 Olympic Games held in Mexico City (2380 m) experienced difficulties in reaching their maximum performance. However, athletes who have trained at moderate or high altitudes, or who are normally resident there, may show superior performance (at low or at high altitudes), particularly in middle- or long- distance running events. Sherpas are known for their support as porters in mountaineering expeditions: at about 3000 m they can carry loads of about 60 kg, and even at 7000 m can manage loads of about 18 kg. This is more than would be expected from a newly arrived lowlander and is certainly related to a Sherpa's superior capacity to utilise oxygen.

If native highlanders were **genetically adapted** to high altitude, the features which allow normal activity in the hypoxic conditions at high-altitudes would be inherited and irreversible. There is little evidence to support this, nor is there evidence that the features would persist if the highlander went to live at sea level. It is preferable to consider both the natives and the visitors as showing different degrees of **acclimatisation** to the high altitudes. Acclimatisation implies that reversible non-inherited changes are shown in structural and physiological features in the body, in response to the conditions experienced at high altitude, and these changes help survival. Thus the features which allow success in hypoxic conditions are of the same kind in both native highlanders and lowland visitors. However, in the native these features probably developed during normal growth and in the early years of childhood, so appear to be permanent, whereas in the visitor the features are acquired over a period of time after arrival at high-altitude.

A few examples (discussed earlier in this chapter) will help to illustrate this:
- Native highlanders show an increased lung volume which contributes to a higher pulmonary diffusing capacity. This would develop in childhood, though perhaps the feature is partly a response to the physical exercise associated with the lifestyle in mountainous regions. A lowland visitor is unlikely to show an increase in lung volume.
- A 'barrel-shaped' chest with large lung volume is a characteristic of Quechua Indians. (This feature may have already existed and given this group an advantage when they first colonised the high regions in the Andes.)
- Both highlanders and visitors hyperventilate, though the visitor maintains a higher rate for a longer period than the native highlander.

- At high altitudes both groups show a higher red blood cell count and higher haemoglobin concentration similar with comparable groups living at sea level. On descent to sea level, several of the features apparent at high altitude, including hyperventilation, haemoglobin level and cardiac output, fall to a normal level. Native highlanders and visitors to high-altitudes are affected in a similar way.

Certain species of mammals, including yaks and llamas, live permanently at high altitudes, so it may help to look at their features in comparison to native highlanders to understand the extent to which each is adapted or acclimatised. Comparing llamas with Quechua Indians at similar altitudes, llamas do not hyperventilate, nor do they show as much increase in haemoglobin concentration. However, llamas are efficient in their extraction of oxygen by the tissues at low partial pressures and show a shift to the left of the oxygen-dissociation curves. Sherpas show a similar shift, but the Quechua Indians do not. The llama is considered to be adapted, the Sherpas are on their way to being fully adapted rather than acclimatised and the Quechua Indians can best be described as acclimatised. The degree of adaptation is relative and related to the length of time the group has lived at high altitude. The lowland visitor is always at a disadvantage during the first few days at high altitude, though becomes at least partially acclimatised after a period of time.

Turning your theory into practice – some practical questions!

Some university students studying biology are planning an expedition in the Himalayan mountains. They will do some botanical work, surveying distribution of plants and studying their ecology, and also aim to climb some of the peaks in the area. They expect to spend about six weeks at altitudes between 2000 m and 6000 m.

Write about a page of 'advice' which can be circulated to the students intending to join the expedition. The advice should include:
- an indication of the conditions expected as they reach the higher altitudes;
- how the body is likely to react at the higher altitudes and over the period of time spent there;
- a biological explanation of the changes taking place in response to the higher altitudes;
- how best to prepare for the journey and how to take care of themselves during the expedition.

Don't forget it can be cold at high-altitudes, so include reference to suitable clothing.

Then list some simple measurements they could take or observations they could make during the expedition to monitor the responses of the body to higher altitudes.

Remember that the students will be carrying most of their own baggage so will not be able to take heavy or complex equipment with them, but you can include some measurements which could be taken before they depart and again on their return.

Circadian rhythms

We tend to develop clear patterns of activities in our daily lives. For some people, the daily routine becomes obsessive, to the extent that if meals or trains are but a few minutes late, the person shows signs of considerable distress. The most obvious pattern is that of being awake followed by a period of sleep, of activity alternating with rest. This appears to coincide with the natural pattern of daylight and darkness, though we often extend daylength artificially and continue our wakeful activities into the hours of natural darkness. In many rural communities, the daily routine is tied closely to waking at sunrise, going out to the fields with the animals to tend the crops, perhaps sleeping during the heat after midday, and returning to the home village in the evening. For others the routine may be governed more closely by the imposed hours of office work, or by the timetable in a school or college. Even the pattern of our leisure hours may be influenced strongly by the timing of programmes on radio or television.

Figure 10.23 Daily activity patterns in rural communities linked closely to the natural daylight hours:
(a) Kutchis and their camels in Afghanistan;
(b) daily movements of sheep in Afghanistan

If we look more closely at body activities other than sleep, we can detect similar rhythmic patterns which follow a cycle of approximately 24 hours. Daily variations in body temperature are described earlier in this chapter (temperature variations). In most people, body temperature rises rapidly after waking, reaching a peak after midday, then falls during the evening with the lowest temperatures occurring during the hours after midnight. This 24-hour repeating cycle is known as a **circadian rhythm** (in Latin *circa* = about, *diem* = day). A rhythm or part of a cycle associated with daylight hours is described as **diurnal**, in contrast to the **nocturnal** part which is associated with the night or darkness.

Circadian rhythms are shown by a number of body functions. The kidney controls elimination of water in urine and excretion of various solutes from the blood, and its activity shows rhythmic fluctuations. Most people sleep about 8 hours at night without needing to pass urine, whereas during the day the bladder is emptied at much more frequent intervals. The composition of the urine also varies over a 24-hour period. The rates of excretion of sodium, potassium and chloride ions are relatively low during the night, and show a rise soon after waking to a maximum after midday. The peaks come in phases, with potassium followed by chloride then sodium. Certain hormones show circadian rhythms with respect to their presence in the blood plasma. As examples, cortisol (linked to carbohydrate metabolism) shows high values early in the morning, just before waking, falling to a minimum towards bed-time, whereas growth hormone has its peak during the early hours of sleep.

Figure 10.24 Examples of circadian rhythms, as shown by (a) body temperature, (b) heart rate, (c) blood pressure and (d) excretion of potassium ions. The measurements were made on people who slept at night between midnight and 08.00 hours, and ate, drank and were active during the day

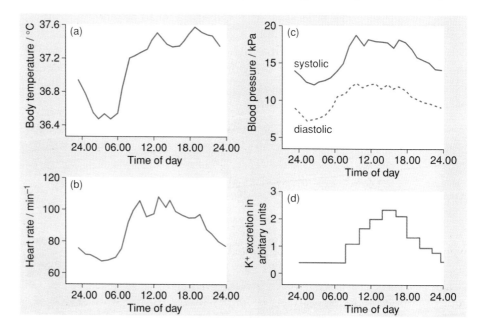

Nature and control of circadian rhythms – endogenous or exogenous?

It is quite difficult to investigate the physiological rhythms shown by an individual person. You cannot easily isolate people and study them away from the influence of the environment and particularly from the activities of the people they live with and the social organisation of our society. It is naturally more convenient to be awake, to eat and work during daylight hours, then to sleep during the hours of darkness. It is reasonable to assume that body temperature is likely to be higher as a consequence of physical and metabolic activities when awake and to fall when resting or asleep. Similarly, the rhythms shown by kidney function may appear to be linked directly to changes in blood composition related to food intake and other activities while we are awake.

Environments where continuous light or continuous dark may be experienced, such as polar regions or deep caves, offer the possibility of studying natural body rhythms away from the influence of alternating daylight and darkness. However, even here, any social effects due to the presence of other people still tend to impose a regular regime on each individual and the inevitable onset of fatigue prevents the individual living freely in a 'timeless' situation. Some useful research has been done in artificial environments, in which there is close control of the normal timing influences such as lighting, temperature, food intake and noise. The volunteer is submitted to a 'constant routine' with controlled alternation of light and dark, and given identical snacks of food at perhaps hourly intervals. In such investigations, one set of volunteers might be kept awake during a full 24 hours and another set allowed to sleep for 8 hours (midnight to 8.00 am). Measurements are taken of a range of body functions, such as core temperature, heart rate, blood pressure and potassium excretion in the urine.

A number of features became clear from these investigations. Firstly, the rhythms do not disappear during the constant routine. There appears to be an innate timing mechanism, or internal 'clock', which maintains at least some of the rhythms. Rhythmic patterns which appear to originate within the body and continue in the absence of external changes are described as **endogenous** rhythms. However, certain rhythms do show variations when the volunteer is allowed the conventional pattern of wakefulness and sleep compared with that during a contrived routine. Such differences may, for example, be shown by higher maximum and lower minimum values of the measurements taken. This suggests that the rhythm can be influenced by the environment and activities which form part of the daily routine of the person. This external influence is described as **exogenous** and may be superimposed upon an endogenous rhythm or, in some cases, may be the sole controller of the rhythm.

Generally the endogenous and exogenous rhythms coincide or are synchronised so that the exogenous or external influences reinforce the endogenous or internal rhythms. When the body is allowed to exist in a truly time-free environment, the natural circadian rhythm is nearer to a 25-hour cycle than to 24 hours. But in practice, the combination of lifestyle, daily activity patterns, social interactions and the regular alternation of light and darkness all drag us into an approximately 24-hour cycle. Perhaps there is justification for the 'Monday morning feeling' as the body struggles to stretch to the full 25-hour cycle.

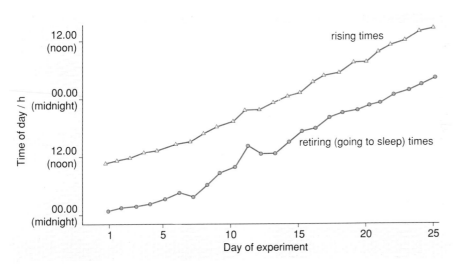

Figure 10.25 Should we really have a 25-hour day?
Walking and retiring times for an individual kept isolated from all knowledge of the passage of time, over a period of 25 days. How many "days" did this person experience?

Further investigations can be carried out, again in artificial environments, to see the relative strengths of the endogenous and exogenous influences on particular rhythms. 'Volunteers' can be trained to live in a 28-hour day, so that the normal routine of a 24-hour day (including eating, sleeping and alternation of light and dark) is spread over an artificially longer time period. In such a situation, the **heart rate** quickly adjusts to the 28-hour day but **oral temperature** continues to follow the previous 24-hour cycle for several days before adjusting. These two circadian cycles thus come out of phase. This suggests that heart rate is influenced mainly by exogenous factors whereas body temperature is controlled by endogenous factors.

Figure 10.26 How a 28-hour routine affects circadian rhythms – measurements of oral temperature and heart rate of a person living through an artificial 28-hour routine. The real times of day are given on the horizontal axis and the dashed vertical lines show the end of the artificial 28-hour days. The coloured bars show when the person was asleep. The oral temperature rhythm continues to follow the pattern of the 'real' days whereas the heartbeat rhythm follows that of the artificial day

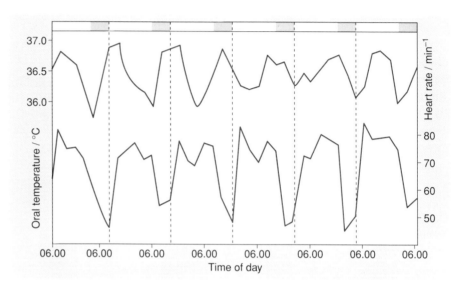

Blood pressure also shows a circadian rhythm, with higher values during the day and lower values at night, falling to remarkably low levels during some phases of sleep. However, like heart rate, blood pressure is strongly influenced by other factors such as stress, metabolic demands, physical activities and emotional states of the person. The exogenous factors are thus relatively more important than the endogenous factors in controlling the rhythms shown by the heart and blood system.

Mental performance

Mental performance, general alertness and vigilance of individual people appear to vary during the day (or waking hours). It may be crucial in some situations that people are performing effectively and avoid making errors in the tasks being carried out. An analysis was made of records over a 20 year period at a Swedish gas works. The people in the gas works worked in 8 hour shifts. Every hour, they took readings from meters, carried out some simple arithmetic then entered the information on a record sheet. All the tasks were double-checked. Mistakes were most frequently detected in the records made at 02.00 and 03.00 hours. If the mistakes had been the result of fatigue, say towards the end of a shift, we would expect three groups of mistakes during the 24-hour period. The study suggests there was a regular circadian rhythm in performance of the individuals and that the rhythm was not affected by the alteration in time pattern imposed by shift working.

It is not easy to define precisely nor to measure the rhythms relating to mental performance in an individual because of the interaction of numerous different influences, including apparent fatigue and outside stimuli. Some people (known as 'larks') claim to be at their most alert soon after waking in the morning while others appear to show better mental performance in the evening towards their normal time for sleeping ('owls'). During the afternoon many people show reduced mental activity or experience an overwhelming desire to sleep, and while it may often be attributed to the after-effects of a midday meal, these feelings of drowsiness and a tendency for dullness in mental activity occur even when only a little food has been eaten. There appears to be some synchronisation of temperature variations with mental performance ('larks' have relatively higher morning body temperatures and 'owls' have relatively higher evening temperatures), but it is not clear whether this is coincidence or an indication of a direct relationship.

Attempts to measure mental performance are usually made by using simple tests. Some examples are listed below:

- reaction time – extinguish a red light with the right hand and a green light with the left hand when they are flashed in random order;
- sort cards in a pack into colours or suits;
- cross out all the letter 'e's in a page of prose;
- search for particular sequences in strings of random letters (e.g. BX or DJBHBIO);
- transcribe symbols into letters in a coded sequence;
- perform mental arithmetic;
- copy irregular drawings viewed in a mirror.

Administration and interpretation of such tests need to be carried out with care. Relatively simple tasks are chosen to help with standardisation but the tests may not represent realistic situations in relation to mental performance. The tests are simple, so soon become repetitive and boring. In attempting to establish the nature of any rhythm over a 24-hour period, inevitably fatigue becomes a factor to consider as well as the performance by individuals in the few minutes just after waking. However, existence of a circadian rhythm in mental performance is accepted, though it may be easier to use other rhythms, such as body temperature, as markers to indicate likely levels of mental performance. Remember though, that the post-midday dip in mental performance is not reflected in the temperature fluctuations. It is generally established that, with respect to the relative importance of the endogenous and exogenous influences, the rhythm for mental performance lies between that for body temperature (strongly endogenous) and heart rate (strongly exogenous).

Disruption of circadian rhythms – jet lag and shift work

We saw earlier how artificially altering the length of a 'day' from 24 hours to 28 hours upsets the circadian rhythms of body temperature and heart rate. At first, in the new daylength regime (of alternating light and darkness), these two rhythms are out of phase. Heart rate (influenced mainly by exogenous factors) starts to adapt first to the new daylength and does so more quickly than body temperature (strongly endogenous control). After about a week (or 6 'days' of the new regime), both rhythms have adjusted and are again synchronised, with each other and with the 28-hour day cycle. Other physiological circadian rhythms are affected in a similar way.

With increasing fast air travel across time zones from west to east or from east to west, many people have experienced the effects of disruption to circadian rhythms. After, say, a 12-hour flight a person may arrive at a destination where the local time (and the pattern of light and darkness) is several hours ahead of or behind the home-time at the start of the flight. There is a mismatch between the internal body clock and external influences related to lifestyle which are imposed by the new local time. During a journey in a west to east direction the traveller has a much shorter day, and for a flight in an east to west direction the traveller has in interminably long day. The familiar symptoms of jet lag are general tiredness with irritability and a desire to sleep or be awake at the 'wrong' time of day. There also tends to be a loss of appetite, and the habitual bowel movements and urinary rhythms are also out of phase with the new or imposed sleep pattern. A person is often disturbed and wakes in the middle of the 'new' night because of a full bladder. Symptoms of jet lag are not experienced by travellers on long flights in a north-south (or south-north) direction, emphasising the fact that the effects are related to the disruption of circadian rhythms rather than being due to anxiety linked to the journey being undertaken.

For eastward travellers, it may take up to a week for the body rhythms to adjust and become synchronised, in a similar way to the adjustment noted in the study with the day experimentally lengthened to 28 hours. Adjustment after westward travel occurs more quickly, perhaps because the internal clock is naturally longer than 24 hours, so the lengthened day is more easily accommodated. It is not clear whether the cause of the jet-lag syndrome is the result of the rhythms being out of phase with the local time or because they become out of phase with each other during their adjustment period. It is, however, of importance to realise that mental activity may be impaired on arrival at the destination, so business travellers should bear this in mind when arranging their schedule of appointments after a long flight. A morning meeting with important negotiations might be disastrous after an eastward flight as the body struggles to overcome the natural desire to sleep.

Shift work has an effect on circadian rhythms similar to that described for fast air travel. Symptoms experienced by many shift workers resemble those of jet lag. A particular difficulty with shift work is that the routine may change fairly frequently, so that the various rhythms that have been disturbed do not have a chance to adjust and settle down permanently. Workers who are on night shifts over prolonged periods may feel deprived of normal social activity during the real daytime. They often sleep less well during daylight hours, partly because of the light and likely noise, but also because of the strong endogenous influence of the internal body clock on, for example, the patterns of sleep, body temperature and need to urinate. Shift workers tend to have a relatively high level of complaints about digestive and gut disorders, though this may be a reflection of their taking irregular or poor quality meals rather than a direct result of disruption of the circadian rhythms.

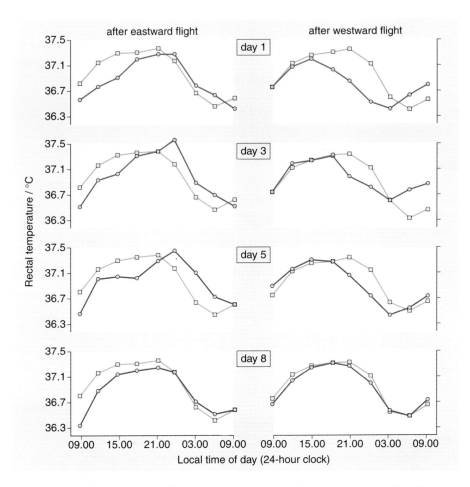

Figure 10.27 Flight across time zones and how it affects circadian rhythms. Measurements of rectal temperature (means for 8 people) after flights across 6 time zones in an eastward and a westward direction. The 'normal' pattern (before the flight, blue line) is given so you can see the effect on successive days as the body adjusts to the 'new' time of day.

Adjustment of circadian rhythms in response to shift work tends to be slower than adjustment after crossing time zones. The jet traveller arrives in a destination where all the external influences are in harmony and combine to reinforce the new local time. Shift workers, however, experience a conflict between their artificially contrived daily pattern as opposed to the stable external influences of alternation of daylight and darkness together with the social activities of other people in the community who are not doing night work. Some rhythms, including body temperature, remain incompletely adjusted even after 21 days of continuous night work. Shift workers and their circadian rhythms appear to adjust back to normal time very rapidly on days off work or when they work day shifts instead of nights.

Weekly shift systems may be particularly difficult for the body to cope with. During the first 5 days the person is likely to be incompletely adjusted to the new daily pattern, then any adjustment achieved is quickly lost by a 'normal' weekend pattern. Some organisations prefer a rapid rotation of shifts – morning / evening / night – so that each shift is worked only once, or at the most twice before moving on to the next in the sequence. Supporters of this system believe that the changes are too rapid to produce any significant adjustment in the circadian rhythms, thus causing less apparent disruption of the body functions or activity patterns of the person. It is also likely that some people adjust more easily than others both to time-zone changes and to shift work, and that adjustment of circadian rhythms may improve with experience.

Causes and consequences of circadian rhythms

There is little doubt that the development of an internal biological clock and the related circadian rhythms are associated with the daily alternation of light and dark. For each rotation of the Earth, there is a period when we are overwhelmed with a desire to sleep, and the pattern of our daily lives normally allows us to do so. There is some evidence that the seat of the biological clock is in the pineal gland in the brain and there are also strong links between observed circadian rhythms and the activities of the hypothalamus. The pineal gland secretes **melatonin** and this is stimulated by darkness and inhibited by light. Perception of light and dark is through the retina of the eye and it is found that some blind people have difficulty in synchronising their circadian rhythms with the rest of the community. The precise mechanism of control is not fully understood though fortunately the internal clock can be 'reset' when transferring to a different time zone.

The natural rhythms appear to be close to a 25-hour cycle rather than a 24-hour cycle, but external influences **entrain** or synchronise the rhythms to fit into the 24-hour day. These external factors of the environment are called **timekeepers** (or **zeitgebers**, to use a term borrowed from the German language). Light is certainly an important timekeeper, but social interactions and our own daily lifestyle routines are also factors which help to adjust the rhythms to the 24-hour cycle. In newly born babies, at first the sleep/wake pattern is about four hours, and the fully adult pattern is established by about the age of four.

If we take a long-term evolutionary view, it would appear to be an advantage for the activity/sleep patterns of people to coincide with the light and dark phases of the 24-hour day. It allows a saving of the body's energy during the resting period and utilisation of light for carrying out activities during the day. During the evening, various internal factors (such as mental activity and body temperature) decline in readiness for sleep. Similarly, from about 4.00 am onwards, levels of body temperature, blood pressure and certain hormones rise, preparing the body for the expected activities after waking. Different patterns shown in other animals are also linked to light and dark, for example, as a means of escape from predators (in some nocturnal animals), or avoidance of high temperatures (in some desert reptiles).

Knowledge of circadian rhythms has a number of practical consequences. Because of the fluctuations in some physiological factors, there are medical implications for both diagnosis and treatment of certain conditions. If levels in the blood of, say, a hormone, vary at different times of day, clinical diagnosis of a condition related to such levels would need to be interpreted in relation to the rhythm. As an example, Cushing's disease is related to overproduction of the hormone cortisol, but since cortisol levels fluctuate in a circadian manner, the condition is best detected in blood samples taken in the evening when cortisol levels are at their lowest. Similar care should be taken in assessing levels of substances excreted in the urine at different times of day (or night). Effectiveness of drugs used in treatment may be critical in relation to the circadian rhythms in metabolic or other physiological activities.

Administration of certain steroids, used to overcome low levels of production, may be adjusted in time and quantity to coincide with the circadian rhythm of their normal activity in the body. Pain and sensitivity thresholds also vary: sensitivity to pain in teeth is at its lowest during the afternoon. Like other shift workers, doctors and nurses carrying out medical checks are more likely to make errors during the real night. This has particular implications for maternity units where more births take place during the night compared with daylight hours. A number of people have difficulty in adjusting their internal clock to the 24-hour day. Some blind people thus naturally follow a 25-hour cycle and go through bouts of insomnia followed by daytime sleepiness. The cycle repeats itself approximately every 25 days, due to the failure of the natural 25-hour cycle to adjust to the social norm of 24 hours.

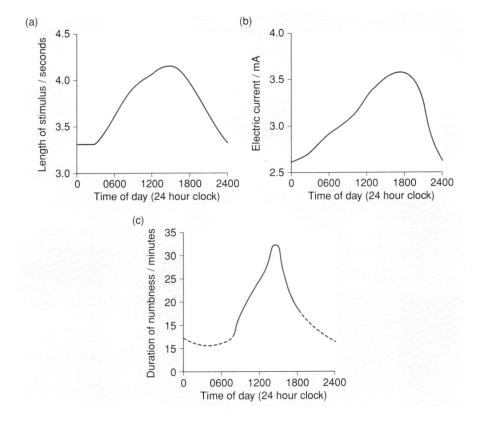

Figure 10.28 Circadian rhythms and your visit to the dentist
(a) threshold of tooth pain to cold stimulus
(b) threshold of tooth pain to electric stimulus
(c) duration of numbness from a local anaesthetic (lignocain)

Some questions to wrestle with!

Your dentist is probably least alert during the afternoon (post midday dip).
Your tolerance of pain is highest during the afternoon.
Numbness from anaesthesia (if used by your dentist when working on your teeth) lasts several times longer during the afternoon than at other times of day.

1   What is the 'best' time of day to visit your dentist?

Assume Delhi time is 5 hours ahead of London time.

2   What would be the best time to visit your dentist in London the day after a flight from Delhi?

3   What would be the best time to visit your dentist in Delhi the day after a flight from London?

Assume you are doing shift work, with three shifts of 8 hours during 24 hours, each for several weeks at a time.

4   What would be the best time to visit your dentist after working on the 8.00 am to 4.00 pm shift? . . . . after the 4.00 pm to midnight shift? . . . . after the midnight to 8.00 am shift?

5   How can you find out the best time of day to answer these questions correctly?

6   Would your answers be the same if you were living north of the Arctic circle during midsummer?

People and populations

Human populations

The world population has grown very slowly for most of the half million years of human existence, but in the last two hundred and fifty years there has been an overwhelming rise in population numbers (see Figure 6.1, page 99). To use rough figures, there were about one billion people in 1830, two billion in 1930, and four billion in 1975. In the 1990s, the world population has already exceeded five billion and predictions suggest there will be six billion soon after the turn of the century with a likely eight billion by 2030. Early figures, long before records were kept, can only be estimates whereas recent figures, based on surveys and censuses, are more reliable. The unmistakable trend is of increased population on a world scale and this brings with it serious questions regarding the future with respect to pressure on space and availability of food and the other resources required to support the demands of an expanding population.

The first census of a national population took place in Sweden in the 18th century, followed by other countries in Europe during the 19th century. In the 20th century some less developed countries still lack accurate figures.

- List some of the practical problems to be faced when doing an accurate census of population numbers. Think about nomadic people, homeless people, literacy, suspicion and so on.
- What sort of evidence can be used to estimate world population before records were kept?

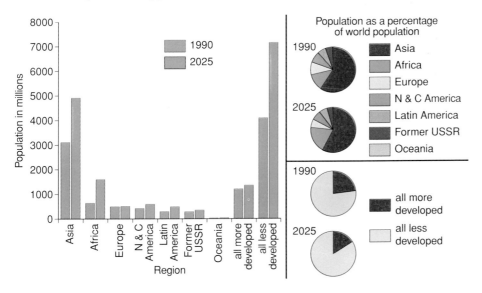

Figure 11.1 Estimates of world population by region in 1990, with predictions for the year 2025.

Table 11.1 *Population densities in major regions of the world in 1990. Note that these data do not reflect the enormous differences in density of population within each region*

Region	Population density in 1990 / km²
Asia (excl. former USSR)	113.0
Africa	21.7
Europe (excl. former USSR)	101.4
North America	12.9
Latin America	21.7
Former USSR	13.0
Oceania	3.1
World	39.2

Work out an estimate for population densities in 2025 in Africa, in Asia and in Europe. Use the population numbers given in Figure 11.1 and the figures for population density in Table 11.1. Which region is likely to show the biggest increase in population density?

Present day distribution of world population shows there are regions which are very sparsely populated and others with extremely high **population density**. The low density areas are likely to be those which are inhospitable for humans, due to extremes of climate or altitude (see Chapter 10) or because of difficult terrain, which is unsuitable for growing crops, such as steep rocky mountains, ice sheets or deserts. At the other extreme, population density exceeds 20 000 individuals per square kilometre in large cities (as diverse as Paris, Manila and Cairo), and Calcutta reached over 88 000 per square kilometre in the late 1980s. Population densities for the major regions of the world in 1990 are summarised in Table 11.1.

Population growth

As an introduction to the factors which affect population growth, we can look at the relatively simple situation of a population of yeast cells growing in a glucose solution in a flask. Figure 11.2 shows how the number of yeast cells increases slowly at first, then rises through the exponential phase to reach a stationary phase followed by a decline and death phase. If we transfer this model to an animal population introduced into a defined area, we would expect a similar initial pattern of growth, with an increase up to a maximum size, which may then fluctuate over a period of time (Figure 11.3). A variety of environmental factors, known collectively as **environmental resistance**, exert pressures which limit the size of the population and prevent it continuing to grow indefinitely. Factors which exert such pressure include availability of food, competition, predation and disease. The level of population reached is a measure of the **carrying capacity** of the area. Overall, numbers of this hypothetical animal population are controlled by:

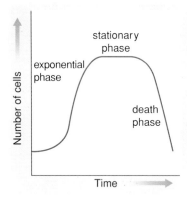

Figure 11.2 Growth curve for a population of yeast cells in a flask. In the exponential phase, the yeast cells reproduce frequently; during the stationary phase, fewer cells divide and some die. The decline in population occurs mainly because the nutrients are used up and also due to the production of toxic substances

- birth rate – higher birth rate can lead to increase in population numbers;
- death rate – lower death rate can lead to increase in population numbers;
- immigration – influx of individuals to the area increases the population numbers;
- emigration – individuals leaving the area decrease the population numbers.

In the case of human populations, social and cultural practices must be included with factors which contribute to 'environmental resistance' and which influence the overall growth of a population.

If we look more closely at the growth of world human population over the last 350 years, we can tease out different patterns in different regions. As a broad generalisation, the more developed countries were the first to show the beginnings of a rapid population increase followed in succession by a number of the less developed countries (Figure 11.4). Growth in the developed countries coincided with the start of the Industrial Revolution. In 19th century Europe, the Industrial Revolution produced an increase in economic activity with development of technological industries and changes in agricultural practices. At the same time there were fundamental changes in social structures and way of life as a greater proportion of people became wage-earners and thus had less direct dependence on the land and what it could produce. The combined effect of these circumstances undoubtedly contributed to the noticeable increase in growth rate of the population. A similar, but not identical pattern of change has occurred in less developed countries during the 20th century.

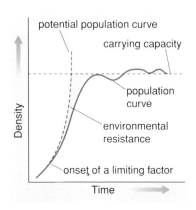

Figure 11.3 Relationship between population growth, environmental resistance and carrying capacity. The carrying capacity is the maximum size of a population that the particular area can support

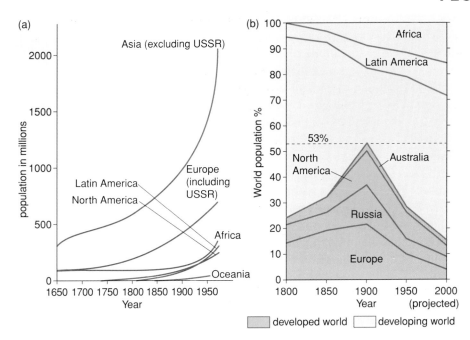

*Figure 11.4 Comparing population
growth in developed and developing
countries. (a) Population growth in
major regions of the world, from 1650
to the late 20th century; (b) Share of
world population between developed
and developing countries, from 1800
to the late 20th century. Developed
countries showed an earlier increase
in population growth (during the
19th century) whereas in developing
countries similar increases have
occurred in the 20th century. Diagram
(b) shows how the developed world
expanded rapidly between 1800 and
1900 but since that time the
population in the less-developed world
has markedly increased in numbers
and at the same time increased its
share of the world population*

Crude birth rate:

$$\frac{total\ number\ of\ live\ births\ in\ one\ year \times 1000}{total\ mid\text{-}year\ population}$$

Crude death rate:

$$\frac{total\ number\ of\ deaths\ in\ one\ year \times 1000}{total\ mid\text{-}year\ population}$$

Attempts to describe and interpret patterns of population growth have led to a number of models. The **demographic transition model** (Figure 11.5), has been used to represent changes in recent European populations. This model identifies four main stages, related to birth rate and death rate. The term **crude birth rate** refers to the number of births per 1000 persons in a year and **crude death rate** is expressed as the number of deaths per 1000 persons per year. In this chapter, these will be referred to as 'birth rate' and 'death rate' respectively.

- **Stage 1** – births approximately equal deaths in a year. Birth and death rates are both high and the population remains steady (with some fluctuations) at a low level.
- **Stage 2** – shows the start of a rapidly expanding population, linked mainly to a decline in death rate. During this stage, life expectancy increases. In European countries in the 18th and 19th centuries this was linked at first to improved agricultural production and nutrition and later to better housing and medical knowledge as well as improved personal hygiene.
- **Stage 3** – shows a decline in birth rate. This stage is associated with an increasingly urban society in which the economic value of a large family becomes less important.
- **Stage 4** – has low birth rate and low death rate, giving a second relatively stable phase.

The demographic transition model should be treated as a generalisation, which does not necessarily apply to all nations but gives a useful overall summary of trends in population growth. A similar pattern is recognised in some developing countries in the 20th century.

Demographic changes in England and Wales since 1700 are summarised in Figure 3.6 where stages of the demographic transition model can be recognised. Stage 1 ended around the 1740s and showed typical high and fluctuating birth and death rates with a long-term slow increase in the

How far do you think events occurring at the time of the Industrial Revolution became a means of increasing the carrying capacity of the land?

Why might large families be considered an advantage in rural populations especially if there is little mechanisation in the agricultural practices?

population. Stage 2 lasted from the 1740s through to the 1880s. Research has suggested that this population increase is accounted for mainly by changes in fertility linked to the timing of marriage. When economic conditions were favourable and wages rose, there were more marriages, people tended to marry at a younger age and to produce larger families, hence have more births per family. Stage 3 is characterised by a fall in both death and birth rates. The latter may be linked to continued improvement in living standards together with improved methods of birth control which gave parents greater freedom to restrict the size of their families if they wished. The noticeable drop in death rate in the late 19th century was due to a decrease in infectious diseases, such as typhus, smallpox and diphtheria. Because of improved nutrition and increased prosperity, people showed greater resistance to disease. Better hygiene and the provision of urban water supplies and drainage systems contributed to the reduction of disease. Towards the end of Stage 3, the effect of the First World War on both death and birth rates can be seen. Since the 1920s, low but fluctuating birth and death rates have led to a slowing down of population growth. A low birth rate during the 'depression' years in the 1930s is contrasted with a post-war 'bulge'. In Figure 11.7, patterns of birth and death rates in three other European countries over this same period, can be compared with that of England and Wales.

Figure 11.5 The demographic transition model of population growth. The terms 'high stationary phase' and 'low stationary phase' refer to the birth and death rates rather than to the numbers in the population

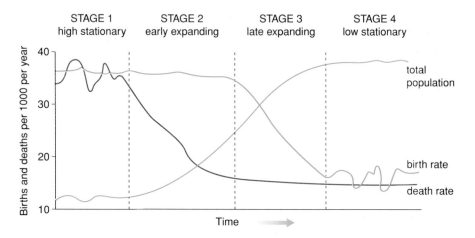

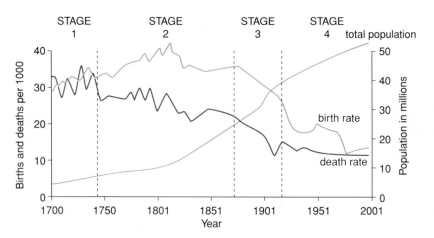

Figure 11.6 Demographic evolution in England and Wales since 1700. Stages 1, 2, 3 and 4 of the demographic transition model are recognisable

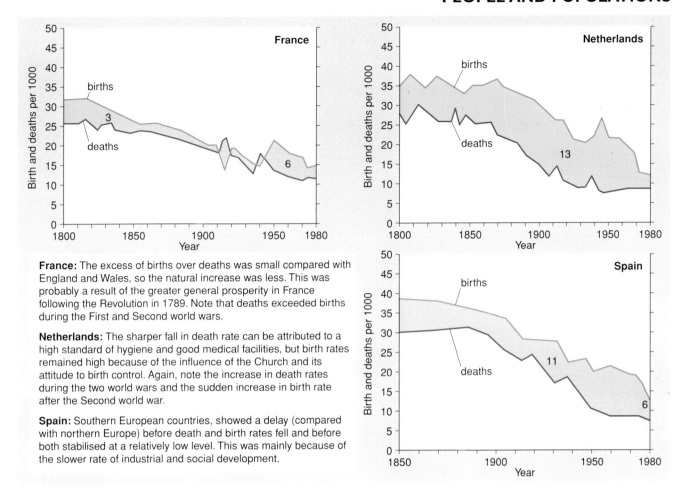

France: The excess of births over deaths was small compared with England and Wales, so the natural increase was less. This was probably a result of the greater general prosperity in France following the Revolution in 1789. Note that deaths exceeded births during the First and Second world wars.

Netherlands: The sharper fall in death rate can be attributed to a high standard of hygiene and good medical facilities, but birth rates remained high because of the influence of the Church and its attitude to birth control. Again, note the increase in death rates during the two world wars and the sudden increase in birth rate after the Second world war.

Spain: Southern European countries, showed a delay (compared with northern Europe) before death and birth rates fell and before both stabilised at a relatively low level. This was mainly because of the slower rate of industrial and social development.

Figure 11.7 Natural population increase in France and the Netherlands from 1800 and Spain from 1850. Compare these countries with Figure 11.6 for England and Wales

The **natural increase** of the population in an area is given by the excess of births over deaths: the wider the gap, the greater the increase. Table 11.2 gives the rate of natural increase in 1990 for the top 20 countries. The natural increase contributes to the growth rate of the country, but **migration** of people into and out of a country (immigration and emigration respectively) must also be taken into consideration. Migration is discussed in more detail on page 253. For world population, migration is not relevant.

List some of the reasons why, in the late 20th century in Britain, couples may wish to limit the size of their families.

Population pyramids

The age and sex structure of the population of a particular nation can be represented by a **population pyramid**. These are histograms, turned on their side and placed back to back, with males on the left-hand side and females on the right. The population is generally presented in 5-year age groups, except for the oldest (say 85+ years), and each group is expressed as a percentage of the total population. New births are added on at the base of the pyramid and each age group moves up as time passes. Interpretation of population pyramids allows us to trace events in the recent history of different age groups within the population and also to make predictions about the likely growth in the future. The actual numbers reflect births and deaths and are also affected by migration.

PEOPLE AND POPULATIONS

Table 11.2 *Rate of natural increase in population in 1990 in the top 20 countries*

	Population estimate mid-1990 in millions	Total population as % of world total 1990	Rate of natural increase 1990	Population projection to 2020 in millions	Total population as % world total 2020
China	1119.9	21.0	1.4	1496.3	18.2
India	853.4	16.0	2.1	1374.5	16.7
Former USSR	291.0	5.5	0.9	355.0	4.3
USA	251.4	4.7	0.8	294.4	3.6
Indonesia	189.4	3.6	1.8	287.3	3.5
Brazil	150.4	2.8	1.9	233.8	2.8
Japan	123.6	2.3	0.4	124.2	1.5
Nigeria	118.8	2.2	2.9	273.2	3.3
Bangladesh	114.8	2.2	2.5	201.5	2.4
Pakistan	114.6	2.2	3.0	251.3	3.1
Mexico	88.6	1.7	2.4	142.1	1.7
Germany*	79.5	1.5	0.0	77.3	0.9
Vietnam	70.2	1.3	2.5	119.5	1.5
Philippines	66.1	1.2	2.6	117.5	1.4
Italy	57.7	1.1	0.1	56.1	0.7
UK	57.4	1.1	0.2	60.8	0.7
Turkey	56.7	1.1	2.1	93.8	1.1
France	56.4	1.1	0.4	58.7	0.7
Thailand	55.7	1.0	1.5	78.1	0.9
Iran	55.6	1.0	3.6	130.2	1.6

* Data for Germany combines that for the former Federal Republic of Germany and German Democratic Republic.

Source: 1990 World Population Data Sheet, Population Reference Bureau, Inc.

Compare the rates of natural increase of population in countries in Europe, Africa and Asia (Table 11.2). What generalisations can you make?

Population pyramids for England and Wales, in 1881, 1931 and 1986 are shown in Figure 11.8. These illustrate different situations with respect to structure of the population and can be related to the demographic transition occurring over these years. The 1881 pyramid is typical of a population with a high death rate, or at an early stage of demographic evolution. It shows a young population, with 70% under the age of 35. The 1931 pyramid shows a population with a reduced birth rate and reflects an older population with increased life expectancy. In the 1986 pyramid, the narrow base shows a continuing reduction in birth rate and the higher numbers over 65 years (particularly of females) indicate an increasing life expectancy. In the 1986 pyramid, there are two bulges, one in the 35 to 40 age range and also in the 20 to 25 age range. The first group was born in the late 1940s to early 1950s, in what is known as the post-war 'baby boom'. The second group was born when the earlier 'baby boom' group reached reproductive age.

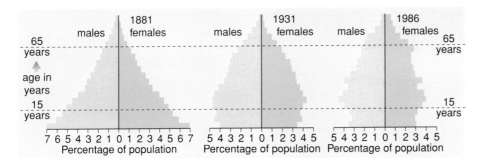

Figure 11.8 Population pyramids for England and Wales in 1881, 1931 and 1986. The data are arranged in 5-year age groups (vertical axis) and represented as a percentage of the total population (horizontal axis). The oldest age group is an exception as it contains all over the age of 85. Other population pyramids in this chapter are represented in a similar way

Some questions about populations

Look at the population pyramids in Fig 11.9 for Kenya, Peninsular Malaysia, Jamaica, France, Sweden and Japan, all in the 1980s. From them we can work out a number of different things about the population of each country.

- Which countries are at an early stage of demographic transition? Are these 'developed' or 'developing' countries?

- Which pyramid is most likely to represent an expanding population?

- Which country is least likely to have an effective birth-control policy?

- Which country has the lowest life expectancy? Which has the highest life expectancy?

- Which countries show a clear difference in life expectancy between males and females?

- In the Japan pyramid, there is a noticeable bulge midway between the 15- and 65-year-old lines. What age are these people? Suggest a reason for this bulge. There is a second bulge just below the 15-year-old line. What does this suggest about the age at which people have children?

- In 1985, what percentage of the population of Kenya was under the age of 35?

- What deductions can you make about birth rate in relation to the economic status or social systems of the country? Suggest some reasons for these relationships.

- What differences can you see in the balance between males and females in different age groups – around birth, up to 15 years, 15 to 49, over 65?

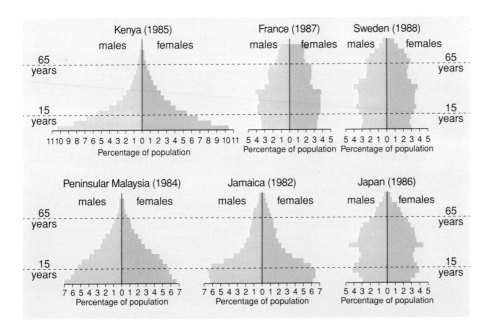

Figure 11.9 Some questions about populations (see text) – Some population pyramids in the 1980s (actual data is given in brackets). See caption for Figure 11.8 for explanation, but note that the oldest age group for the Kenya pyramid includes all over the age of 80

Fertility and birth rates

The term **fertility** can be used generally in relation to aspects of human reproduction, but more specifically it refers to the actual number of live births achieved in a population over a period of time. The term should be distinguished from **fecundity** which refers to the physiological ability of a woman (or of a population) to conceive and bear children, otherwise known as the **reproductive potential**.

Crude birth rate, as defined on page 237, relates the number of births in a year to the whole population. This includes people who cannot give birth, notably males, and females who are outside the normal reproductive age (generally taken to be between 15 and 49 years) as well as those unlikely to do so (because they are single or infertile, for example). Figure 11.8 and 11.9 show that there are considerable variations in the sex and age structure of a population, so other measures of fertility are used which relate the birth rate more specifically to women of child-bearing age, or to particular cohorts. Such measures can be useful when comparing demographic trends in different populations.

- **General fertility rate** relates the number of births in a year to the number of women between 15 and 49 years and is expressed by the formula

$$\frac{\text{number of live births} \times 1000}{\text{number of women aged 15–49 years}}$$

- **Total fertility rate** (TFR) indicates the average number of children that would be born to 1000 women in the current population assuming that each lives to the end of her child-bearing age (15–49 years) and produces children at the same rate as women of these ages did in the year of calculation. It can be interpreted as the average number of live-born children which women would have if they experienced the same age-specific fertility rates of that particular year throughout their lives. A TFR of 2.1 to 2.5 is considered to be the level needed for natural replacement of the population, though this assumes no net migration.

- **Gross reproduction rate** is a modification of total fertility rate by giving it in terms of female children only.

- **Net reproduction rate** is a further modification of gross reproduction rate and takes account of women who die before reaching the end of the fertile age range. A net reproduction rate of 1.0 indicates that the population is replacing itself exactly. If less than 1.0 the population is failing to replace itself. If greater than 1.0 the population is likely to increase because the number of potential mothers in the next generation is increasing.

Attempts to explain past patterns of fertility and to predict future trends with any certainty are highly complex because of the interaction of a large number of variables concerning the lives of people. These include social, cultural and religious practices, economic conditions, the level of education and the extent of government controls or intervention. The way that these factors influence fertility will be different in different countries and inevitably change over time.

Fertility is also affected by the consequences of war and peace and of hardship or plenty in terms of food production. Examples outlined in the next section should be considered only as illustrations of how fertility rates change rather than as representing a comprehensive overview.

Proportion of women in marriage

In the 16th century in western Europe, between 40 and 50% of women in the age range 15 to 44 (and so potentially able to give birth), were married whereas at the same time in eastern Europe, Africa and Asia an estimated 60 to 70% of women were married. Countries in these latter regions would be likely to have a higher fertility (or birth) rate. Cultural practices relating to birth in or outside marriage vary in different countries and are changing. In Britain, for example, the number of births outside marriage has increased notably in recent years, from 9.8% in 1977 to 34.8% in 1994. In considering trends in birth rates it is appropriate to use the term 'marriage' to include those in stable sexual partnerships as well as formal marriage.

Age of marriage

In some primitive societies, marriage may occur soon after menarche (first menstruation) giving the potential for many years of child bearing. Some valuable demographic information was obtained in the 1960s for the !Kung Bushmen, living in the Kalahari desert in southern Africa and who still practised a hunting and gathering lifestyle. Marriage was likely at about 16 years though the first birth was often delayed until about 19 years due to there being a delay of ovulation after the onset of menstruation (known as adolescent infertility). By contrast, in northwestern Europe in the 16th century, the estimated average age for first marriage for women was relatively late, between 25 and 27 years. This was, perhaps, a reflection on the inheritance customs and the need to delay marriage until suitable security could be offered, say in terms of land (for peasants) or tools (for skilled manual workers). In the late 20th century, there is a trend in all countries for later marriage. This is doubtless linked to there being a higher proportion of educated women: they would need to take a career break to have children and often have a wide range of interests outside the raising of children within a family. Later marriage is likely to delay the birth of the first child and reduce the number of children born within the child-bearing age. Table 11.3 shows the increasing age of mothers in England and Wales, from 1964 to 1994. In the 1970s, the government in China took positive steps to delay the age of marriage: at first there was pressure on couples not to marry before 25 for males and 23 for females, and in 1980 this was formalised as the legal age of marriage at 22 and 20 respectively.

Number of children within a marriage (family size)

In traditional societies, particularly those that are rural with an agricultural subsistence-based way of life, high fertility is advantageous to, and becomes a form of security for, the family unit. From a young age, children provide services such as caring for livestock, weeding the crops, sweeping, carrying wood and water and looking after younger children. Older children provide labour on the land and offer gifts even after they have left the family home. Adult children care for parents in old age, which is important when there is no state social security

Figure 11.10 (a) Young children used to care for younger children: two sisters in Sichuan, China
(b) Young children tending livestock: Bedouin girl, Jordan

provision. There exists an extended family of mutual obligations, and high fertility is linked to status. But in recent decades in many developing countries, there has been mass provision of primary schools, so that younger children are no longer available for general agricultural and household chores. New agricultural technology prefers weedkillers to child labour and the practices of freely grazing animals and collection of dung or firewood for fuel are disappearing. Children grow up with different expectations and become alienated from the traditional family way of life. When they reach adulthood they are likely to leave the family home and support an independent family. A reduction in family size has been witnessed widely both in developed and developing countries and often occurs sooner in urban than in rural communities.

Table 11.3 *Mean age of mothers at childbirth in England and Wales (1964 to 1994). The general trend for first births is for mothers to be older, but there is less difference for later births*

Year	All births	First births	Second births	Third births	Fourth births
1964	27.2	23.9	26.8	29.3	31.0
1977	26.5	24.4	26.8	28.8	30.6
1984	27.0	24.7	27.3	29.2	30.7
1989	27.3	25.3	27.6	29.5	30.9
1994	28.4	26.5	28.6	30.3	31.2

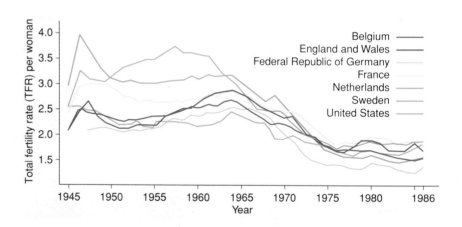

Figure 11.11 Changes in total fertility rate in selected countries, from 1945 to 1986. Note the post-war baby boom in most countries and the downward trends towards the 1980s

Table 11.4 *Total fertility rates in selected countries, comparing rural and urban families and comparing mothers with different education. Data derived from births reported in the year 1986 to 1987*

Country	Rural	Urban	Education of mother		
			None	Primary	Secondary
Latin America					
Brazil	5.0	3.0	6.5	5.1	3.0
Columbia	4.9	2.8	5.4	4.2	2.3
Peru	6.6	3.2	7.0	5.1	2.9
Asia					
Thailand	2.6	1.7	3.5	2.5	1.5
Africa					
Liberia	6.8	6.1	6.7	7.0	4.8
Senegal	7.1	5.5	6.8	5.7	3.8
Burundi	6.9	5.3	6.8	7.2	5.5

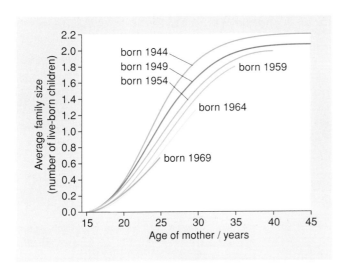

Figure 11.12 *Changes in family size in England and Wales, for various cohorts of fertile women born 1944 to 1969. Note the trend to smaller families as well as the later age at which mothers give birth and complete the family – for mothers born in 1944, at the age of 25 years the average family size was 1.2 but 2.1 when they were 35, whereas for mothers born in 1959 the average sizes were 0.75 and 1.75 respectively*

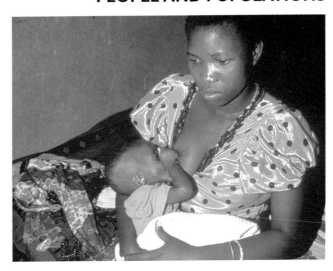

Figure 11.13 *Woman breast feeding baby in Tanzania*

Control of family size

Control or limitation of family size is practised through deliberate measures taken before and after conception, including abstinence, contraception and termination of pregnancy. Conception usually does not occur during lactation, since breast-feeding suppresses ovulation making pregnancy significantly less likely. This is known as **lactational amenorrhoea** and, particularly in primitive societies where breast-feeding may be prolonged, it can be a valuable way of delaying pregnancy and spacing out the family. It is probably important in nomadic people, such as the !Kung Bushmen of southern Africa, where a certain amount of mobility is essential in their hunting and gathering activities - a mother with more than one young child on her back or hips would be at a considerable disadvantage.

Modern family planning (birth control) is effected in a number of ways:
- rhythm method (limiting intercourse to the days in the menstrual cycle when conception is least likely);
- barriers which prevent fertilisation (condom, diaphragm, cervical cap);
- prevention of implantation (intra-uterine device);
- hormonal prevention of ovulation (most frequently administered as the 'pill');
- sterilisation (females by surgical removal of the uterus, ovaries or oviducts or by closing the oviducts; males usually by vasectomy);
- induced abortion, resulting in deliberate termination of pregnancy.

The extent to which modern contraceptives are used varies considerably throughout the world (Table 11.5). Even within these major regions there are differences, related to factors such as religion, level of income and of education, and how far there are appropriate government-backed family-planning programmes. These often include dissemination of advice as well as making available the contraceptive devices. Where contraceptives are not used, abortion may be the alternative, but often carried out in unhygienic conditions and therefore dangerous to the mother.

Table 11.5 *Use of contraceptives in different parts of the world in mid 1980s. Contraceptive prevalence rate is given as a percentage of married women of childbearing age who are using (or whose partners are using) some form of contraceptive, modern or traditional. Estimates are also given of the total fertility rate (TFR) and annual population growth rate (1960 to 1990) in the same areas*

Group of countries	Contraceptive prevalence rate	TFR	Population growth rate
Industrialised countries	70	1.9	0.8
All developing countries	52	3.9	2.3
Least developed	19	6.1	2.5
Sub-Saharan Africa	15	6.5	2.8
World	**56**	**3.5**	**1.8**

Table 11.6 *Use of contraceptives in some countries in eastern Europe, late 1980s*

Country	Married women using contraceptives / %
Bulgaria	8
Czechoslovakia	25
Hungary	62
Poland	26
Romania	6

Success in terms of use of contraceptives is generally linked with the level of development of the country and the extent to which women are educated. In poor countries, cost may be a deterrent. Resistance to the use of contraceptives comes from followers of certain religions, notably Islam and Roman Catholicism. Government family planning policies generally encourage limitation of family size, as a means of controlling growth of the population and so raising the standard of living. However, a declining population can lead to a diminished labour force resulting in reduced economic potential, so some governments take an opposite view and have a **pro-natalist** policy, encouraging larger families. Some examples will illustrate the effect of government intervention in promoting family planning policies.

Table 11.7 *Family size in relation to religion – Roman Catholic and non-Roman Catholic areas of Belfast, Northern Ireland. Data refers to married women, aged 45 to 49, in 1983*

Region	Number of children in the family	
	Roman Catholic	Non-Roman Catholic
Belfast	4.0	2.4
Belfast suburbs	3.7	2.6
Belfast fringes	3.3	2.2

An extreme example of attempts to limit family size (**anti-natalist policy**) is seen in government policies in China since 1949. Between 1949 and 1990 the population more than doubled, from 540 million to 1134 million. During the 1950s, there was little restraint on family size, the optimistic implication being that the population could be fed, however rapid the increase. Widespread famine in the early 1960s led to increased mortality and had a major effect on lowering fertility, though there was a brief 'baby boom' from 1963 to 1965 when food supplies returned to normal. From the late 1960s through the 1970s there was strong government encouragement to reduce family size. Community-based family planning services were expanded and there were campaigns to stress the beneficial effects of a small family in terms of health of mother and children as well as the economic advantages. There were slogans and social pressures from the community to have 'later marriage, longer intervals between births and fewer children'. By 1979 this had been stepped up to the 'one child per family' policy. This introduced incentives for couples to have just one child but penalties for those having more than two. Success of the policy has been greater in urban than rural communities.

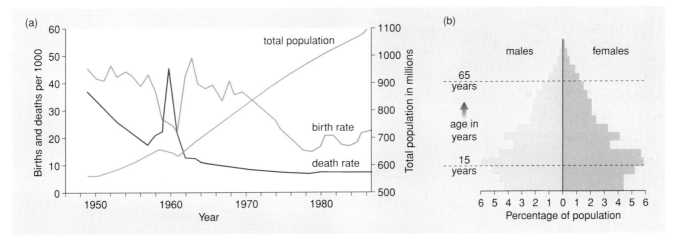

Figure 11.14 (a) Demographic transition in China, 1949 to 1987, an example of a government with an anti-natalist policy; (b) Population pyramid for China (1988). Link information in the pyramid with the demographic trends shown in the graph

Figure 11.15 (a) Family planning advice available in remote rural areas - Wolong, Sichuan, China; (b) Poster promoting China's one-child policy

European countries, particularly those in eastern Europe, have faced declining birth rates and populations, and some governments have adopted pro-natalist policies. In eastern Europe, during the 1970s, there was relatively little attempt to encourage the use of modern contraceptives. Because of low wages, couples found difficulty in surviving on a single income so the pressure for women to remain in employment contributed to the falling birth rate. Incentives to increase family size were implemented through social welfare programmes aimed at helping mothers in work to maintain a family as well. These included maternity leave, birth payments, income tax reductions, priority subsidised housing and provision of nursery places for children under 3 years old. Extreme measures were taken in Romania when, in 1966, abortions were made illegal except in exceptional circumstances, and the import, manufacture and sale of most contraceptives were prohibited. In the late 1980s, all women up to the age of 40 were subjected to frequent examination to ensure that pregnancies were not terminated and voluntarily childless couples were taxed heavily. Data in Figure 11.16 suggest these measures in Romania did have some effect in increasing birth rates, compared with neighbouring eastern European countries.

Figure 11.16 The effect of government pro-natalist policies. Birth rates in Romania (1957 to 1985) are compared with the average of equivalent data for Bulgaria, Czechoslovakia, East Germany, Hungary and Poland

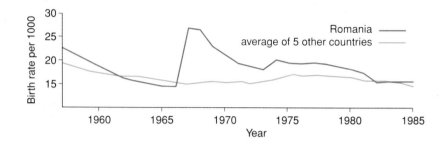

In China, in the late 1980s, up to 90% of births in some urban areas were the first-born in the family whereas in rural areas the equivalent figure was as low as 40%.

• What does this suggest about the success of the 'one-child' policy or the difficulties of implementing it?

• Many of the ethnic minority peoples live in rural areas. Suggest reasons for promoting a pro-natalist policy among such minority people.

What problems might be encountered in attempting to increase the use of contraceptives in a developing country? Would you anticipate a better response in rural or in urban areas?

Effects of infertility

In some couples, one or other partner might be infertile for physiological or genetic reasons. Causes of infertility in females include failure to ovulate, blockage of the oviducts or disease of the lining of the uterus (endometrium). In males, infertility may be due to low numbers or complete absence of sperms, or to reduced motility of sperm. Medical technology has opened up ways to help childless couples overcome infertility through, for example, drugs to stimulate ovulation, *in vitro* fertilisation and artificial insemination. While important to the individuals affected, infertility and its treatment does not have a significant effect on fertility trends in the population as a whole.

Mortality and death rates

Mortality or death rates vary within a population in relation to age, sex, environmental conditions, social factors and disease. Death rates also vary between different countries. Changes in death rates influence the demographic trends or growth of a population. When making comparisons, it must be realised that some of the data is unreliable. For example, earlier records in England and Wales were taken from parish registrations of baptisms and probably discounted infant deaths between birth and baptism. Similarly, early deaths of unwanted children are sometimes obscured and even now in some countries registration of deaths may be incomplete.

Crude death rate is defined (page 237) as the number of deaths per 1000 in the population in a given year, usually taken as the mid-year population. However, as with crude birth rates, these 'crude' death rates do not allow for differences in composition of the population, particularly with respect to age and sex, both of which affect mortality probabilities. Crude death rates for a number of countries are given in Table 11.8. Generally, crude death rates are higher in developing than in developed countries. It may, however, seem anomalous that certain developed countries (e.g. Sweden) have higher death rates than some developing countries (e.g. Brazil). While this appears to conflict with the data relating to life expectancy which is higher in Europe than in these same developing countries, it can be explained by looking at the age structure of the population. There is a higher proportion of older people in the developed countries, which accounts for the higher overall death rate. In most countries, death rate of males is higher than that of females, though the biological basis for this difference is not understood. Because of these variations, death rates may be adjusted and expressed, say, as age-specific death rates, or related to males or females.

Table 11.8 *Estimates of death rates, infant mortality rate and expectation of life in selected countries around 1990*

Country	Death rate per 1000 per year	Infant mortality rate per 1000 live births	Expectation of life at birth / years	
			Male	Female
Africa				
Central African Republic	17	100	44	47
Egypt	10	61	59	61
Mali	20	164	46	50
Zimbabwe	10	61	57	60
Americas				
Brazil	8	60	62	68
Cuba	7	11	73	76
Jamaica	6	16	71	77
USA	9	10	71	78
Europe				
Poland	10	16	67	76
Portugal	10	13	68	75
Sweden	12	6	74	80
UK	12	9	72	78
Asia				
Bangladesh	15	114	57	56
China	7	30	68	71
India	11	94	52	52
Japan	7	4	75	81
Sri Lanka	6	26	68	72
USSR	10	25	65	74

Table 11.9 *Age specific death rates for males and females, in England and Wales compared with Mauritius (mid 1980s)*

Age / years	Age specific death rate per 1000			
	England and Wales		Mauritius	
	Male	Female	Male	Female
Under 1	10.4	7.9	30.8	23.1
1–4	0.5	0.4	1.2	1.3
5–14	0.2	0.1	0.5	0.4
15–24	0.7	0.3	1.1	1.0
25–34	0.9	0.5	2.1	1.4
35–44	1.7	1.1	4.5	2.3
45–54	5.0	3.2	11.7	5.0
55–64	15.7	9.0	27.8	13.7
65–74	41.7	22.8	57.3	39.6
75–84	98.8	60.7	123.0	83.5
85+	212.7	168.6	249.5	192.5

PEOPLE AND POPULATIONS

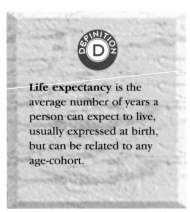
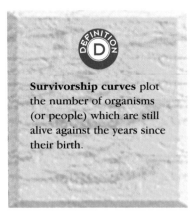
Life expectancy can be used to give an indication of the patterns of mortality in a population. The life expectancy at birth is usually less than that at 1 year old because of relatively high mortality of infants soon after birth. Figure 11.17 shows how expectation of life in England and Wales has altered over the 140 years from 1841 to 1981. Note the great improvement in life expectancy at birth and link this with the stages in demographic transition for England and Wales (Figure 11.6).

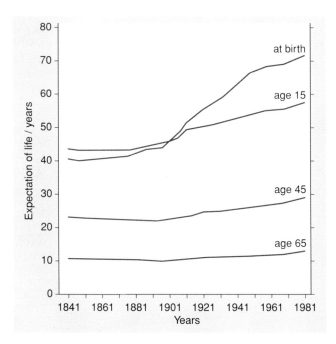

Figure 11.17 Expectation of remaining years of life at birth and other ages in England and Wales, 1841 to 1981

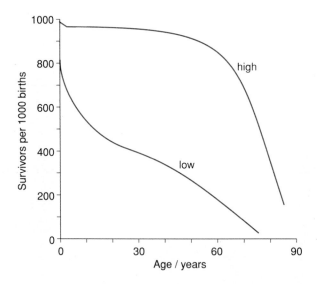

Figure 11.18 Survivorship curves – comparing populations with high and low expectations of life

Table 11.10 *Comparing expectation of life for males and females at various ages, in UK and Bangladesh (late 1980s). Note that life expectancy is lower in Bangladesh (a developing country) in the early years of childhood because of high mortality rates, but after the age of 30, life expectancy in the two countries is relatively close*

Suggest reasons for the trend in all countries towards lowering of infant mortality rates.

Sex	Country	Expectation of life at age / years					
		0	1	5	10	30	50
Male	Bangladesh	57	64	63	59	41	23
	UK	72	72	68	63	43	25
Female	Bangladesh	56	61	61	57	40	24
	UK	78	77	73	69	49	30

Causes of mortality – patterns and trends

Patterns in the causes of mortality have changed over time and continue to show variations in different regions of the world, depending on local conditions or sometimes on isolated events. When reviewing data, it must be appreciated that reporting of deaths may be incomplete and that diagnosis of illnesses may be vague or inaccurate, particularly where medical knowledge is inadequate compared with the best known in the late 20th century. In England and Wales, death certificates were first issued in 1850 and this led to greater reliability in the data, but even today in some countries, there is no complete system for recording deaths. Changes in the causes of death in England and Wales over the 140 year period from 1851 to 1990 are given in Figure 11.19. These data show that deaths from infectious and parasitic diseases and from tuberculosis have undergone a marked decline since 1851 whereas there is a substantial increase in deaths from heart disease, strokes and cancer. You can link these changes with the stages in demographic transition represented in Figure 11.6 (page 238).

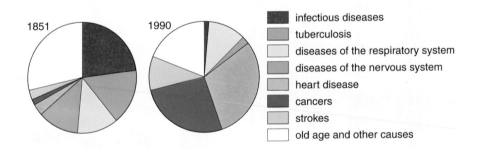

Figure 11.19 Causes of deaths in England and Wales, comparing 1851 with 1990

In looking for explanations for the decline in infectious diseases, we can consider a number of factors:

- **biological** – such as a reduction in virulence of a microorganism or an increase in natural resistance in the population. These may have occurred in some pathogens, but can be discounted as the main reason for a decline in infectious diseases since they are unlikely to have occurred simultaneously in a range of microorganisms.
- **environmental and social** – improvements in living conditions (especially in housing, clean water supplies and nutrition), and better understanding of hygiene (including food preparation and removal of

Figure 11.20 Public health advice in rural Bhutan - to keep water supplies clean

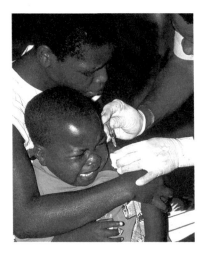

Figure 11.21 Immunisation programme - Tanzanian baby being vaccinated against malaria

refuse and sewage) leading to less exposure to infectious pathogens. These factors are best summarised as 'improved standard of living'.

- **medical intervention** – including vaccination and immunisation programmes encouraging prevention, as well as understanding of diseases and medication or treatment which improves chances of curing them.
- **public health advice and legislation** – to ensure appropriate education of the people and implementation of measures to control or reduce disease.

The prevalence of strokes, heart disease and cancers as the main causes of death in developed countries in the late 20th century represents a shift towards chronic or deteriorative diseases, associated with older adulthood and a more affluent lifestyle. To put it in simple terms, this can be seen as a consequence of there being fewer deaths from infectious diseases. We should, however, remember that many of the less developed countries are situated in warm latitudes where the climate encourages spread of diseases such as yellow fever, malaria and bilharzia. In these countries, improvements in housing, sanitation and general standard of living together with public health and improved medical facilities are likely to bring about a similar reduction in mortality.

Trends in mortality can be related to life expectancy and to the stage of development of a country. This link is represented in Figure 11.23 which gives the likely distribution of causes of deaths in relation to life expectancy. Group 1 diseases (infectious, parasitic and respiratory) account for nearly 50% of deaths in countries with low life expectancy and, except for pneumonia and influenza in older people, are relatively unimportant in developed countries. Diseases in Groups 2 and 3 become more important in countries where the population structure supports an older age range. Group 4 deaths include those from motor vehicle accidents and show a fairly constant proportion of the pattern of mortality.

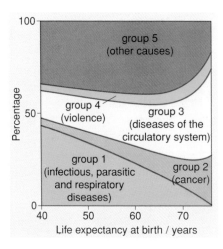

Figure 11.22 Likely causes of death (in a standard population) in relation to life expectancy. Causes of death are collected into five groups and their distribution is shown as a percentage against levels of life expectancy at birth ranging from 40 to 76 years

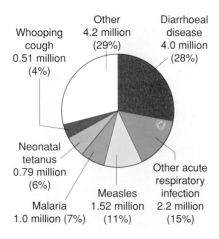

Figure 11.23 Causes of infant mortality in developing countries, around 1990

Migration

We have seen how population pyramids (Figure 11.9) can be used to represent the age and sex structure of a nation's population. They also reflect events over a period of time. A bulge, for example, can be interpreted as a 'baby boom'; a constriction as high mortality due, say, to famine or war, or the result of lowered birth rate for a variety of reasons. But one factor not yet considered is that of **migration** across international boundaries. Migration within a nation state (for example from rural to urban communities, or nomadic peoples moving between lowlands and highlands) has no effect on the overall population statistics of a nation and migration has no relevance to global population.

If we look again at the population pyramid for Jamaica (Figure 11.9) and try to interpret its shape, we should include the effect of migration as well as births and deaths, because migration was an important factor affecting the size of this population. Employment problems in the 1950s stimulated large-scale emigration, the people being attracted by opportunities elsewhere. Eventually immigration controls in the destination countries had a limiting effect on the emigration from Jamaica. Those who emigrated were mainly young adults, giving rise 30 years on to the constriction in the 45+ age group. The narrow base shows it is partly the effect of birth-control policies, but also due to there being fewer people of reproductive age because of the earlier emigration.

There have been repeated examples of movements of people from one area to another. Such migrations started perhaps 100 000 years ago with the earliest populations of *Homo sapiens*. Possible dispersal routes taken by our ancestors as they spread from their origin in Africa to all continents of the globe are shown in Figure 11.24. Nomadic hunter-gatherers certainly would have moved short distances as they exploited the resources of their surrounding environment. Indeed, our estimates of early population numbers are based on the carrying capacity of the land, or the maximum population density that could be sustained by the way of life of the people in the area. As resources diminished, it is likely that the people moved elsewhere. Transition to settled agriculture in the Neolithic Age inevitably increased the potential productivity of the land for the people and was probably associated with a surge in population numbers. At that time, there were no national or international boundaries, but as nation states became identified, we can trace large-scale migrations of people which have had noticeable effects on the population numbers of a particular area or country.

Questions about hygiene and disease
A cholera epidemic in London in the mid 1850s was linked to a water pump in Soho.

- What steps are taken now to ensure clean water supplies? List other hygiene measures which are likely to have led to a reduction in infectious diseases.

- Find out when the first smallpox vaccinations were carried out and the date when this disease was eradicated on a world-wide scale.

- What methods are used today for control of malaria and why are some preventative measures becoming less successful?

PEOPLE AND POPULATIONS

Figure 11.24 Possible dispersal routes of Homo sapiens, *from Africa around the globe over the last 100 000 years. The map has been reconstructed using evidence from fossils and from genes*

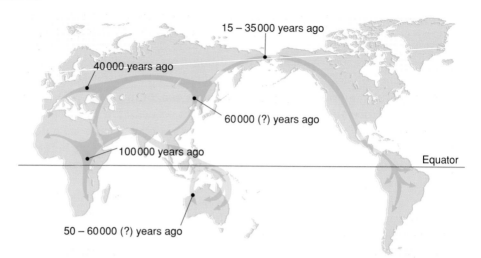

The impetus to migrate is usually linked to pressure at home, often triggered by increasing population, crop failure and scarcity of food. A glimpse into history highlights some mass movements of people over the centuries. Over 2000 years ago, the expanding Greek empire spread into Egypt, Europe, Turkey and beyond. The Romans later invaded territories as far west as Britain, bringing people as well as the influence of their civilisation and culture. From the 16th century onwards we can see the expansion of Europeans into distant lands. One attraction was for tropical and sub-tropical regions because of the potential offered by production of exotic crops in a warm, humid climate. Later colonisers turned to temperate zones that were sparsely populated. The Spaniards invaded the Americas, followed by other Europeans who acquired new land in the United States of America, Canada, Argentina, Australia and New Zealand. These settlers brought with them an expertise in new agricultural methods and spread the influence of the Industrial Revolution. They were thus able to exploit the huge potential of these new territories, particularly their agricultural productivity. Establishment of commercial plantations of coffee, sugar, tobacco and cotton in the Americas and Caribbean brought the need for cheap labour and led to the start of the slave trade, with the resulting importation of large numbers of people from Africa. The European migrants were pushed by the rapidly growing population of Europe, and pulled by the lure of new lands and an escape from hunger and poverty. Many were very successful, but often at the expense of the local inhabitants. During the 19th century in New Zealand, for example, the native Maori population was reduced from over 150 000 to about 40 000, due mainly to the introduction of 'foreign' infectious diseases (such as measles and venereal disease) but also by the social collapse of their own society in the face of competition from the colonisers.

In the twentieth century, population numbers in different countries have continued to be affected by emigration and immigration. Between 1933 and 1938 anti-Semitic legislation in Germany exerted extreme pressure on Jews to emigrate. Since the 1950s, high numbers of foreign workers have migrated into Europe – probably in excess of 1.5 million Turkish workers to West Germany alone (*see* Table 11.12). In 1956, Hungarian refugees settled in Britain after the invasion of Hungary by Soviet armed forces; in 1972, Asians were expelled from Uganda and many migrated to Britain. Forced migrations in the 1990s are seen in movements of Kurds from Iraq into Turkey and Iran, Afghans into Pakistan and as far as Vladivostoc, and Hutus from Rwanda into Zaire in central Africa.

Table 11.11 *Origins of immigrants from some countries to the United States, between 1820 and 1930. Numbers are in 1000s. Note, for example, the high numbers who left Ireland in the late 1850s following the famine resulting from the disastrous potato crops, affected by potato blight (Phytophthora infestans)*

Years	All countries	England and Wales and Scotland	Ireland	Germany	Scandinavia	Russia and Baltic States	Italy
1820–29	127	26	50	4	–	–	–
1830–39	537	73	170	124	2	–	1
1840–49	1427	217	655	384	13	–	–
1850–59	2814	444	1029	975	25	–	8
1860–69	2081	532	427	723	96	1	9
1870–79	2741	577	422	750	208	34	45
1880–89	5247	810	673	1444	672	182	267
1890–99	3693	328	404	578	391	449	602
1900–09	8202	469	344	327	488	1500	1930
1920–19	6346	371	166	174	238	1106	1229
1920–29	4294	337	206	386	205	87	527

Table 11.12 *Origins of immigrant workers in selected countries in Europe (Belgium, France, West Germany, Sweden and Switzerland), in the mid 1980s. For each of the countries listed, the three leading origins of immigrants are given. Note the high number of immigrants from Turkey to West Germany. Numbers are in 1000s*

Origin	Belgium	France	West Germany	Sweden	Switzerland
Finland	–	–	–	131	–
Italy	252	–	544	–	385
Portugal	–	751	–	–	–
Spain	–	–	–	–	113
Turkey	76	–	1481	22	–
Yugoslavia	–	–	598	39	87
Algeria	–	821	–	–	–
Morocco	126	516	–	–	–
Total	853	3752	4630	401	979
As % of total population	8.6	6.8	7.6	4.8	15.0

Figure 11.25 Afghan refugees in Vladivostoc

Implications of world population trends

Graphs of population numbers (Figure 6.1, page 99) suggest that the human population is now in the exponential phase of growth (Figure 11.2) and, in the late 1990s, there is little sign that we are approaching the stationary, let alone the declining phase. The growth rate in some developed countries (Table 11.2) is relatively low but it remains higher in many developing countries. Even if extreme measures (such as China's 'one-child' policy) were taken on a global scale in an attempt to limit population growth, with the numbers present in the existing population, continuation at a very low growth rate would still add huge numbers to the world population over the next few decades. Predictions which suggest the world population will reach 8 billion in 2030 assume that 'environmental resistance' will not exert sufficient effect to limit its growth.

Availability of food is one of the factors which controls or limits the growth of a population, leading towards the stationary or decline phase when supplies of nutrients become low. This applies to a yeast population growing in glucose in a flask, or indeed to any animal population in a given area. In humans, the population size of early hunter-gatherers must have been limited by the area within which they could hunt for food and on the availability of natural food sources in that area. The transition of these early human hunter-gatherers to settled agriculture in the Neolithic Age was almost certainly associated with a surge in population numbers. The practice of growing crops, harvesting and storing excess through a lean season, together with provision of a reliable source of meat and other products from domesticated animals, offered enormous potential for increasing the food available in a given area.

Over the centuries, developments in agriculture have been a means of increasing the carrying capacity of the land with respect to human demands. Selection of favourable characteristics in crops has resulted in improved varieties with higher yields and tolerance of a wide range of environmental conditions. Similarly, breeds of domesticated animals have been developed and provide humans with a rich range of meat (and eggs and milk) in far greater abundance than our Neolithic ancestors could have imagined. Gradual introduction of mechanisation, now seen in the form of tractors, machinery for irrigation and every stage of crop cultivation through to harvesters, has given a further boost to the potential yield that farmers can get from the land.

In the late 20th century, pressure on land is becoming acute. While there is still some scope for bringing fresh land into cultivation, parts of the land surface across the globe are unsuitable for cultivation because of the topography, the nature of the soil or the climate. In some regions, particularly Europe and Asia, a very high proportion of potential arable land is already under cultivation. Social factors may discourage people from cultivating land in remote areas. As populations expand, often the best agricultural land is swallowed up in housing and the supporting infrastructure of urbanisation.

Intensive farming methods strive to maximise productivity and increase the yield from agricultural land. Artificial fertilisers are used to increase the nutrient content of the soil (or replace that which was lost when crops were removed);

pesticides help to reduce crop losses from pests and disease. Control of the environment by cultivation in glasshouses is a way of extending the growing season and allows the harvest to be more predictable in terms of quality, quantity and timing. In rearing animals for meat, production has increased through control of diets, manipulation of reproduction and dense stocking in housing where environmental conditions can be kept at their optimum. There is further potential for increasing production by culture of microorganisms. Protein or other organic molecules derived from these sources can be introduced into novel foods to become part of the human diet. Application of gene technology gives scope for being very specific about the precise characteristics to be introduced into the organism being produced for food. Modern methods of packaging and storage of foods have contributed to a considerable reduction in the wastage of food after harvest through deterioration and spoilage.

(a)   (b)

Figure 11.26 (a) Intensive glasshouse culture - tomatoes grown by hydroponics (Kent, UK)
(b) Housing for livestock - use of protective housing at lambing time (in Suffolk, UK)

Despite the success of applying science and technology, there is a global imbalance in the availability and distribution of food and there is concern whether production can continue to keep pace with the increasing number of mouths to feed in the 21st century. There is little doubt that in affluent, well developed societies, there is an abundance of high quality food while at the same time, in other places, there are people living on minimal diets, suffering from hunger and malnutrition with some close to starvation. This may be the result of crop failure in subsistence areas, or perhaps due to the effect of wars or simply a consequence of poverty within an otherwise prosperous society. As income per head increases in developing countries, there is a corresponding demand for more food, in terms of quality, quantity and variety. This puts further pressure on the capacity of the land to fulfil these demands.

Some of the consequences of human activities on the environment in the context of increasing world population are discussed in Chapter 6. Consideration is given to changes in land use, the impact on natural ecosystems and limitations of natural resources including sources of energy. There are also effects of pollution as a result of human activity. In terms of feeding the world population, the emphasis must be to find ways of increasing food production in a sustainable way that maintain long-term ecological stability. While 'progress' and changes are inevitable, the ultimate resource is people and we must be optimistic that, as in the past, human ingenuity will find ways that will continue to provide for and support the increasing demands of the human population in the future.

Figure 11.27 *Food production and the increasing population. (a) Total world production of food more than doubled between 1950 and 1980; (b) Comparison of developed and developing countries indicates that the developing countries showed a larger percentage increase than developed countries. The year 1970/71 is given an arbitrary value of 100*

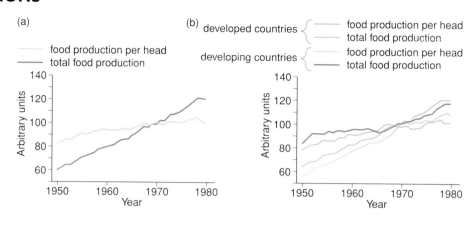

The following poem 'The technologist's reply to a conservationist's lament' was written by an American economist, Kenneth Boulding:

Man's potential is quite terrific
You can't go back to the
 Neolithic
The cream is there for us to
 skim it
Knowledge is power, the sky's
 the limit.
Every mouth has hands to
 feed it.
Food is found when people
 need it.
All we need is found in
 granite
Once we have the men to plan
 it.
Yeast and algae give us meat,
Soil is almost obsolete.
Men can grow to pastures
 greener
Till all the earth is Pasadena.

Questions about food production

- List some ways that people can *increase* the carrying capacity of the land for human populations. Think about practices in modern agriculture (scientific and technological applications), biotechnology (old and new) and possible future developments.

- Then list some human activities that effectively *decrease* the potential of land to produce food.

- What sacrifices are made in the drive for increased food production? Some points to consider include animal welfare, quality, taste, diminished gene pool, loss of natural habitats, use of energy.

As well as there being more mouths to feed in the future, the composition of populations is also changing, particularly with respect to the age structure.

- What are the consequences of an increasingly older population in developed societies?

- As medical facilities become more widely used in less developed countries, what effects will this have on the age structure – now, in 10 years time, and in 50 years time? What further demands will these countries make on world resources, such as food and energy?

Geographical variation

In different geographic regions around the world, the native or indigenous human populations show a number of distinct differences in their features, even though all humans belong to the same species (*Homo sapiens*). There is variation within the populations and there are differences between populations. Since the origin of early humans, believed to be in Africa nearly half a million years ago, human populations have expanded and migrated into new territories (Figure 11.24). As with any animal population, there is variation within the gene pool. The combined effects of isolation and selection pressure are likely to have led to some divergence in characteristics shown by the populations as they moved apart. Some features may have evolved in a way that is advantageous to the population, allowing the people to adapt or live successfully in the particular region. We easily notice differences in features like height, hair and skin colour and general body form, but there are other internal characters, such as blood groups and enzyme systems, that are also associated with certain geographical regions.

Generally there are considered to be four main geographical groups: the **Caucasoids**, **Negroids**, **Mongoloids** and **Australoids**. For each group we can recognise certain types, but because no human population is totally isolated from all others, and because people mix and migrate, there is overlap between the groups and the descriptions given cannot be taken as rigid definitions. These groups and their subgroups are further characterised by the development of different languages and cultural activities. The main visible distinctive features are summarised below, though it must be appreciated that there is considerable variation within each of these main groups.

- **Caucasoids** are indigenous to Europe, north Africa and the Middle East through to the Indian subcontinent. The skin has relatively little pigment, ranging from fair through pink to light brown or olive; the hair is usually fine and either wavy or straight, ranging in colour from blond or red through to dark brown. The hair is usually fairly well developed on the body. The face is narrow with a prominent nose and thin lips. The eye colour varies from pale blue through green to brown.
- **Negroids** are indigenous to Africa south of the Sahara. The pigment is dense producing dark brown to black skin. The hair is usually black and tightly curled; the face is rather short with dark eyes, thick lips and a broad nose.
- **Mongoloids** occur in eastern Asia (China, Korea and Japan), northeast Asia and include the American Indians. The skin is brown to light; the hair is coarse, black and usually straight, but sparse on the body and face. The face is broad and flat, with dark eyes and in some regions the eyelid is partly covered by an internal fold of skin (the epicanthic fold) which gives the characteristic slant-eyed appearance.
- **Australoids** are the aboriginals of Australia and Melanesia. The skin is dark and the hair ranges from wavy (Australia) to frizzy (Melanesia). The head is long and narrow, with projecting jaw and sloping forehead with prominent brow ridges.

Climate is a possible selective force in determining body form. People indigenous to tropical regions tend to be tall and slim, with long limbs. In central Africa, the Tutsi people of Rwanda and the Masai people of northern Kenya are typical of this body shape. Since heat is lost from the body surface, mostly by sweating, exposure of a large surface area in relation to volume favours heat loss in these hot climates. By contrast, those living in cold climates, such as the Inuit (Eskimos) of northern Canada and Alaska, have a short, squat body with relatively short limbs, which is well adapted to conserve heat.

Figure 11.28 Faces of main geographical groups: (top) Caucasoids; (upper centre) Negroids; (lower centre) Mongoloids; (bottom) Australoids

Figure 11.29 A mixed group of students

PEOPLE AND POPULATIONS

Figure 11.30 Variations in body form with climate: (left) tall, slim Masai tribesmen in Kenya; (right) Inuits from North West Territories, Canada

It is harder to rationalise differences in skin pigmentation with respect to heat exchange. Dark-skinned people are associated with warmer regions and light-skinned people with the colder regions of the globe. Dark-coloured objects gain heat faster in the sun so the dark skin would appear to be less well adapted to hot climates. There may, however, have been some natural selection with respect to vitamin D balance and skin pigmentation. Synthesis of vitamin D occurs by the action of ultraviolet (UV) light on a derivative of cholesterol, but the dark pigment (melanin) reduces the UV penetration. It may be an advantage to those living in less sunny climates to have paler skins, thus giving a higher level of vitamin D synthesis.

The shape of the nose is another feature which shows some adaptation to climate. Broad, short noses are characteristic of people living in hot tropical climates whereas long, thin noses are associated with people in dry and in colder climates. It is likely that the long nose allows incoming air to be warmed and made moist before reaching the lungs. This may be an important means of conserving water in dry climates and at high altitude.

The distribution of blood groups shows marked differences around the world. While these patterns must reflect something of the evolutionary history, there is no known selective advantage of one group over another. The three maps in Figure 11.31 show the world-wide frequencies in the A, B and O alleles. Note, for example, how Central and South America have low frequencies of both the A and B alleles (less than 5% for each) but more than 90% of the population carry the O allele. Compare this with Australia, where parts have a relatively high frequency of the A allele and correspondingly lower frequencies of the B and O alleles. In northwest Europe, B is relatively low and A and O frequencies are higher.

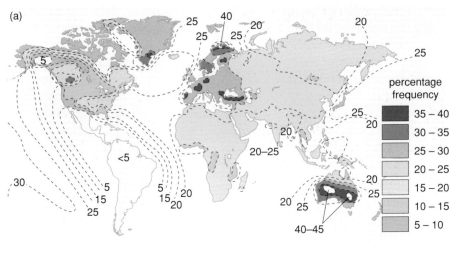

Figure 11.31 Maps of global distribution of the ABO blood group system. (a) frequency of the A allele; (b) frequency of the B allele; (c) frequency of the O allele

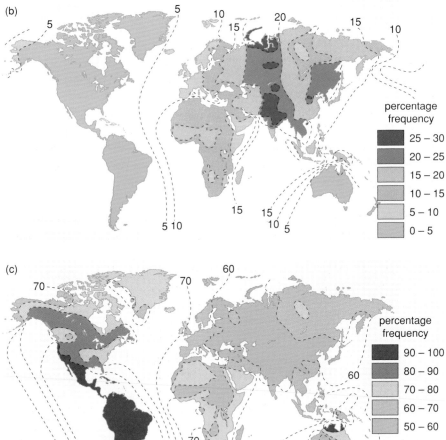

In adults, the global distribution of the enzyme lactase, which breaks down lactose (the sugar in milk) into glucose and galactose, shows an interesting pattern. In most ethnic groups, adults show loss of lactase after puberty which means they cannot digest fresh milk containing lactose. Undigested lactose in the gut of these lactose-intolerant people is likely to be fermented by bacteria in

PEOPLE AND POPULATIONS

Figure 11.32 Convergence of human populations – marriage between Negroid and Caucasian

the large intestine, resulting in nausea, abdominal pain and diarrhoea. Lactase is lacking in an estimated 90% of Chinese and lactose intolerance is also found in Africans, Arabs and some people from the Indian subcontinent. However, in northern and western Europe lactase is found in about 95% of adults. Within some of the lactose-intolerant areas there are pockets of people who have traditionally been pastoralists and so continued to drink milk in adulthood; these people do have the enzyme. Lactose intolerance doubtless accounts for the very low usage of fresh milk in traditional Asian foods. In fermented milk products, such as yoghurt and cheese, the lactose is reduced or eliminated, thus making them digestible and acceptable to lactose-intolerant people.

Some features have been described which show variation on a geographical basis and we can see how certain features may have had some selective advantage to people living in different regions of the world. Divergence of the genetic make-up of populations would have resulted from barriers of distance, as well as oceans and rivers or mountains and deserts, all of which reduce the likelihood of people mixing and marrying. These geographical barriers are reinforced by language and cultural differences which have evolved in the different human populations. Languages can, in fact, be useful in tracing evolutionary history of different groups and show close associations with the genetic relationships (Figure 11.33). The Basques in northern Spain, for example, speak a different language and are genetically distinct from their neighbours, reflecting their different origin. Perhaps with greater global communication and easier means of travel, as we approach the 21st century, the future trend will be for increased convergence of human populations and less distinctive geographical variation in their features.

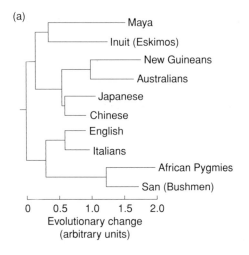

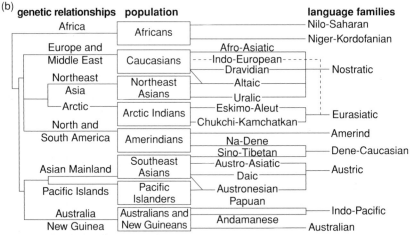

Figure 11.33 Tracing patterns of genetic relatedness in human populations. Such evidence may give clues about the migrations of human populations. (a) A tree of evolutionary relatedness based on 58 blood group genes. You can identify the four main geographical groups and see that this evidence suggests a west-east divide – with Europeans closer to Africans and Australoids to Mongoloids; (b) Human populations and language families – The rate of change of languages over time has been used as an evolutionary time clock in looking at relatedness between groups. In this diagram, the genetic relationships are based on evidence from blood groups, enzymes and other genes. You can identify the four main geographical groups and compare the relatedness suggested by language families with that from the genes

Examination questions

Chapter 1

1   Read through the following account of photosynthesis, then write on the dotted lines the most appropriate word or words to complete the account.

Photosynthesis is a type of .... nutrition, involving the synthesis of organic molecules from inorganic materials. The process involves two types of reactions, light-dependent and light-independent.

In the light-dependent reactions, light energy is absorbed by chlorophyll molecules located on the .... of the chloroplasts; .... and .... are produced and oxygen gas is given off as a by product.

In the light-independent reactions, ... accepts molecules of carbon dioxide, which together with the products of the light-dependent reactions, results in the formation of .... This compound can be converted to .... or used to regenerate the carbon dioxide acceptor molecule.

(Total 7 marks)

2   The apparatus shown below can be used to study mineral nutrition in flowering plants.

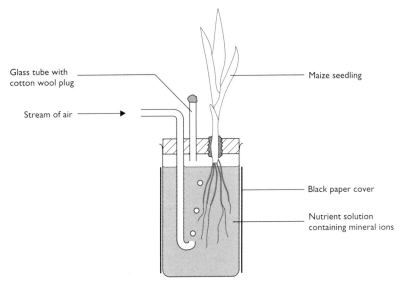

Glass tube with cotton wool plug

Stream of air

Maize seedling

Black paper cover

Nutrient solution containing mineral ions

(a)   Suggest the function of each of the following.
      (i)   The stream of air.   (2 marks)
      (ii)   The black paper cover.   (2 marks)

(b)   The nutrient solution contains various mineral ions including
      magnesium and phosphate. Give *one* reason why each of these
      ions is essential to the plant.   (2 marks)
                                                         (Total 6 marks)

Chapter 2

1   Read through the following account of carbohydrate digestion then write
    on the dotted lines the most appropriate word or words to complete the
    account.

    Digestion of starch starts in the .... where it is

    hydrolysed to .... by the enzyme

    .... This process is halted in the stomach but

    continues in the duodenum, catalysed by an enzyme secreted by the

    .... Also in the duodenum, sucrose is hydrolysed to

    .... and .... by enzymes

    produced by secretory cells in the duodenum wall.

                                                         (Total 6 marks)

2   The diagram below shows the structure of a villus as seen in longitudinal
    section.

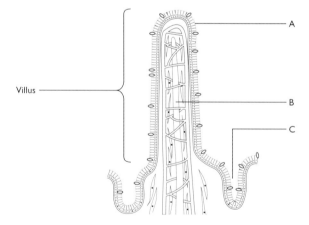

(a)   Name the parts labelled A, B and C.   (3 marks)
(b)   Describe *one* way in which the structure of the villus is adapted for
      the functions it performs.   (2 marks)
                                                         (Total 5 marks)

Chapter 3

1   Read through the following passage on the water potential of plants cells, then write the most appropriate word or words to complete the account.
    The water potential of a cell (Ψ_{cell}) is determined by its solute potential (Ψ_s) and its pressure potential (Ψ_p). The relationship can be summarised in the following equation.

$$\Psi_{cell} = \Psi_s + \Psi_p$$

When a cell is plasmolysed, Ψ_p is ...., Ψ_{cell} is equal to

.... and the cell is said to be ....,

If this cell is placed in pure water, its water potential is

....than that of the external solution so water

....by osmosis. The Ψ_p...

until the cell becomes ....   (Total 7 marks)

2   The diagram below shows the structure of a freshwater protozoan.
    (*a*)   On the diagram label the contractile vacuole.   (1 mark)

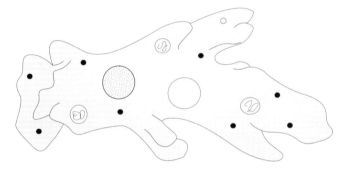

    (*b*)   Explain how the contractile vacuole functions in the regulation of the internal environment of the protozoan.   (3 marks)
                                                        (Total 4 marks)

Chapter 4

1   The table below refers to descriptions of some major groups of organisms. State the name of each group described in the spaces provided.
                                                        (Total 6 marks)

Features	Group
Heterotrophic organisms with rigid cell wall; body structure normally a mass of hyphae	
Photosynthetic organisms, with no flowers and no roots or stems or leaves, usually aquatic	
Photosynthetic non-flowering plants, with roots, stems and leaves	
Photosynthetic flowering plants, seeds enclosed in a fruit	
Heterotrophic, radially symmetrical, multicellular animals with nematocysts	
Heterotrophic, multicellular animals possessing a post-anal tail and pharyngeal clefts at some stage of their life	

2   The diagram below shows the distribution of daisy plants in a lawn. Each dot represents one plant.

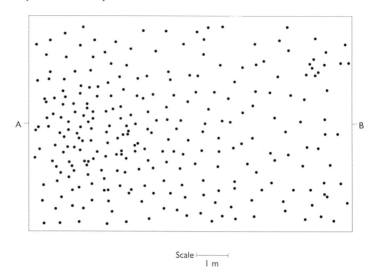

Scale ⊢————⊣
1 m

(*a*)   Describe how you would use a quadrat to obtain an accurate estimation of the density of daisy plants in this lawn.   (5 marks)

(*b*)   Suggest *one* limitation of this method.   (1 mark)

(*c*)   Describe how you would investigate changes in plant populations between points A and B.   (2 marks)

(Total 8 marks)

Chapter 5

1   Explain the meaning of each of the following ecological terms.

(*a*)   Productivity   (2 marks)

(*b*)   Pyramid of numbers   (2 marks)

(*c*)   Community   (2 marks)

(Total 6 marks)

2   The diagram below shows the energy flow for part of a large pond. All values are given in kJ m^{-2} yr^{-1}.

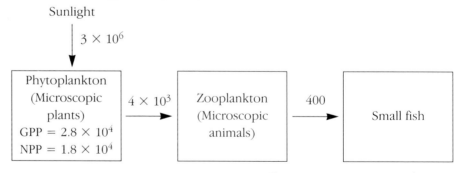

GPP = gross primary productivity
NPP = net primary productivity

(*a*)   Calculate the percentage energy from sunlight which is fixed as GPP by phytoplankton. Show your working.   (2 marks)

(*b*)   Suggest *two* reasons why not all of the incident sunlight is utilised in photosynthesis.   (2 marks)

(Total 4 marks)

Chapter 6

1   The diagram below shows a biogas plant as used by a family in China.

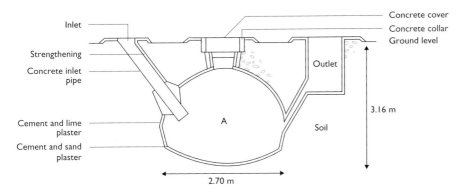

(a)   Name *two* constituents of biogas.   (2 marks)

(b)   (i)   Describe the processes which occur in A   (2 marks)

       (ii) Explain why the pH within the biogas plant falls during the
             first stages of gas production.   (1 mark)

(c)   Explain why the biogas plant is normally kept sealed.   (1 mark)

(d)   Suggest why this biogas plant is built underground.   (1 mark)

                                                              (Total 7 marks)

2   Give an account of the use of biomass for fuels.   (Total 10 marks)

Chapter 7

1   Give an account of the ways in which human activities lead to the
    pollution of water.   (Total 10 marks)

Chapter 8

1   Diagrams A–D below show the heads of four of the ten species of finch
    inhabiting a volcanic island in the Galapagos group, about 600 miles from
    the mainland of Ecuador, South America.

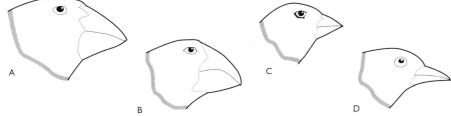

(a)   (i)   What major difference between the four species is shown in
             the diagrams?   (1 mark)

       (ii) Suggest how this difference may be related to the way of life
             of these finches.   (2 marks)

(b)   These finches probably evolved from a common ancestral species.
       The different species do not interbreed. What might this indicate?

                                                                           (2 marks)

(c)   All the species of finch on the island are dull-coloured. Suggest
       two ways in which recognition may occur between sexes of the
       same species of finch.   (2 marks)

                                                              (Total 7 marks)

Chapter 9

1   The table below refers to cultural developments during the evolution of the genus *Homo*. Three species of *Homo* are listed in the table.

If the statement is correct, place a tick ✓ in the appropriate box and if the statement is incorrect, place a cross ✗ in the appropriate box.

Statement	*H. habilis*	*H. erectus*	*H. sapiens*
Use of fire			
Oldowan (pebble tool) culture present			
Acheulian (hand-axe) culture present			
Production of cave art			
Burial of the dead			
Development of speech			

(Total 6 marks)

2   DNA hybridisation is a method of comparing DNA from different species. In this technique, purified DNA samples from two different species are heated to separate the strands, then mixed and allowed to cool together. As the separated strands cool they re-combine into double helices. Some of the new double helices are hybrid: that is, they consist of one strand from each species. This process is shown in the diagram below.

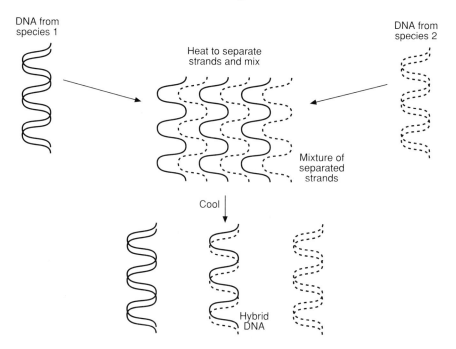

(a)   The strands in the hybrid DNA separate at a lower temperature than those in DNA from a single species. The more closely related the two species, the lower the difference between the separation temperatures of hybrid DNA and single-species DNA.

(i)   Suggest why the strands in hybrid DNA separate at a lower temperature than those in DNA from a single species. (1)

(ii) Suggest why the separation temperature of hybrid DNA from distantly related species is lower than that of hybrid DNA from closely related species. (2)

(b) The table below shows the difference between the separation temperatures of hybrid DNA and single-species DNA for a number of pairs of primate species or groups. Each figure is a mean of many tests using DNA from different individuals.

Sources of hybrid DNA	Difference in separation temperature / °C
Human / chimpanzee	1.8
Gorilla / chimpanzee	2.2
Human / gorilla	2.3
Human / orang-utan	3.6
Gorilla / orang-utan	3.6
Chimpanzee / orang-utan	3.6
Gibbon / other apes	4.8
Old World monkeys / apes	7.2

It is assumed that the difference in separation temperature is directly proportional to the time since the evolutionary lines of the two groups diverged.

(i) The evolutionary lines of Old World monkeys are thought to have diverged 30 million years ago. Use this figure to calculate the time represented by a difference of 1 °C in separation temperature of hybrid DNA. Show your working. (2)

(ii) Calculate how long ago the evolutionary lines of humans and chimpanzees diverged, according to DNA hybridisation studies. Show your working. (2)

(c) The diagram below is an incomplete family tree of primate evolution based on the table in part (b).

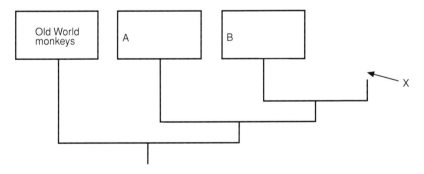

(i) Fill in the names of the missing species of groups in boxes A and B. (2)

(ii) Complete the family tree from the branching point marked X. (2)

(d) The relationship and divergence times calculated using DNA hybridisation data have been the subject of considerable debate and disagreement among students of primate evolution. In particular, it has been claimed that the difference in separation temperature between hybrid DNA and single-species DNA may not be directly proportional to the time since the evolutionary lines of the two groups diverged.

Give two alternative sources of evidence concerning relationships and divergence times in primate evolution. (2)

(Total 13 marks)

Chapter 10

1 The table below shows the mean voluntary energy intake per day of soldiers stationed in climates with different local mean temperatures.

Local mean temperature / °C	Voluntary energy intake per day / kJ
+35	13 000
+15	15 000
+5	16 800
−5	18 000
−20	20 000
−30	21 000

(a) Discuss the physiological significance of the energy intake in relation to local mean temperature. (4)

(b) State *three* ways in which the body acclimatises to high temperatures. (3)

(c) (i) Compare the tolerance of the body to increases and decreases in core temperature. (2)
    (ii) Describe the effects of cold stress. (2)

(Total 11 marks)

2 Reduced atmospheric pressure and low oxygen partial pressures may have important physiological effects on mountain climbers, who may suffer from high altitude stress and mountain sickness if they are not acclimatised.

The table below shows the effects of increasing altitude on atmospheric pressure.

Altitude / m	0	3 000	6 000	9 000	12 000
Atmospheric pressure / kPa	101.3	69.7	46.5	30.1	18.8

The graph opposite shows the effects of increasing altitude on the partial pressures of oxygen in the atmosphere and in the alveoli.

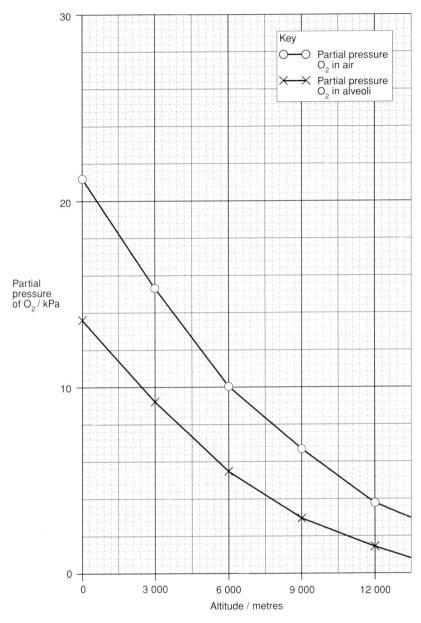

(a) Calculate the difference between the partial pressures of oxygen in the atmosphere and in the alveoli when a person is at a height of 4500 metres above sea level. Show your working. (2)

(b) Comment on the similarities and differences between the curves on the graph. (3)

(c) Describe the symptoms of mountain sickness. (3)

(d) The summit of Mount Everest is approximately 9000 m above sea level. A climber who is not fully acclimatised would have to breathe pure oxygen to survive at the top of Mount Everest, but members of the Sherpa group, who normally live at very high altitudes, can survive at 9000 m for short periods without using extra oxygen.

Describe *three* long term physiological changes which may have occurred in the Sherpas to adapt them to high altitudes. (3)

(Total 11 marks)

3   Read through the following passage on biological rhythms, then write on the dotted lines the most appropriate word or words to complete the passage.

Patterns of physiological activity which follow a cycle of about 24 hours are referred to as .... rhythms. Rhythms which follow a spontaneous internal cycle are called .... rhythms. A good example of a spontaneous internal cycle is .... , while blood pressure changes are an example of .... control. Travel across time zones can disrupt the rhythms of the body, an effect commonly known as .... The hormone .... , produced by the pineal gland, is involved in the control of body rhythms and may be able to relieve some of the effects of this condition.

(Total 6 marks)

Chapter 11

1   During the period 1964 to 1969, the Shipibo tribe of the Peruvian Amazon region had one of the highest documented fertilities in a human group. The gross reproduction rate (defined as the number of female children born to each woman) was 4.9.

(a)   If there are 1.05 male births to each female birth, calculate the mean number of births to each woman when there is a gross reproduction rate of 4.9.

(2)

(b)   Monogamy describes the practice of each man or woman having only one spouse. When polygyny is practised, one man has several wives.

Polygamy is common but not universal amongst the Shipibo. It has been found that the fertility of the wives of polygynous men is lower than that of the wives of monogamous men. Suggest *two* reasons why this occurs.

(2)

(c)   Polygyny reduces the fertility rate yet the overall rate is very high. Suggest an explanation for this.

(1)

(d)   Contact with western society has increased not decreased the gross fertility. Suggest an explanation for this.

(1)

(Total 6 marks)

2   *(a)*   The graph below shows the mortality rate, expressed as deaths per thousand of population, for males and females in the United States for the year 1940.

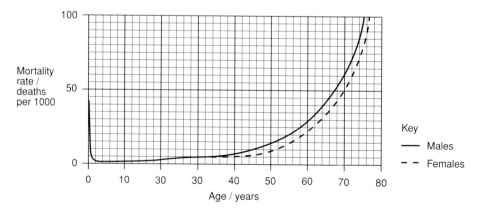

(i) From the graph, determine the mortality rate in 1940 for males aged 55. (1)

(ii) Comment on the mortality rates in the following age ranges: 10 to 30 years; 40 to 60 years. (4)

(b) The graph below shows two survival curves for Europeans, one from the middle of the eighteenth century and the other from the middle of the twentieth century.

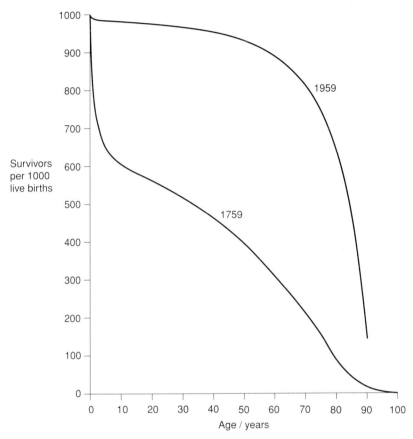

(i) Compare the survival rates from 0 to 10 years in 1759 with the survival rates from 0 to 10 years in 1959. (2)

(ii) Suggest two reasons for the differences you have described in (b) (i). (2)

(iii) Comment on the differences in survival patterns, as shown by the curves, for Europeans after the age of 10. (3)

(Total 12 marks)

3 Give an account of the factors which influence the size of human populations.

(Total 12 marks)

Mark schemes

In the mark schemes, the following symbols are used.
;   indicates separate marking points
/   indicates alternative marking points
eq   means correct equivalent points are accepted

Chapter 1

1   autotrophic ; lamellae / thylakoids / grana ; NADPH / reduced hydrogen carriers ; ATP ; ribulose bisphosphate / RuBP ; phosphoglyceric acid / glycerate 3-phosphate; triose / hexose / glucose / etc ;

(Total 7 marks)

2   (*a*)   (i)   provides oxygen ; for respiration ; for active uptake / active transport of mineral ions ;   (2 marks)

             (ii)   excludes light ; prevents the growth of algae ; which compete for mineral ions ;   (2 marks)

    (*b*)   Magnesium: chlorophyll / enzyme cofactor ;
             Phosphate: phospholipids / ATP / nucleotides / nucleic acids ;

(2 marks)
(Total 6 marks)

Chapter 2

1   mouth / buccal cavity ; maltose / dextrins ; amylase ; pancreas ; glucose ; fructose ;   (Total 6 marks)

2   (*a*)   A = columnar epithelium / brush border ;
             B = lacteal / lymphatic vessel ;
             C = crypt of Lieberkuhn / intestinal gland ;   (3 marks)

    (*b*)   long and thin ; increases surface area / dense capillary network ; for transport of digestion products / lacteal ; transport of products of fat digestion / presence of microvilli ; increases surface area / simple epithelium ; short distance for uptake of nutrients / large surface area ; increases uptake / muscle cells ; move villus to keep in contact with contents of intestine ;   (2 marks)

(Total 5 marks)

Chapter 3

1   zero ; Ψs ; flaccid ; lower / more negative ; enters ; increases ; turgid ;

(Total 7 marks)

2   (*a*)   contractile vacuole correctly labelled ;   (1 mark)

    (*b*)   water enters protozoan by osmosis ; because the cell has a lower water potential than freshwater ; water enters the contractile vacuole ; vacuole fuses with cell membrane ; water is pumped out;

reference to involvement of ATP / mitochondria / contractile
vacuole maintains water potential / prevents cell lysis / eq ; (3 marks)
(Total 4 marks)

Chapter 4

1   Fungi ; Algae ; Ferns / Filicinophyta ; Angiospermophyta ; Cnidaria ;
Chordata ;   (Total 6 marks)

2   (*a*)   use tables of random numbers ; pairs of numbers used as
coordinates ; use a quadrat of stated size (up to 1 m^2) ; count the
number of daisies per quadrat ; repeat at least ten times ;
density = number of plants per unit area ;   (5 marks)

(*b*)   may miss an area / sample same area twice / variation in
distribution ;   (1 mark)

(*c*)   use a line transect / line of quadrats ; record the numbers of each
species at regular intervals ;   (2 marks)
(Total 8 marks)

Chapter 5

1   (*a*)   energy stored / fixed by plants ; production of biomass ; per unit
area per year ; NPP = GPP − R ;   (2 marks)

(*b*)   represents the numbers of organisms ; at each trophic level ;
producers at the bottom ; area of each bar is proportional to
numbers ;   (2 marks)

(*c*)   group of several populations of different species ; in a habitat ;
at one time ; interacting through energy flow / mineral cycles ;
(2 marks)
(Total 6 marks)

2   (*a*) $(2.8 \times 10^4 \div 3 \times 10^6) \times 100$; = 0.9% ;   (2 marks)

(*b*)   not absorbed by chlorophyll ; some absorbed by water ; some light
is reflected ; lost as heat ; another limiting factor / carbon dioxide
limiting ;   (2 marks)
(Total 4 marks)

Chapter 6

1   (*a*)   methane ; carbon dioxide ;   (2 marks)

(*b*)   (i)   fermentation / anaerobic respiration occurs ; dung / waste /
named organic matter digested ; by activity of bacteria /
methanogens / eq ; complex molecules are converted to
simple carbon compounds / eq ;   (2 marks)

(ii)   acidic gases / carbon dioxide / acids produced ;   (1 mark)

(*c*)   to keep the conditions anaerobic / maintain the temperature /
prevent smell / store the gas / prevent leakage / eq ;   (1 mark)

(*d*)   heat insulation / temperature reference / safety reference /
aesthetic reference / easier to construct / fill / easier to seal / saves
space ;   (1 mark)
(Total 7 marks)

2   photosynthesis results in the formation of organic compounds ; direct burning of biomass releases heat energy ; results in the formation of carbon dioxide and water as wastes ; use of wood ; examples, such as tree thinnings from managed woodland ref: short rotation coppicing ; use of straw bales in central heating boilers / burning of dried dung / eq ; use of domestic or agricultural wastes ; in production of biogas ; details of biogas fermenter ; reference to landfill capping in relation to methane ; production of ethanol from sugar ; mixed with petrol / gasohol reference ; use of fossil fuels and example ; production of greenhouses gases / comment on global warming / eq ; sulphur dioxide from burning oil / coal ; comment on acid rain eq ; hydrocarbons / nitrogen oxides from burning oil ;   (Total 10 marks)

Chapter 7

1   sewage / waste organic matter from poor treatment / agricultural / industrial spillage; nitrates from agricultural fertilisers ; leaching into run-off water ; oil spillage ; less light so less photosynthesis ; effects on plumage / skin / fur ; warm water from industry / power stations ; some organisms grow faster / less well / die due to temperature change ; oxygen is less soluble in water at higher temperatures ; detergents from domestic use / industrial use / from oil treatment ; acid rain from combustion of fossil fuels ; reduces pH of lakes ; increases aluminium concentration; pesticides from agricultural use / heavy metals from industrial processes ; radioactive material from nuclear reactions ; oestrogenic compounds from surfactants / plastics / pesticides ; abnormal development in fish ; (organic matter) utilised by bacteria which increase in number ; increased respiration uses up more oxygen / increases BOD ; eutrophication / death of other organisms ; toxic effect of substance on organisms ; bioaccumulation reference ; disruption of food webs / algal bloom as a consequence of fertilisers / sewage / phosphates in detergents ;   (Total 10 marks)

Chapter 8

1   (*a*)   (i)   there is variation in the beaks ;   (1 mark)
             (ii)   may be related to the diet of the finches ; A and B may eat seeds ; C and D may eat insects ;   (2 marks)
       (*b*)   speciation has occurred ; due to reproductive isolation / eq ;
                                                                             (2 marks)
       (*c*)   feather patterns ; song ; behaviour pattern / eq ;   (2 marks)
                                                                      (Total 7 marks)

Chapter 1

1   ✗   ✓   ✓;
    ✓   ✓   ✗;
    ✗   ✓   ✗;
    ✗   ✗   ✓;
    ✗   ✗   ✓;
    ✓   ✓   ✓;
                                                                      (Total 6 marks)

2 *(a)* (i) fewer hydrogen bonds / base pairings between strands / eq ;
                                                                                (1)

       (ii) fewer base pairs in common between distantly related species /
             more unmatched / unpaired areas / converse closely related ;
             fewer hydrogen bonds between strands / eq ; fewer gene /
             cistron similarities ;   (2)

    (b) (i) 30 000 000 ÷ 7.2 ; = 4.2 million years / 4.2×10^6 years ;
                                                                                (2)

       (ii) $4.2 \times 10^6 \times 1.8$; = 7.5 million years / eq ;   (2)

    (c) (i) A = gibbons ; B = orang-utan ;   (2)

       (ii) *name and correct branching needed*
             gorilla ;   chimpanzee / human   human / chimpanzee ;
                                                                                (2)

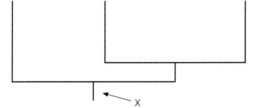

    (d) fossils / palaeontology ; chemical comparison (*one or two from*)
           albumen / fibrinopeptides transferrin / haemoglobin / cytochrome
           oxidase / eq / immunology / amino acid sequence ;   (2)
                                                              (Total 13 marks)

Chapter 10

1 *(a)* lower mean temperature causes more heat loss ; thus more energy intake
          needed ; lower external temperature stimulates skin receptors ; impulses
          to hypothalamus ; thyroid / adrenal medulla ; stimulates adrenaline /
          thyroxine secretion ; ref increase in metabolic rate / increased heat
          production ; stimulation of appetite centre ; caused greater food intake ;
          more heat from outside means fewer kJ needed ;   (4)

    (b) lower heart rate ; less salt in sweat ; increased blood volume ;
           increased number of sweat glands ; 'thermostat' re-set ;   (3)

    (c) (i) moderate / high tolerance to decrease ; low tolerance to increase ;
                figures for one / both ;   (2)
       (ii) hypothermia described ; necrosis in frostbite ; 'trench foot'
             described ;   (2)
                                                              (Total 11 marks)

2 *(a)* air 12.7 ± 0.2, alveoli ± 0.1 ; 5.1 – 5.7 kPa ;   (2)

(b) both curves fall (with increased altitude) ; difference between curves greater at lower altitudes / converse / converge at high altitudes / diverge at low ; rate of decrease in partial pressure falls with increasing altitude ; any use made of the figures / manipulation of figures ; partial pressure in air always greater than that in alveoli / converse ; oxygen in air a steeper gradient ;　　　　　　　　　　　　　　　　　　(3)

(c) hyperventilation / deeper breathing / faster breathing / shortness of breath ; nausea / dizziness / like drunkenness / blurred vision ; carbon dioxide blown off reducing blood acidity / alkalaemia / greater pH ; increased secretion of ADH / more water reabsorbed from urine to blood / less urine produced ; redistribution of body fluids / oedema / water retention ; impaired mental functions / hallucinations / eq ;

　　　　　　　　　　　　　　　　　　　　　　　　　　　　　(3)

(d) increased lung volume / thoracic volume / alveolar density / number / pulmonary diffusing capacity / more pulmonary capillaries ; increased red cell count / haemoglobin concentration / myoglobin in muscle / greater affinity of haemoglobin for oxygen ; increased cardiac output / pumping effectiveness improved / blood flow faster ;　　　　(3)

　　　　　　　　　　　　　　　　　　　　　　　　　　(Total 11 marks)

3　circadian ; endogenous ; body temperature / REM sleep / ACTH level / cortisol level / ion level / urination / bowel movements / mental performance / menstrual cycle / ovulation ; exogenous ; jet lag ; melatonin ;

　　　　　　　　　　　　　　　　　　　　　　　　　　(TOTAL 6 marks)

Chapter 11

1　*(a)*　$4.9 \times 1.05 = 5.145$; $4.9 + 5.145 = 10.045$;　　　　　　(2)

(b) longer sexual abstinence postpartum / after birth ; lower frequency of coitus / intercourse per woman ; preference for sexual activity with one wife / some wives have little coitus / eq ;　　　　　　(2)

(c) no effective contraception ;　　　　　　　　　　　　　　(1)

(d) decreased polygyny rate ;　　　　　　　　　　　　　　(1)

　　　　　　　　　　　　　　　　　　　　　　　　　(TOTAL 6 marks)

2　*(a)*　(i)　accept 20 to 22 per 1000 ;　　　　　　　　　　　(1)
　　　　(ii)　*(10 to 30 years)* low / stated rate ; almost constant ; no difference between males and females ;　　　　　(2)
　　　　　　(40 to 60 years) rate increases ; higher in males ; exponential / accelerating rate ;　　　　　　　(2)

(b)　(i)　1759 - number of survivors decreases steeply / converse for 1959 ; decreases from 900 to 600 but little change in 1959 ; decrease slows down after first few years / little change in 1959 ;

　　　　　　　　　　　　　　　　　　　　　　　　　(2)

(ii) hospital / medically attended birth ; better hygiene ; water supply / sewerage ; immunisation / medication ; better diet ; nutrition ; less overcrowding / better housing ; (2)

(iii) (1759) decrease greater / converse for 1959 ; steady / constant decrease to 82 to 85 years ;

(1959) steady / constant decrease to 55 to 55 years ; steeper decrease after 60/70 years ; ref increased heart disease / eq ;

(3)

(Total 12 marks)

3 birth rate ; proportion of women bearing children ; effects of polygyny ; factors affecting family size ; effect of contraceptive practices ; ref pro-natalist / anti-natalist policies ; ref environmental / social factors ; examples / figures ; death rate ; effects of disease / parasitism ; effects of hygiene / sanitation / medicine ; effects of war ; female / selective infanticide ; examples / figures ; immigration ; emigration ; credit suitable examples ; ref carrying capacity ; environmental resistance ; overcrowding / availability of living space ; food availability ; effects of agricultural / green revolution / intensive agriculture / eq ; limitations on hunter-gatherers ; predation in early populations ; etc ;

(Total 10 marks)

Index

INDEX